VOLUME ONE HUNDRED AND NINETY TWO

PROGRESS IN MOLECULAR BIOLOGY AND TRANSLATIONAL SCIENCE

Human Microbiome in Health and Disease - Part B

VOLUME ONE HUNDRED AND NINETY TWO

PROGRESS IN MOLECULAR BIOLOGY AND TRANSLATIONAL SCIENCE

Human Microbiome in Health and Disease - Part B

Edited by

BHABATOSH DAS

Molecular Genetics Laboratory, Infection and Immunology Division, Translational Health Science and Technology Institute, Faridabad, Haryana, India

VIJAI SINGH

Department of Biosciences, School of Science, Indrashil University, Rajpur, Mehsana, India

ELSEVIER

ACADEMIC PRESS

An imprint of Elsevier

Academic Press is an imprint of Elsevier
50 Hampshire Street, 5th Floor, Cambridge, MA 02139, United States
525 B Street, Suite 1650, San Diego, CA 92101, United States
The Boulevard, Langford Lane, Kidlington, Oxford OX5 1GB, United Kingdom
125 London Wall, London EC2Y 5AS, United Kingdom

First edition 2022

ISBN: 978-0-323-91210-5
ISSN: 1877-1173

For information on all Academic Press publications
visit our website at https://www.elsevier.com/books-and-journals

Publisher: Zoe Kruze
Acquisitions Editor: Leticia Lima
Developmental Editor: Jhon Michael Peñano
Production Project Manager: James Selvam
Cover Designer: Matt Limbert

Typeset by STRAIVE, India

Contents

Contributors

Ramesh Agarwal
Department of Pediatrics, All India Institute of Medical Sciences, New Delhi, India

Swati Aggarwal
Regional Centre for Biotechnology, Faridabad, Haryana, India

Taruna Ahrodia
Molecular Genetics Laboratory, Infection and Immunology Division, Translational Health Science and Technology Institute, NCR Biotech Science Cluster, Faridabad, India

Vineet Ahuja
Department of Gastroenterology and Human Nutrition, All India Institute of Medical Sciences, New Delhi, India

Md Jahangir Alam
Department of Biotechnology, National Institute of Pharmaceutical Education and Research (NIPER), Guwahati, Assam, India

Krishnamohan Atmakuri
Bacterial Pathogenesis Lab, Infection and Immunology Division, Translational Health Science and Technology Institute, Faridabad, India

Sanjay K. Banerjee
Department of Biotechnology, National Institute of Pharmaceutical Education and Research (NIPER), Guwahati, Assam, India

Krishna Singh Bisht
Regional Centre for Biotechnology, Faridabad, Haryana, India

Santanu Chattopadhyay
Rajiv Gandhi Centre for Biotechnology, Trivandrum, Kerala, India

Meenal Chawla
Molecular Genetics Laboratory, Infection and Immunology Division, Translational Health Science and Technology Institute, NCR Biotech Science Cluster, Faridabad, India

Bhabatosh Das
Molecular Genetics Laboratory, Infection and Immunology Division, Translational Health Science and Technology Institute, NCR Biotech Science Cluster, Faridabad, India

Amit Ghosh
ICMR-National Institute of Cholera and Enteric Diseases, Kolkata, India

Aeshna Gupta
School of Biology and Environmental Science, University College Dublin, Dublin, Ireland

Rashi Gupta
Department of Microbiology, Institute of Home Economics, University of Delhi, India

Bharti Kandiyal
Molecular Genetics Laboratory, Infection and Immunology Division, Translational Health Science and Technology Institute, NCR Biotech Science Cluster, Faridabad, India

Shashi Kumari
DBT-Translational Health Science and Technology Institute, Faridabad, India

Shruti Lal
Molecular Genetics Laboratory, Infection and Immunology Division, Translational Health Science and Technology Institute, NCR Biotech Science Cluster, Faridabad, India

Tushar Kanti Maiti
Regional Centre for Biotechnology, Faridabad, Haryana, India

Indra Mani
Department of Microbiology, Gargi College, University of Delhi, New Delhi, India

G. Balakrish Nair
Rajiv Gandhi Centre for Biotechnology, Trivandrum, Kerala, India

Lekshmi Narendrakumar
Molecular Genetics Laboratory, Centre for Human Microbial Ecology, Translational Health Science and Technology Institute, Faridabad, India

Angitha N. Nath
Rajiv Gandhi Centre for Biotechnology, Trivandrum, Kerala, India

Archana Pant
Regional Centre for Biotechnology, Faridabad, Haryana, India

Deepjyoti Paul
Molecular Genetics Laboratory, Infection and Immunology Division, Translational Health Science and Technology Institute, NCR Biotech Science Cluster, Faridabad, India

Vaishnavi Puppala
Department of Biotechnology, National Institute of Pharmaceutical Education and Research (NIPER), Guwahati, Assam, India

Thandavarayan Ramamurthy
ICMR-National Institute of Cholera and Enteric Diseases, Kolkata, India

Animesh Ray
Department of Medicine, All India Institute of Medical Sciences, New Delhi, India

R.J. Retnakumar
Rajiv Gandhi Centre for Biotechnology, Trivandrum, Kerala; Manipal Academy of Higher Education, Karnataka, India

M. Jeeva Sankar
Department of Pediatrics, All India Institute of Medical Sciences, New Delhi, India

Vijai Singh
Department of Biosciences, School of Science, Indrashil University, Rajpur, Mehsana, Gujarat, India

Kiyoshi Takeda
Laboratory of Immune Regulation, Department of Microbiology and Immunology, Graduate School of Medicine, Osaka University, Suita, Japan

Shravan K. Uppulapu
Department of Biotechnology, National Institute of Pharmaceutical Education and Research (NIPER), Guwahati, Assam, India

Jyoti Verma
Molecular Genetics Laboratory, Infection and Immunology Division, Translational Health Science and Technology Institute, NCR Biotech Science Cluster, Faridabad, India

J.R. Yodhaanjali
Molecular Genetics Laboratory, Infection and Immunology Division, Translational Health Science and Technology Institute, NCR Biotech Science Cluster, Faridabad, India

Preface

Advances in culturomics, DNA sequencing technologies, and computational biology have revealed that trillions of microbes inhabit our body with body site–specific, distinct microbial communities tremendously contributing to human physiology. Autochthonous microbiota associated with the human body provides metabolic functions, resistance against pathogen colonization, and signaling molecules that modulate a wide range of cellular processes and immune maturation. However, dysbioses in the compositions or functions of the human microbiome may lead to several health disorders, including malnutrition, obesity, cancer, diabetes, gastroenterologic disorders, sepsis, cardiovascular diseases, neurologic disorders, respiratory diseases, and adverse birth outcomes. In addition, transiently colonized microbiota and horizontally acquired functions of the autochthonous microbiota affect the efficacy, permeability, stability, and bioavailability of therapeutics and pose an additional burden to healthcare management.

We abstracted this volume for providing concise and updated information on human microbiome–associated communicable and noncommunicable diseases. In this volume, we have included cancer, metabolic diseases, and nonalcoholic fatty liver disease. We have also included a chapter covering the role of the gut microbiome in the emergence of drug-resistant bacterial pathogens. The concluding chapter covers our recent discovery of the role of the vaginal microbiome in preterm birth delivery.

An improved understanding of causality, mechanistic of microbiome-associated disease, and disease-specific microbial taxa or functions will help in diagnostics and therapeutics discovery and preventive strategies. We believe that this volume will be an excellent primer in which scientific knowledge would grow and widen in the field of microbiome biology in health and disease. We hope that the volume will appeal to a wide readership from research scientists, clinicians, pharmacologists, and students.

Bhabatosh Das
Vijai Singh

CHAPTER ONE

Gut microbiome in the emergence of antibiotic-resistant bacterial pathogens

Deepjyoti Paul and Bhabatosh Das*
Molecular Genetics Laboratory, Infection and Immunology Division, Translational Health Science and Technology Institute, NCR Biotech Science Cluster, Faridabad, India
*Corresponding author: e-mail address: bhabatosh@thsti.res.in

Contents

Abstract

The human gastrointestinal tract is home to a complex and dynamic community of microorganisms known as gut microbiota, which provide the host with important metabolic, signaling, and immunomodulatory functions. Both the commensal and pathogenic members of the gut microbiome serve as reservoirs of antimicrobial-resistance genes (ARG), which can cause potential health threats to the host and can transfer

Progress in Molecular Biology and Translational Science, Volume 192
ISSN 1877-1173
https://doi.org/10.1016/bs.pmbts.2022.07.009

the ARGs to the susceptible microbes and into the environment. Antimicrobial resistance is becoming a major burden on human health and is widely recognized as a global challenge. The diversity and abundance of ARGs in the gut microbiome are variable and depend on the exposure to healthcare-associated antibiotics, usage of antibiotics in veterinary and agriculture, and the migration of the population. The transfer frequency of the ARGs through horizontal gene transfer (HGT) with the help of mobile genetic elements (MGEs) like plasmids, transposons, or phages is much higher among bacteria living in the GI tract compared to other microbial ecosystems. HGT in gut bacteria is facilitated through multiple gene transfer mechanisms, including transformation, conjugation, transduction, and vesicle fusion. It is the need of the hour to implement strict policies to limit indiscriminate antibiotic usage when needed. Developing rapid diagnostic tests for resistance determination and alternatives to antibiotics like vaccination, probiotics, and bacteriophage therapy should have the highest priority in the research and development sectors. Collective actions for sustainable development against resistant pathogens by promoting endogenous gut microbial growth and diversity through interdisciplinary research and findings are key to overcoming the current antimicrobial resistance crisis.

Abbreviations

AMR antimicrobial resistance
ARDB antibiotic resistance gene database
ARGO antibiotic resistance gene online
ARGs antimicrobial-resistance genes
CADRD comprehensive antibiotic resistance database
ESBLs extended-spectrum beta-lactamases
HGT horizontal gene transfer
MDR multidrug resistance
MGEs mobile genetic elements

1. Introduction

The human microbiome is considered a complex ecosystem of microbial communities where multiple organisms comprising bacteria, archaea, viruses, and protists reside mostly on the environmentally exposed surfaces of the human body. Microbial communities in the human body are more dynamic and diverse, and the interactions among the microbes may be symbiotic or pathogenic, as observed in healthy individuals and patients suffering from microbiome associated health disorders. The balance among these microorganisms within the microbiome is complicated and is continually developing defense mechanisms against each other, leading to a real "arms race."[1] The term "microbiome" was originally described as the ecological community of commensal, symbiotic, and pathogenic microorganisms harboring within the body.[2,3] Microorganisms colonize various sites in the human body, including the skin, mucosa, respiratory tract, urogenital tract, mammary gland, and gastrointestinal tract, and this microbial group is important in a variety of activities that keep the host healthy. The study of the microbiome is considered to be a very broad area, and the microbes usually differ based on different body sites such as skin, gut, or genital microbiomes, and here in this chapter, we will discuss the gut microbiome as well as the factors and parameters that influence the colonization and dissemination of multi-drug resistant bacteria in the human intestinal microbiome.

The largest collection of the microbial community is observed in the human gut, known as the gut microbiota, and the microbes within the gut play a key role in maintaining and sustaining the health of humans. The human gut microbiome comprises of various networks of microorganisms, inclusively known as microbiota, which play a supreme role in maintaining host well-being by affecting gut maturation, microbial resistance, nutrition, and also in causing diseases. The human gastrointestinal tract includes the stomach, small intestine, cecum, large intestine, and rectum, and the environmental conditions like pH and oxygen concentration vary along the tract.[4] Within the gastrointestinal tract, the large intestine carries the highest microbial load compared to the small intestine, where the pH is low. The gut microbiota exhibits dissimilarity in the distribution and heterogeneity of microbiota according to physiological conditions of the microbiome. The gut microbiota is beneficial for the host in several ways, like the healthy growth of the intestinal tract, supplying crucial nutrients, synthesizing vitamins, helping in the digestion of undigested foods, and also facilitating the growth of the nervous system. Apart from this, the intestinal microbiota also plays a significant role in protecting the host against pathogenic microorganisms by preventing the microbes from invading the gastrointestinal tract through a complex set of events, viz. colonization resistance. A healthy microbiome has intense effects on the development of gut-associated lymphoid tissue, understanding the variation of gut immune cells, and production of various immune mediators like IgA and microbial defense peptides.[5] However, in some adverse situations, the gut microbiota becomes compromised and no longer provides protection against pathogenic bacteria. As a result, the pathogens get colonized and start proliferating in the gastrointestinal tract, and thus it serves as an imperative reservoir for various groups of bacterial pathogens. These pathogens can cause many infections in healthy and immunocompromised patients, which may be responsible for the wider spread of resistance determinants.

The gut microbiome is referred to as a "metabolically active organ" as it is linked to multiple functions.[6,7] A number of projects, such as the Human Microbiome Project, My New Gut, and Meta-HIT, have been undertaken to study and understand the functional potential of the gut microbiome and also to search for conceivable strategies to help the host through the modification of the gut microbiome.[8] Microbes can break down the nutrients that are not accessible to the host, and in exchange, the host provides the raw materials and shelter or protection to the microbiome. The gut microbiome is unique for every individual and also very dynamic, and the changes in the microbiota associated with various health issues are called dysbiosis.[9] The human microbiome project has demonstrated that the changes in the immune environment may affect the dysbiotic flora of the gut. This perturbation in the microbiota is linked to several life-threatening conditions such as cancer, cardiovascular disease, inflammatory bowel disease, vaginosis, obesity, and infections caused by resistant bacteria.[10] Also, recent studies have illustrated the influence of intestinal microbes on host energy metabolism, intestinal epithelial proliferation, and immune responses. The indigenous microbiota within an individual develops immediately after the rupture of amniotic membranes, and subsequently, the bacterial load and diversity increase as the host matures. During the first days of life, bacteria started colonizing in the infant's gut from the mother's birth canal, the surrounding environment, or from other persons handling the infant. Normally borne infants are colonized first by maternal fecal and vaginal bacteria, whereas surgically delivered babies are primarily exposed to bacteria from the hospital setting or surrounding environment, as well as healthcare workers.[11,12] Hence, multiple features like the environment during birth, prematurity, and hygiene are very significant and influence the neonatal gut microbiota.

2. Community structure of gut-microbiota

The human microbiome consists of approximately 40 trillion bacterial cells and the intestinal microbiota institutes a diverse ecosystem that comprises thousands of different microbial species and many of them are contributing to several functions to the hosts.[13,14] The diversity of the gut microbiome differs from one individual's microbiota to another but a major population of healthy adults share a "core microbiota" in the gut and most of the microbiota primarily belong to four phyla viz. Bacteroidetes, Firmicutes, Actinobacteria, and Proteobacteria. Among them, the predominant phyla present within the normal human gut microbiome includes Bacteroidetes and Firmicutes which are found in maximum number in the colon, followed by *Actinobacteria* and *Proteobacteria* phyla in low abundance.[15,16] The predominant microbial genera in the gut microbiome include *Bifidobacterium*, *Streptococcus*, *Enterococcus*, *Clostridium*, and *Lactobacillus* which can be identified in stool whereas *Clostridium*, *Lactobacillus*, and *Enterococcus* are predominantly mucus associated isolates and are identified in the mucus layer of the small intestine. The development of gut microbiota start immediately after a few hours of birth and the composition of microbiota differs in normally delivered infants to the infants delivered by caesarean section. The gastrointestinal microbiota of vaginally delivered infants contains a higher percentage of lactobacilli and the majority of their fecal microbiota is alike to their mothers in comparison to the C-section delivered infants. The diversity of gut microbiota in the initial year of lifespan is mainly limited to Actinobacteria and Proteobacteria phyla whereas Firmicutes and Bacteroidetes but the majority of the gut inhabitants when the child reaches the age of 3–4 years. After the early years, the structure and composition of microbiota bear a resemblance to adult microbiota which remains constant but depends on the factors like age, lifestyle, socio-economic cultural factors, environment, dietary habits, illness or use of probiotics, prebiotics, and antibiotics, the composition of an individual's microbiota may change.

In the human gut microbiome, the Firmicutes to Bacteroides ratio is considered to be significant as high ratio of Firmicutes and low Bacteroides usually correlate with a healthy diverse microbiome and reflect a largely plant-based diet whereas the reverse is considered as an unhealthy microbiome. The change in the structure and composition of the microbiome has a potential impact in the individual's health and happiness. A balanced microbiota has a significant role in human health, however, an alteration in the human microbiota also plays a pivotal role in causing an extensive range of disorders.

2.1 Phyla bacteroidetes

This phylum comprises Gram-negative aerobic and anaerobic bacteria colonizing different parts of the intestine. Among them, the Bacteroides are found to be the most predominant group and is highly beneficial to the human host as they can digest those complex polysaccharides that remain unaffected to human digestive enzymes and also contributes to other important metabolic functions. But surprisingly, these microorganisms maintain this beneficial relationship until they are retained in the intestinal lumen of the host, whereas few members can become pathogenic if they disperse or disseminate from the original place. The Enterotoxigenic *B. fragilis* is known to release a toxin *B. fragilis* toxin (BFT) which can cause colitis and also been linked with inducing colon tumorigenesis.[17,18] Apart from this, the *Bacteroides* spp. has been found to carry multiple resistance mechanisms showing resistance towards many antibiotics.

2.2 Phyla Firmicutes

Most members of this phyla belong to Gram-positive anaerobic bacteria comprised of *Lactobacillus*, *Bacillus*, *Clostridium*, and Enterococcus genera. Several clusters of *Clostridium* (XIV and IV) are found inhabiting the intestinal tract and recognized as one of the predominant groups of bacteria in this phylum. This group of bacteria has been implicating several beneficial roles like fostering the host immune homeostasis and helping in maintaining the intestinal epithelial cells by colonizing between mucosal folds and also the release of butyrate as an end-product of fermentation promotes intestinal epithelial health.[4] Apart from the beneficial part, the members of clostridia such as *C. perfringens*, *C. tetani*, and *C. difficile* are also important human pathogens causing multiple infections. Bacilli, also belong to phyla Firmicutes, and comprises clinically significant and pathogenic bacteria *Enterococcus* and *Streptococcus* spp. which are normally found in less numbers in the intestine, however, *Lactobacillus* spp. which are facultative anaerobes, are known to be the permanent residents of the human gut.

2.3 Phyla actinobacteria

This phylum embraces aerobic and anaerobic Gram-positive bacteria and among them, *Bifidobacteria* spp. is the predominant bacteria residing in the intestinal tract.[19] This group of bacteria has many beneficial effects like *B. longum*, known to have probiotic effects and also can perform many essential functions such as digestion and staving off harmful bacteria which normally live in the intestines. In the absence of this bacteria, members of Proteobacteriaceae, Bacteroidaceae, Staphylococcaceae, Clostridiaceae are known to be the main reservoirs of clinically significant resistant genes that predominate in the breastfed infant gut. This indicates that the existence of Bifidobacteria in the human intestine reduces the abundance and rate of antimicrobial resistant genes.

2.4 Phyla proteobacteria

This phylum includes a number of facultative anaerobic Gram-negative bacteria unlike most of the obligate anaerobic microbes in the gastro-intestinal tract. The increasing prevalence of proteobacteria in a microbial community may be an indication of dysbiosis and encourage the threat of infection.[20] This dysbiosis increases potential enterobacterial pathogens like *Escherichia coli*, *Klebsiella* spp., and *Serratia* that are normally less in numbers and a majority of these organism are linked with gastrointestinal infections and carry specific adhesins, which aid their adhesion to the intestinal mucosa.[21,22]

3. Gut microbiome is a potential reservoir of antibiotic-resistant genes

Antimicrobial resistance (AMR) is an imperative public health threat that can tremble the foundation of the basic health care system and also add a major cost to the country's economic condition. It has been stated that millions of people develop antibiotic-resistant infections each year and the majority of the bacteria accountable for healthcare-associated infections are either resistant to first-line therapy or at least one of the antibiotics used worldwide.[23] The gut microbiome is known to have a significant role in the host's fitness and well-being, but it is also an important source of many antibiotic-resistant genes. The gut microbiome is a significant reservoir of multi-drug resistant bacteria, which can cause potential health threats to the host and can indirectly transfer ARGs into the environment.[24] In comparison to other environments, the human gut microbiome harbors a significantly greater number of ARGs. The MDR determinants can integrate with the gut microbiome through two different mechanisms, i.e., either the exogenous MDR bacteria can be acquired by the host followed by the colonization of the intestinal epithelium or the existing susceptible bacteria may show a resistant phenotype through the acquisition of AMR genes due to the selection pressure of antibiotics or by horizontal transfer of genes.[24] The diversity and abundance of ARGs in the gut microbiome vary depending on antibiotic exposure, population density in a country or region, and antibiotic use in animal feed. However, the understanding of the influence of antibiotic exposure on the stability of the microbiome as well as the progression of antibiotic resistance is important. The application of antibiotics has a direct link to the occurrence of a higher number of ARGs, and a previous study revealed that the resistant genes in the gut microbiome are less influenced by factors like age, gender, height, or weight ratio compared to usage of antibiotics or consumption of antibiotic-treated foods.[25,26] The collection of resistance genes within an individual is known as a resistome, and they persist for a minimum of 1 year at the individual level. This gut resistome is very diverse among the culturable and non-culturable microbes, and the ARGs within the resistome not only vary genetically but also in abundance within the ARG pools.[24]

As the human gut microbiota is rich in diverse bacterial species and home to numerous commensals, the imbalance caused by antibiotics leads to the enrichment of resistant organisms in the microbiome. These commensals can acquire ARGs from resistant microbes in the gut and can cause many serious infections. For instance, a report by[27] mentioned that a high abundance of ARGs showed non-susceptibility towards the drugs that have been used for both humans and animals for a prolonged time.[27] It is also suggested in this study, that the majority of ARGs are correlated with the practice of antibiotics in different nations, and it is also confirmed that increased exposure to antibiotics leads to the acquisition of resistance genes by the gut microbiota. ARGs are more likely to transfer and spread from one ecological niche and habitat to another, and especially the isolates originating from farm animals, were found to share ARGs with human isolates. The relatedness among different microbiomes, viz. human, animal, and environmental, is an essential factor for the selection and spread of antibiotic resistance.

Surprisingly, not only do adults have ARGs in their gut microbiota, but the elderly, children, and infants also have a resistance reservoir within their gut. The infant gut microbiota is very active and the development of this susceptible microbiota depends upon several factors, like host genetic makeup, nutrition, and environment. At this stage, dysbiosis of the gut microbiota could have significant effects on the metabolic and immune systems. The gut microbiota of a new-born baby harbors a diverse range of resistance genes even without any antibiotic treatment. However, antibiotic treatments facilitate the increase in the abundance of pathogenic enterobacterial isolates and lower the number of healthy microbiota like *Bifidobacteriaceae* and *Lactobacillales* spp., which are the essential constituents for the maturation and growth of the infant gut microbiome and are known to originate from the maternal microbiome.[25,28]

The usage of prophylactic antibiotics during pregnancy to stop streptococcal infections in the new-born has a potential impact on the microbiome of the new-born baby and affects not only the taxonomy of microbiota but also the ARG content at the beginning of the gut microbiome. The ARGs harboring within maternal gut microbiomes are transferred to the new-born during or shortly after birth, and these ARGs can easily be disseminated between commensal and pathogenic bacteria among individuals. As per a recent report by Gosalbes et al.[29], several ARGs have been identified in meconium, and these ARGs are not only related to the individual who has taken antibiotic treatments, but their presence has also been observed in human populations who have never taken any antibiotics. It suggests that a resistance gene can stably persist in the human gut without the presence of any antibiotic pressure and, similarly, in many natural environments, with a minimum exposure to antibiotics. The maternal gut microbiome can be conveyed to the baby, so the resistance may also be vertically inherited to the infant from the mother before the birth of the child. Multiple research studies have identified the presence of shared resistance determinants in the fecal samples of mother and offspring and they have also been observed in meconium or colostrum.[30,31] Apart from the shared ones, other ARGs found in infants are absent in mothers and are most probably acquired from the hospital environment or from other sources.

However, the occurrence of ARGs in the gut microbiota of human populations staying in remote areas may be due to resistant genes being ancestral even prior to the widespread distribution of resistance due to exaggerated use of antibiotics, or they might have been horizontally acquired from a resistant pathogen from different locations. Similar kinds of resistance genes were observed in the gut microbiome as well as among the pathogenic strains causing disease in humans, which indicates that the resistance determinants in the gut microbiota can be horizontally disseminated from one organism to another and is a serious matter of concern.

4. Microbiome: Accumulated effects of antibiotic exposure

Antibiotics are powerful medicines used for the prophylactic and curative treatment for certain life-threatening bacterial infections and have the ability to save lives against many severe infections. But, the prolonged application of some antibiotics leads to unexpected consequences in the intestinal microbiota as they can alter the proportion of microbes that can cause various physiological effects in absorption of nutrients. Further, absorption of antibiotics in the intestinal lumen depends on the specific transport mechanism, as well as the integrity of the intestinal membrane.

Also, the consequences of antibiotics on intestinal microbiota depends on the members of the microbiota, intestinal absorption of antibiotics as well as the specific properties of the antibiotics such as the spectrum of activity, class, potency, regimen, and administration route. The consumption of antibiotics may contribute to alterations in the host's indigenous microbiota or it may cause dysbiosis, i.e., interference in composition and function. The perturbation of microbiota due to the consumption of antibiotics can change the gut microbial composition, eventually leading to the loss of colonization resistance and as result, the host becomes susceptible to colonization by pathogens. This is a serious concern as many nosocomial infections emerges from gastrointestinal colonization.

It has been reported that antibiotics having higher spectrum of activity can affect the gut microbiota, and may substantially reduce bacterial diversity, and evenness. Gut microbiota alterations may result in severe health problems such as increased susceptibility to intestinal infections, or antibiotic-associated diarrhoeas mainly caused by nosocomial pathogens viz. *K. pneumoniae, Staphylococcus aureus,* and *Clostridium difficile.* The dysbiosis of gut microbiota may also increase the risk for infections and may affect basic immune homeostasis, atopic, inflammatory, and autoimmune diseases. Further, the unwarranted usage of antibiotics leads to the expansion of multidrug resistant bacteria within commensals. The early exposure to antibiotics affects the gut microbiota of an infant as it reduces the microbial diversity of the infant's microbiota and alters its composition by increasing the number of proteobacteria and depleting *Bifidobacterium* in the gut. A similar alteration in the gut microbiota of the infants is observed, whose mothers received antibiotics during pregnancy or delivery. Franzosa et al., in the year 2015 mentioned that the use of antibiotics also affects the gene expression, protein activity, metabolism, and physiological state of the gut microbiota apart from altering the bacterial taxonomic composition.[32]

Several factors such as the indiscriminate use of antibiotics, misapplication or inappropriate use of antibiotics may facilitate the horizontal transfer of resistant genes.[33] The two important concepts designed to understand the influence of a drug on the gut microbiota are overall fecal microbiota structure, i.e., α-diversity, and compositional dissimilarity or β-diversity. The interpretation of perturbation of microorganisms within the gut of an active person may be complex due to the lack of a well-defined healthy or abnormal fecal microbial community.[34]

The implementation of antimicrobial stewardship and necessary consumption of antibiotics may significantly limit the effect on the gut microbiota. The impact of antibiotics in the gut microbiome mainly depends on the class and spectrum of the antibiotic. β-Lactams are the most commonly prescribed antibiotics and the excessive use of antibiotics in clinical practice may contribute to the human gut resistome and also aid in the transfer of AMR genes. There is a direct relation between reduction in broad-spectrum antibiotic consumption and decrease in AMR. However, total knock-down of antibiotics for infected or hospitalized patients may increase the risk of treatment failure, hence judicious use of antibiotics is very important along with revision of national antimicrobial policies.

Antibiotic use has various effects on gut microbiota and among them, the most significant one is that it reduces the abundance of gut microbiota and on the other side increases resistance genes. As per the earlier studies, macrolides administration results in the decrease in *Actinobacteria* with an increase in Bacteroides and Proteobacterial population,[35,36] whereas the administration of clindamycin reduces the relative abundance of anaerobic bacteria and in contrast increases the abundance of Enterobacteriaceae in infected adults and also, increases efflux pump expression. A low dosage of antibiotics from food and the environment has also been associated with gut dysbiosis which negatively influence the state of the health of the host.

5. Factors affecting gut resistome and spread of ARGs

It is necessary to recognize the factors influencing the dissemination of resistant genes in the gut resistome to combat and tackle the problem. The diversity of gut microbiota and the type of resistance gene varies as per the lifestyle, travel, and location of people, diet, health status, and environment.

5.1 Application of antibiotics in farm animals

Antibiotics are often administered to the farming animals and this may enrich the ARGs in the gut microbiome of animals and as a result number of resistance genes targeting efflux pumps, coding antibiotic degrading enzymes have been detected in animal farms. Tetracycline antibiotic is highly used in farm animals and also the tetracycline resistance gene was predominantly reported in large-scale human gut microbiome analysis which may be due to the possible transfer of this resistant mechanism from animal to human.[37] The usage of meat from these animals for human consumption may result in exposure of antibiotics to the gut microbiota and also eventually leading to increasing selective pressure of antibiotics in the environment. Thus, the intestinal microbiota of farming animals may act as a reservoir of resistance gene which can be acquired by both bacterial and human foodborne pathogens. The use of antibiotics in livestock has a consequential impact on the emergence of resistance genes and as a result, it was found that the higher abundance of different resistance genes in fecal samples encoding higher resistance to those antibiotics which are used in clinical practice for a long time. The association between farm animals and soil also enhances the probability of genetic interchange followed by transfer of novel ARGs to humans via direct contact or by intake of animal-related foods.

In the Indian agricultural system, animal husbandry plays significant importance in covering the world's livestock population. The spread of AMR genes between animals and humans is facilitated mainly due to the random use of antibiotics in the livestock farming and food manufacturers. As per a report by,[33] the use of antibiotics in poultry animals enriches the human gut resistome more than due to the antibiotic use in clinical practice. Further, the study states that the antibiotics excreted from animal's guts also leads to the development of the resistome of the human gut and the environment. Hence, antibiotics use in veterinary hospital is considered to be one of the significant reasons for contributing the resistance burden in humans. The increased use of antibiotics in animals is may be due to the limited access to veterinary services in the rural areas and also the effortless availability of antibiotics through over-the-counter.

5.2 Diet and its consequence on resistome

Specific diet of an individual may configure the diversity of bacteria in gastrointestinal tract of humans and a difference in food may alter the pattern of gut microbiota. The gut microbiota also differs based on the food habit of the people of different continents and consumption of contaminated or uncooked foods may cause several disorders in the gut microbiome. The acquisition of resistant genes can make a susceptible commensal phenotype to a drug-resistant strain. The application of chemicals, metals or biocides in agriculture or foods produces a concoction of selective pressures which may induce the transfer and acquisition of resistance determinants. Hence, to understand the spread of new AMR genes, we must keep track to the gut microbe. The consumption of raw foods like milk or fish are potential risk factors as they can carry resistant bacteria. Surprisingly, the colistin-resistant gene *mcr-1* was first described in an infected patient from China and sequencing analysis revealed that the dissemination of this *mcr-1* gene through the food chain.[38,39] Gut resistome may be reduced using certain dietary interference such as the use of whole grains, foods, and prebiotics as they alter the fermentation pathways from protein to carbohydrate. As a consequence, the bacteria with increased capability to utilize carbohydrates like *Bifidobacterium* are enriched, and minimize the burden of AMR genes among bacteria.[40] The gastrointestinal microbiota of a new-born also depends on the diet and the gut flora of breast-fed infants carry increased number of *Bifidobacterium* whereas infants receiving artificial alimentation carry a minimal number of *Bifidobacterium* and lower microbial diversity.

5.3 AMR genes in waste and effluent

The AMR may also develop from the selective pressure generated through the hospital and urban household effluent, antibiotic contaminated water, or wastewater treatment plants which are known to be the significant reservoirs of MDR bacteria. These MDR organisms show resistance to different groups of antibiotics like β-lactams, quinolones, and aminoglycosides, and also aid in the dissemination of resistance genes to human populations. The antibiotic residues released from various research centers and drug manufacturers can also contribute to the emergence of AMR in the environment and eventually, these may gain access to our system.[41] Hence, proper sanitization protocols should be maintained for these waste effluents to minimize the AMR loads for the benefit public health.

5.4 Tourism and migratory birds

The increase in frequency and ease of travel have also accounted for the quick proliferation of resistance determinants from local to global environments. The transfer of resistance genes may also mediate through the gut microbiome among different places of a country or across the countries as in many cases it has been found that the AMR pattern increases after traveling to a new place even when not exposed to any antibiotics. Similarly, another study reported the acquisition of $bla_{\mathrm{CTX\text{-}M}}$ and mutation in GyrA increases the level of cephalosporin and quinolone resistance in the gut resistome of healthy travelers.[42–44] In case of the migratory birds, flying long distances may affect the dissemination of ARGs from one region to another.

6. Gut microbiome: A well-known transporter of antibiotic resistance gene

The gut resistome is considered the large reservoir of antibiotic-resistant genes among pathogenic and non-pathogenic bacteria found in the gut microbiome. The dissemination of antimicrobial-resistant bacterial pathogens is a critical concern and widely recognized as a global challenge. The anatomy of the gastrointestinal tract allows the accommodation of a diverse bacterial flora, serving as a reservoir of AMR genes originating from several sources, such as the food chain, soil, water, animals, and humans.[45] This diversity allows the transfer of genes among the bacterial strains and is also facilitated by the gastrointestinal environment, including temperature, the presence of biofilms and the permanent flow of nutrients.

This genetic interchange of resistance determinants through horizontal gene transfer among bacteria in the human gastrointestinal microbiome is much higher than among bacteria in other microbial ecosystems. The types of AMR genes found in the gut microbiome are either intrinsic or mobile resistance genes. The intrinsic genes are inherited, non-mobile, and exhibit resistance to a specific group of drugs without any prior exposure, but may occasionally get captured into MGEs and turn into mobile-resistant genes. However, the mobile resistomes are encoded in mobile genetic elements (MGEs) like plasmids, transposons, integrons, genomic islands, and phages, and the propagation of resistance genes occurs mainly through these MGEs (Table 1). They play a remarkable role in the horizontal transfer of antibiotic-resistant genes and are increasingly found among Gram-negative bacteria of clinical background. The selective pressure generated due to the higher usage of antibiotics, exposure of disinfectants, and heavy metals may facilitate the acquisition of these genetic elements. MGEs are transferred among related or unrelated bacteria from human enteric pathogens to gut bacteria through the horizontal gene transfer mechanisms including transformation, conjugation, and transduction (Fig. 1). Among them, transduction is considered the crucial vehicle for the spread of resistant genes in the gut microbiome, which may be due to the presence of an equivalent number of bacteria and phages residing in the intestinal tract. MGEs and horizontal transfer play a notable role in the emergence of antibiotic resistance and are found to be more common among the bacterial isolates associated with the gut. This may be due to the compact environment in the gut microbiome, which provides ideal conditions for the exchange of resistant determinants by horizontal gene transfer between transient and residential microbes. The transfer of resistance determinants of similar genetic information has been documented among distantly related organisms in the gut microbiota, and due to this, the gut microbiota is not only known as a reservoir of multiple ARGs but also a habitat where these resistant determinants can disseminate among different species levels. Recent studies have demonstrated the

Table 1 Association of mobile genetic elements with antibiotic resistance genes.

Sl. No.	Antibiotics	Class	Resistance mechanism	Resistance genes	Associated with MGEs	References
1.	Penicillin Cephalosporin Carbapenems Aztreonam	β-Lactam	Extended spectrum β-lactamases (ESBLs)	bla_{CTX-M}	IS*Ecp1*, IS*903B*, IncF, IncX, IncH, IncN, IncR, IncP, ΔIS6, IS*CR1*	46
				bla_{TEM}	Tn1, Tn2, Tn3, Tn801	47,48
				bla_{SHV}	IS*26*	49
				bla_{PER}	Tn1213, IS*CR*	50
				bla_{VEB}	intI1, IS*CR*	51
				bla_{GES}	int1, int3	50,52,53
			ampC β-lactamase	bla_{DHA}	IS*26*, IS3E, Inc. A/C	54
				bla_{ACC}	Inc N, IS*26*,	54
				bla_{CMY}	ISEcp1,	54
			Carbapenemases	bla_{KPC}	ISCR2, ISCR3, IS26, Tn3, Tn1721, Tn4401, Tn5393, IncF, I, A/C, N, X, R, P, U, W, W, L/M and ColE plasmids	55–58
				bla_{IMP}	int1, int3, Tn21, Tn5051, Inc. A/C, IncHI2, Inc. L/M, IncU plasmids	55–58

Continued

Table 1 Association of mobile genetic elements with antibiotic resistance genes.—cont'd

Sl. No.	Antibiotics	Class	Resistance mechanism	Resistance genes	Associated with MGEs	References
				bla_{NDM}	int1, ISAba125, IS26, Tn3, Tn125, IS5 Tn300, IncF, IncFII, IncR, IncX3, Inc. A/C, Inc. K and unknown plasmids	58–64
				bla_{OXA}	IS1999, ISAba1, ISAba2, ISAba3, Tn2006	65–68
				bla_{VIM}	int1, Tn402, Tn5090 and Inc. A/C, IncFI/II, Inc. HI2, IncI1, Inc. L/M, IncN, and IncW plasmids	49,59,60,69,68
2.	Amikacin Gentamicin Tobramycin Neomycin Streptomycin	Aminoglycosides	16S ribosomal RNA methyltransferases	*rmtA*	IS*6100*, Tn*4051*	70,71
				rmtB	Tn*3*	72,73
				rmtC	IS*Ecp1*	74
				rmtD	IS*CR*	73
				armA	IS*26*, Tn*1548*	75,76
			aminoglycoside-modifying enzymes (AMEs)	*aac,*[1] *aac,*[2] *aac,*[3] *aac*[6]	IS1247, ISKpn23, IS1071, Tn3	77,78
				ant,[2] *ant,*[3] *ant(4)*, *ant,*[6] & *ant*[9]	IS*CR6*	78
				aphA1	IS*26*, Tn*4352*, Tn*6020*, IS*50* Tn*5*, IS*903*, Tn*903*, IS*Aba14*	78

3.	Chloramphenicol	Chloramphenicol	Type A chloramphenicol acetyltransferases	*catA1*	Tn*9*	79
				catA2	pRI234, pMR375	79
				catA3	pMVSCS1	80
				catP	pIP401:Tn4451	81
			Type B chloramphenicol acetyltransferases	*catB1*	–	–
				catB2	pNR79:Tn2424, pSp39	79,82
				catB3	pWBH301, IncF1 plasmid	79,83
				catB4	pWBH301, pEKP0787-1	84
				catB5	Tn840	79
4.	Vancomycin	Glycopeptides	Vancomycin resistance operon	*van*A	Tn1546, IncI8 plasmids	85–87
				*van*B	Tn1549, Tn5382-like conjugative transposons (ICEs), pCF10-like plasmid	86,88
				*van*G	ICE	85,89
				*van*M	plasmid	89
				*van*N	plasmid	85,89

Continued

Table 1 Association of mobile genetic elements with antibiotic resistance genes.—cont'd

Sl. No.	Antibiotics	Class	Resistance mechanism	Resistance genes	Associated with MGEs	References
5.	Sulfamethoxazole and trimethoprim	Sulfonamides		*sul1*	IncF, IS*6100*, IS*1006*, IncF, IncY, Inc. K	90
				sul2	IS1294, IS1S, IS26, Inc. F, Inc. K, Inc. I1	
				sul3	IS*1006*, IS*1182*, IS*1081*, IncF, IncFIB, IncK	
				dfrA		
				dfrB		
6.	Tetracycline Doxycycline Minocycline	Tetracyclines	Alteration of ribosomal conformation	*tet (B)*	IS*Vsa5*, IncFII	91
				tet (W)	IS5	92
7.	Azithromycin Clarithromycin Erythromycin	Macrolides		*erm* A, B, C, T, Y	genomic islands, transposons, Tn5253-like composite ICE, plasmids, phage (ΦJN4341-pro)	93–96
				msr	plasmids	96
				mef (A/E)/C	MEGA element, transposons Tn916 composite transposons, ICEst3, other ICEs and IncA/C plasmids	93,95,96
				mphG and *ere*	Plasmids	95,96

8.	Ciprofloxacin Levofloxacin Moxifloxacin Gemifloxacin	Fluroquinolones	Plasmid mediated quinolone resistance	*qnr* genes	SulA integron, IS2, IS26, ISEcI2, Tn*3*, Inc. A/C, H12, F, FII, L/M, N, R, Q, U, I1, X2, ColE1 plasmids & phages	97,47,98–101
				aac (6′)-1b-cr	Inc FII, IncR & IncN plasmids	47,98–100
				qepA	IS29, plasmids	47,98,100
				oqxAB	IS26, Tn3, IncX1, IncHI2, and IncF plasmids	101
9.	Polymyxin B Colistin	Polymyxins		*mcr* genes	IncP-1, IncFII, IncH, IncHI1, IncX4, IncI2 plasmids	102–104
				lpxA, *lpxC*, and *lpxD*	IS*Aba*11	105,106

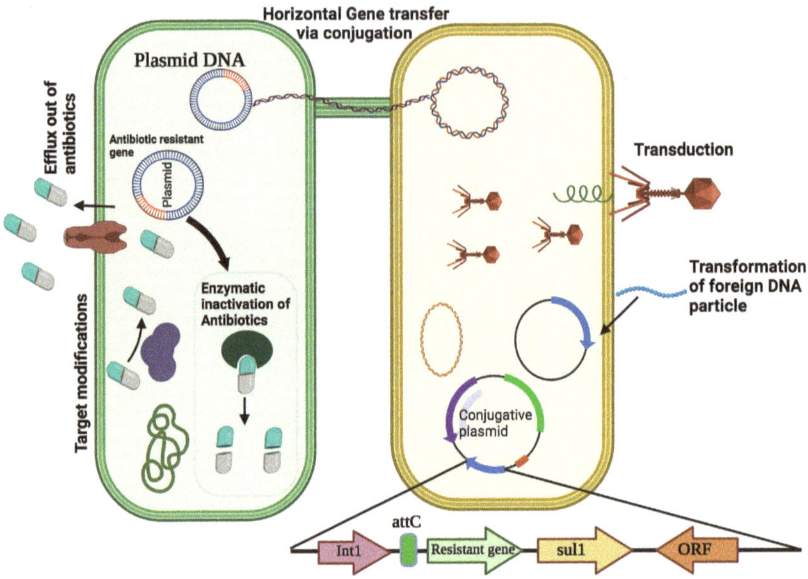

Fig. 1 Antibiotic resistance mechanisms and dissemination resistance genes between bacterial species. Different mobile genetic elements promote dissemination of resistance genes through horizontal gene transfer. The most frequent pathway of horizontal gene transfer between distantly related bacterial species is the transformation.

transfer of ARGs from Actinobacteria to pathogenic proteobacteria and also that the molecular cross-talk between the host and gut pathogens may increase the horizontal transfer of ARGs in the gut.[107] It has also been known that the physiological concentration of some hormones in the gut may increase the horizontal spread of conjugative plasmids between the organisms.[108]

Traveling to different places, regardless of destination, has been known to be an important factor in altering the bacterial resistance pattern, especially in the enterobacterial species residing within the human gut. Proteobacteria naturally exists as a minor fraction of the total human gut microbiota (less than 4%) but it has been found to increase and display higher abundance upon returning from travel, whereas other gut bacteria representing, Actinobacteria, Bacteroidetes, and Firmicutes, remain stable during the travel period. This alteration in the gut microbiome may be due to travel to different places and exposure to different bacterial strains of those regions, and also the changes in food habits contribute to the alteration in the gut microbiome. The enterobacterial strains harboring ESBLs have been reported among travelers around the globe, and the risk factors include the consumption of antibiotics or travel to a region with a high prevalence of antibiotic resistance. The bacterial community in the microbiota is considered a reservoir for ARGs, and a large part of it is associated with travel. The enterobacterial members known to be important human pathogens constitute not more than 1% of the human gut microbiota. The metagenomic study gives a new insight into the changes in the antibiotic resistance potential of the human gut microbiome and found that the core resistome of the high abundance genes remained stable and a fraction of the low abundance genes showed an increase in number during or after travel to a different place. In the case of the resistance pattern of the isolates, notable increases were observed in low abundance genes conferring resistance towards sulphonamides, trimethoprim, and β-lactam groups of antibiotics, leading to the increase in antibiotic resistance burden. Many studies have also found that there is an increase in core resistomes at the population level since the consumption of different antibiotics has also increased.[109,110]

In the gut microbiome, mainly two types of gut resistomes have been recognized, such as resident resistome and transitory resistome. The resident resistome is associated with commensal flora, whereas the transitory resistome is mainly linked with the bacteria that occasionally exist in the gut, which can easily transfer their resistance gene to other commensal flora residing in the same region. It has been noted that the ARGs encoding resistance to potential antibiotics in human pathogens are also found in the microbiome, which suggests that the microbiome is an adequate reservoir for antibiotic resistant bacterial pathogens. Under the selective pressure of antibiotics, bacteriophages, the resident virome, can be swapped between bacteria, and may also contribute to the horizontal transfer of resistance genes. Standard metagenomic and culturomics studies have identified various resistant genes that are predominantly found in the gut microbiome, and these genes encode resistance to antibiotics that are commonly used in clinical practice.[1] The resistance genes found are unique and differ from one microbiome to another. Specifically, the gut microbiota carries a diverse range of genes encoding resistance towards β-lactams, aminoglycosides, tetracyclines, and glycopeptides, followed by chloramphenicol and macrolides. The tetracycline resistance in the commensal flora of Firmicutes and Actinobacteria in the gut microbiota is mainly mediated by major-facilitator super-family (MFS) efflux transporters and ribosomal protection enzymes encoded by the tetQ or tetW genes.[111,112] Resistance to -lactams is primarily caused by the presence of class A, class C, and class D -lactamases enzymes, whereas resistance to vancomycin is primarily caused by the presence of the

VanG and other operon-carrying resistant genes.[24] In the gut microbiome, the acquisition of resistance genes occurs in the early life of a person, and the resistance gene present in a healthy individual appears soon after birth. Also, the types of resistance genes found are similar in the baby and the vagina or gut of the mother, which indicates the influence of the mother in the advent of the gut resistome of the infant.

7. Different approaches to study and understand human gut-resistome

The different methods used for the interpretation of AMR genes include the isolation and identification of resistant bacteria using culture- and non-culture-based approaches and each of the methods is appropriate and beneficial, but for extensive and accurate characterization of the human gut microbiome, an amalgamation of different methods is useful.

7.1 Culture-based approach

Culture-based approaches are the most commonly used method for analyzing resistant gut microbiota using AMR analysis after isolation and identification of specific microorganisms on selective media.[113,24] These culture-based studies are a very effective way to understand the organism's resistant profile and also it is very significant in understanding the relationship within the practice of antibiotics and the susceptibility pattern of gastrointestinal bacteria. The resistant bacteria residing in the gut can be isolated and identified by culturing the stool samples on chromogenic medium, followed by selective medium and antibiotic-containing medium. Apart from this, the resistant bacteria can also be identified by susceptibility testing using the Kirby-Bauer disc diffusion method or by determining the minimum inhibitory concentration of the isolate.[62,64] The limitation of this method is that it relies on the culturing of the organism and is unfeasible for many microbes and also the noble resistance determinants harboring within gut bacteria may be missed.

7.2 Molecular biology-based approach

The molecular approach has gained enormous interest in investigating the gut resistome and identifying the association of resistant microbiota in developing disease. The culture-independent, molecular approach has enabled the identification of many unidentified bacterial species of microbiota and provided novel insights in understanding the resistance pattern, diversity of resistance determinants in the gut microbiome.

7.2.1 Using conventional PCR-based method

This molecular-based method includes the characterization of resistance genes by amplifying them using a polymerase chain reaction followed by sequencing to identify the common resistance mechanisms present in the gut microbiome. The spread of ESBLs, carbapenemase, and other resistance determinants have been investigated among enteric Gram-negative rods by PCR and sequencing analysis.[114–116] These methods also identify the vehicles accountable for the spread of resistance determinants like mobile elements, integrons, and transposons and their function in the spread of resistant genes.[117]

7.2.2 Metagenomics based approach

Metagenomics is a modern method for analyzing the resistome in the gut microbiome, as well as the entire genome of an unknown microorganism or a community of microorganisms. This approach involves the sequencing of the bacterial genome and mapping to a reference database which can identify the presence or distribution of resistant genes within gut microbiome, understanding the mechanisms of acquiring ARGs, and also the dissemination pathways among diverse populations of gut microbiome. The recent development of nanopore technology has also been thoroughly used for analyzing gut resistome. The quick advancement of metagenomics studies through next-generation sequencing technologies has developed a significant amount of data related to the resistome which emerges the gut microbiome as a crucial reservoir for resistant genes. As many of the indigenous bacteria are non-culturable, the metagenomic approach is one of the advanced options to understand and characterize all the complexity including unculturable and rare taxa. The metagenomic approach uses several AMR gene databases which include the antibiotic resistance gene online (ARGO), antibiotic resistance gene database (ARDB), and another bioinformatics tool such as comprehensive antibiotic resistance database (CARD) which is very useful in rapid & easy identification of the existing and new AMR genes along the detailed information. The theme of the ResistoMap is to monitor levels of ARGs and also the comprehensive study of heavy metal resistance genes. This method is mainly based on three different metagenomic approaches such as targeted metagenomics, functional metagenomics, and sequence-based metagenomics.

Targeted metagenomics applies multiple feasible PCR-based methods to identify various resistance genes. This technique also uses the quantitative data generated from the real-time PCR to identify a significant number of genes. The cost-effective nature and high throughput analysis make it a significant tool for analyzing resistome.

Sequence-based metagenomics is another rapidly growing approach to characterize the human gut microbe. This method uses advanced DNA sequencing technologies of the bacterial genome for high throughput analysis of the resistance genes of the gut resistome. Also, it is helpful to characterize the AMR and virulence genes and also it provides significant information on the entire gene contents.

Functional metagenomics includes the cloning of a specific gene of interest into a plasmid vector and analyzing its transcriptional response in different hosts. This approach can identify novel drug resistance genes in the gut microbiota which will help to understand the scenario of resistance mechanism in the gut microbiome.[118] The Emulsion, Paired isolation and concatenation PCR shortly called EPIC-PCR is a technique that can be used to analyze the functional diversity of the gut microbiome.

8. Conclusion and future perspective

The gut microbiome is known to be an essential part of the human body and plays an important role in xenobiotic degradation along with other metabolic functions. The composition and function of the gut microbiome are dynamic and several factors, including diet, environmental exposure, lifestyle, and host genetics, influence their structure. Different members of the gut microbiota contribute differentially to the acquisition, maintenance, and transfer of drug resistance functions and act as a large reservoir of resistance genes in the human body. It has been seen that the wide application of antibiotics in hospital settings, animal husbandry, and the food industry has not only enriched the gut resistome but also caused extended persistence of ARGs in the gut microbiome. This urges the implementation of strict policies to maintain moderate antibiotic usage and also to develop strict approaches to limit the adverse effects of antibiotics on the gut microbiota. This approach, in conjunction with the ecological impact of antibiotics, could help to reduce the burden of ARGs and their transmission in the gut microbiota. The need of the hour is to focus and pay attention to alternative approaches such as vaccination, probiotics, and bacteriophage therapy to avoid the undesirable effects of antibiotics on the gut microbiota. The surge in ARGs indicates that the human gut resistome will be an interesting research area in the near future as further studies are required to comprehend the factors influencing the growth and maintenance of the gut resistome, the mechanisms behind the exchange of resistance genes across different gut bacterial taxa, as well as the impact of the gut microbiome dysbiosis on the community of indigenous microbiota. The in-depth research linking the gut microbiome and the emergence of multi-drug resistant pathogens is of utmost importance, and the key research areas like understanding the association of mobile genetic elements linked with AMR resistance determinants in gut microbiota as well as their carriage in both pathogens and commensals. Further, research work can also be undertaken to develop suitable metagenomic methods for the rapid identification of pathogens which are difficult to grow by traditional microbiological culture methods and also to explore their AMR profiles, which will help to decide early treatment options in clinical settings. The emergence and spread of AMR pose a serious global health threat. Hence, immediate and comprehensive action is required to implement the One Health concept to perform interdisciplinary research in all areas related to the healthcare system, veterinary, and the environment.

Acknowledgment

The authors would like to acknowledge the support of the Translational Health Science and Technology Institute and the Department of Biotechnology (DBT), Govt. of India. Dr. Deepjyoti Paul is the recipient of the DBT-MK Bhan fellowship 2020. The figure incorporated into this chapter was created using the BioRender software.

Conflict of interest

The authors declare that there is no commercial or financial conflict of interest.

References

1. Baron SA, Diene SM, Rolain JM. Human microbiomes and antibiotic resistance. *Hum Microbiome J.* 2018;10:43–52.
2. Berg G, Rybakova D, Fischer D, et al. Microbiome definition re-visited: old concepts and new challenges. *Microbiome.* 2020;8:103.
3. NIH HMP Working Group, Peterson J, Garges S, et al. The NIH human microbiome project. *Genome Res.* 2009;19:2317–2323.
4. Kim S, Covington A, Pamer EG. The intestinal microbiota: antibiotics, colonization resistance, and enteric pathogens. *Immunol Rev.* 2017;279:90–105.
5. Sommer F, Backhed F. The gut microbiota-masters of host development and physiology. *Nat Rev Microbiol.* 2013;11:227–238.
6. Hooper LV, Midtvedt T, Gordon JI. How host-microbial interactions shape the nutrient environment of the mammalian intestine. *Annu Review of Nutr.* 2002;22:283–307.
7. Qi X, Yun C, Pang Y, Qiao J. The impact of the gut microbiota on the reproductive and metabolic endocrine system. *Gut Microbes.* 2021;13:1–21.

8. Roeselers G, Bouwman J, Levin E. The human gut microbiome, diet, and health: "Post hoc non ergo propter hoc". *Trends Food Sci Technol*. 2016;57:302–305.
9. Vijay A, Valdes AM. Role of the gut microbiome in chronic diseases: a narrative review. *Eur J Clin Nutr*. 2022;76:489–501.
10. Patangia DV, Anthony Ryan C, Dempsey E, Paul Ross R, Stanton C. Impact of antibiotics on the human microbiome and consequences for host health. *Microbiology Open*. 2022;11, e1260.
11. Penders J, Thijs C, Vink C, et al. Factors influencing the composition of the intestinal microbiota in early infancy. *Pediatrics*. 2006;118:511–521.
12. Dominguez-Bello MG, Costello EK, Contreras M, et al. Delivery mode shapes the acquisition and structure of the initial microbiota across multiple body habitats in newborns. *Proc Natl Acad Sci U S A*. 2010;107:11971–11975.
13. Sender R, Fuchs S, Milo R. Are we really vastly outnumbered? Revisiting the ratio of bacterial to host cells in humans. *Cell*. 2016;164:337–340.
14. Konstantinidis T, Christina Tsigalou C, Karvelas A, et al. Effects of antibiotics upon the gut microbiome: a review of the literature. *Biomedicine*. 2020;8:502.
15. Almeida A, Mitchell AL, Boland M, et al. A new genomic blueprint of the human gut microbiota. *Nature*. 2019;568:499–504.
16. Forster SC, Kumar N, Anonye BO, et al. A human gut bacterial genome and culture collection for improved metagenomic analyses. *Nat Biotechnol*. 2019;37:186–192.
17. Wexler H. Bacteroides: the good, the bad, and the nitty-gritty. *Clin Microbiol Rev*. 2007;4:593–621.
18. Sears CL. Enterotoxigenic Bacteroides fragilis: a rogue among symbiotes. *Clin Microbiol Rev*. 2009;22:349–369.
19. Rajilic-Stojanovic M, de Vos WM. The first 1000 cultured species of the human gastrointestinal microbiota. *FEMS Microbiol Rev*. 2014;38:996–1047.
20. Shin NR, Whon TW, Bae JW. Proteobacteria: microbial signature of dysbiosis in gut microbiota. *Trends Biotechnol*. 2015;33:496–503.
21. Taur Y, Xavier JB, Lipuma L, et al. Intestinal domination and the risk of bacteremia in patients undergoing allogeneic hematopoietic stem cell transplantation. *Clin Infect Dis*. 2012;55:905–914.
22. Taur Y, Pamer EG. The intestinal microbiota and susceptibility to infection in immunocompromised patients. *Curr Opin Infect Dis*. 2013;26:332–337.
23. Zhang L, Kinkelaar D, Huang Y, et al. Acquired antibiotic resistance: are we born with it? *Appl Environ Microbiol*. 2011;77:7134–7141.
24. Bag S, Ghosh TS, Banerjee S, et al. Molecular insights into antimicrobial resistance traits of commensal human gut microbiota. *Microb Ecol*. 2019;77:546–557.
25. Makino H, Kushiro A, Ishikawa E, et al. Mother-to-infant transmission of intestinal bifidobacterial strains has an impact on the early development of vaginally delivered infant's microbiota. *Plos One*. 2013;8, e78331.
26. Imchen M, Kumavath R. Metagenomics of antimicrobial resistance in gut microbiome. In: Kumavath RN, ed. *Metagenomics for Gut Microbes*. London: IntechOpen; 2018.
27. Forslund K, Shinichi S, Kultima JR, et al. Country-specific antibiotic use practices impact the human gut resistome. *Genome Res*. 2013;23:1163–1169.
28. Palmer C, Bik EM, DiGiulio DB, et al. Development of the human infant intestinal microbiota. *PLoS Biol*. 2007;5:e177.
29. Gosalbes M, Valles Y, Jimenez-Hernandez N, et al. High frequencies of antibiotic resistance genes in infants' meconium and early fecal samples. *J Dev Orig Health Dis*. 2016;7:35–44.
30. Ferretti P, Pasolli E, Tett A, et al. Mother-to-infant microbial transmission from different body sites shapes the developing infant gut microbiome. *Cell Host Microbe*. 2018;24:133–145.

31. Yassour M, Jason E, Hogstrom LJ, et al. Strain-level analysis of mother-to-child bacterial transmission during the first few months of life. *Cell Host Microbe.* 2018;24:146–154.
32. Franzosa EA, Hsu T, Sirota-Madi A, et al. Sequencing and beyond: integrating molecular 'omics' for microbial community profiling. *Nat Rev Microbiol.* 2015;13:360–372.
33. Singh S, Verma N, Taneja N. The human gut resistome: current concepts & future prospects. *Indian J Med Res.* 2019;150:345–358.
34. Backhed F, Fraser CM, Ringel Y, et al. Defining a healthy human gut microbiome: current concepts, future directions, and clinical applications. *Cell Host Microbe.* 2012;12:611–622.
35. Shah T, Baloch Z, Shah Z, et al. The intestinal microbiota: impacts of antibiotics therapy, colonization resistance, and diseases. *Int J Mol Sci.* 2021;22:6597.
36. McDonnell L, Gilkes A, Ashworth M, et al. Association between antibiotics and gut microbiome dysbiosis in children: systematic review and meta-analysis. *Gut Microbes.* 2021;13:1–18.
37. Kyselkova M, Jirout J, Vrchotova N, et al. Spread of tetracycline resistance genes at a conventional dairy farm. *Front Microbiol.* 2015;6:536.
38. Liu YY, Wang Y, Walsh TR, et al. Emergence of plasmid-mediated colistin resistance mechanism MCR-1 in animals and human beings in China: a microbiological and molecular biological study. *Lancet Infect Dis.* 2016;16:161–168.
39. Hu Y, Liu F, Lin IYC, et al. Dissemination of the mcr-1 colistin resistance gene. *Lancet Infect Dis.* 2016;16:146–147.
40. Wu G, Zhang C, Wang J, et al. Diminution of the gut resistome after a gut microbiota-targeted dietary intervention in obese children. *Sci Rep.* 2016;6:24030.
41. Amador PP, Fernandes RM, Prudencio MC, et al. Antibiotic resistance in wastewater: occurrence and fate of Enterobacteriaceae producers of class A and class C β-lactamases. *J Environ Sci Health A Tox Hazard Subst Environ Eng.* 2015;50:26–39.
42. Leangapichart T, Dia NM, Olaitan AO, et al. Acquisition of extended-spectrum β-lactamases by *Escherichia coli* and *Klebsiella pneumoniae* in gut microbiota of pilgrims during the hajj pilgrimage of 2013. *Antimicrob Agents Chemother.* 2016;60:3222–3226.
43. Johnning A, Kristiansson E, Angelin M, et al. Quinolone resistance mutations in the faecal microbiota of Swedish travellers to India. *BMC Microbiol.* 2015;15:235.
44. von Wintersdorff CJH, Penders J, Stobberingh EE, et al. High rates of antimicrobial drug resistance gene acquisition after international travel, the Netherlands. *Emerg Infect Dis.* 2014;20:649–657.
45. Smillie CS, Smith MB, Friedman J, et al. Ecology drives a global network of gene exchange connecting the human microbiome. *Nature.* 2011;480:241–244.
46. Hounmanou YMG, Bortolaia V, Dang STT, et al. ESBL and AmpC β-lactamase encoding genes in E. coli from pig and pig farm workers in Vietnam and their association with Mobile genetic elements. *Front Microbiol.* 2021;12(629139).
47. Poirel L, Naas T, Nordmann P. Genetic support of extended-spectrum b lactamases. *Clin Microbiol Infect.* 2008;14:75–81.
48. Partridge SR, Hall RM. Evolution of transposons containing bla_{TEM} genes. *Antimicrob Agents Chemother.* 2005;49:1267–1268.
49. Miriagou V, Carattoli A, Tzelepi E, et al. IS26-associated In4-type integrons forming multiresistance loci in enterobacterial plasmids. *Antimicrob Agents Chemother.* 2005;49:3541–3543.
50. Poirel L, Cabanne L, Nordmann P. Genetic environment and expression of the extended-spectrum β-lactamase $bla_{PER\text{-}1}$ gene in gram-negatives. *Antimicrob Agents Chemother.* 2005;49:1708–1713.
51. Naas T, Mikami Y, Imai T, et al. Characterization of In53, a class 1 plasmid- and composite transposon-located integron of *Escherichia coli* which carries an unusual array of gene cassettes. *J Bacteriol.* 2001;183:235–249.

52. Dubois V, Poirel L, Marie C, et al. Molecular characterization of a novel class 1 integron containing *bla*$_{GES-1}$ and a fused product of aac3-Ib / aac6-Ib gene cassettes in *Pseudomonas aeruginosa. Antimicrob Agents Chemother.* 2002;46:638–645.
53. Correia M, Boavida F, Grosso F, et al. Molecular characterization of a new class 3 integron in *Klebsiella pneumoniae. Antimicrob Agents Chemother.* 2003;47:2838–2843.
54. Zamorano L, Miro E, Juan C, et al. Mobile genetic elements related to the diffusion of plasmid-mediated AmpC-lactamases or carbapenemases from Enterobacteriaceae: findings from a multicenter study in Spain. *Antimicrob Agents Chemother.* 2015;59: 5260–5266.
55. Mathers AJ, Cox HL, Kitchel B, et al. Molecular dissection of an outbreak of carbapenem-resistant enterobacteriaceae reveals intergenus KPC carbapenemase transmission through a promiscuous plasmid. *MBio.* 2011;2:e00204–e00211.
56. Chen L, Mathema B, Chavda KD, et al. Carbapenemase-producing Klebsiella pneumoniae: molecular and genetic decoding. *Trends Microbiol.* 2014;22:686–696.
57. Bi D, Jiang X, Sheng ZK, et al. Mapping the resistance-associated mobilome of a carbapenem-resistant *Klebsiella pneumoniae* strain reveals insights into factors shaping these regions and facilitates generation of a 'resistance-disarmed' model organism. *J Antimicrob Chemother.* 2015;70:2770–2774.
58. Luo Y, Luo R, Ding H, et al. Characterization of carbapenem-resistant *Escherichia coli* isolates through the whole genome sequencing analysis. *Microb Drug Resist.* 2018; 24:175–180.
59. Paul D, Maurya AP, Chanda DD, et al. Carriage of *bla*$_{NDM-1}$ in *Pseudomonas aeruginosa* through multiple Inc type plasmids in a tertiary referral hospital of northeast India. *Indian J Med Res.* 2016;143:826–829.
60. Paul D, Dhar D, Maurya AP, et al. Occurrence of co-existing *bla*$_{VIM-2}$ and *bla*$_{NDM-1}$ in clinical isolates of *Pseudomonas aeruginosa* from India. *Ann Clin Microbiol Antimicrob.* 2016;15:31.
61. Toleman MA, Spencer J, Jones L, et al. *bla*$_{NDM-1}$ is a chimera likely constructed in *Acinetobacter baumannii. Antimicrob Agents Chemother.* 2012;2773–2776.
62. Paul D, Bhattacharjee A, Ingti B, et al. Occurrence of *bla*$_{NDM-7}$ within IncX3-type plasmid of *Escherichia coli* from India. *J Infect Chemother.* 2017;23:206–210.
63. Potter RF, D'Souza AW, Dantas G. The rapid spread of carbapenem resistant Enterobacteriaceae. *Drug Resist Updat.* 2016;29:30–46.
64. Paul D, Bhattacharjee A, Bhattacharjee D, et al. Transcriptional analysis of *bla*$_{NDM-1}$ and copy number alteration under carbapenem stress. *Antimicrob Resist Infect Control.* 2017;6:26.
65. Poirel L, Nordmann P. Carbapenem resistance in *Acinetobacter baumannii*: mechanisms and epidemiology. *Clin Microbiol Infect.* 2006;12:826–836.
66. Walther-Rasmussen J, Hoiby N. OXA-type carbapenemases. *J Antimicrob Chemother.* 2006;57:373–383.
67. Miriagou V, Cornaglia G, Edelstein M, et al. Acquired carbapenemases in gram-negative bacterial pathogens: detection and surveillance issues. *Clin Microbiol Infect.* 2010;16:112–122.
68. Walsh TR. Emerging carbapenemases: a global perspective. *Int J Antimicrob Agents.* 2010;36:S8–14.
69. Ahmad N, Ali SM, Khan AU. Molecular characterization of novel sequence type of carbapenem-resistant NDM-1 producing *Klebsiella pneumoniae* in NICU of Indian hospital. *Int J Antimicrob Agents.* 2018;53:525–529.
70. Yokoyama K, Doi Y, Yamane K, et al. Acquisition of 16S rRNA methylase gene in *Pseudomonas aeruginosa. Lancet.* 2003;362(1888–93).
71. Yamane K, Wachino J, Doi Y, et al. Global spread of multiple aminoglycoside resistance genes. *Emerg Infect Dis.* 2005;11:951-953.

72. Yan JJ, Wu JJ, Ko WC, et al. Plasmid-mediated 16S rRNA methylases conferring high-level aminoglycoside resistance in *Escherichia coli* and *Klebsiella pneumoniae* isolates from two Taiwanese hospitals. *J Antimicrob Chemother.* 2004;54:1007–1012.
73. Doi Y, Arakawa Y. 16S Ribosomal RNA methylation: emerging resistance mechanism against aminoglycosides. *Clinic Infect Dis.* 2007;45:88–94.
74. Wachino J, Yamane K, Shibayama K, et al. Novel plasmid-mediated 16S rRNA methylase, RmtC, found in a *Proteus mirabilis* isolate demonstrating extraordinary high-level resistance against various aminoglycosides. *Antimicrob Agents Chemother.* 2006;50:178–184.
75. Galimand M, Sabtcheva S, Courvalin P, et al. Worldwide disseminated *armA* aminoglycoside resistance methylase gene is borne by composite transposon Tn1548. *Antimicrob Agents Chemother.* 2005;49:2949–2953.
76. Bogaerts P, Galimand M, Bauraing C, et al. Emergence of *ArmA* and *RmtB* aminoglycoside resistance 16S rRNA methylases in Belgium. *J Antimicrob Chemother.* 2007;59:459–464.
77. Bohm ME, Razavi M, Marathe NP, et al. Discovery of a novel integron-borne aminoglycoside resistance gene present in clinical pathogens by screening environmental bacterial communities. *Microbiome.* 2020;8:41.
78. Partridge SR, Kwong SM, Firth N, et al. Mobile genetic elements associated with antimicrobial resistance. *Clin Microbiol Rev.* 2018;31:e00088–17.
79. Schwarz S, Kehrenberg C, Doublet B, et al. Molecular basis of bacterial resistance to chloramphenicol and florfenicol. *FEMS Microbiol Rev.* 2004;28:519–542.
80. Kehrenberg C, Ojo KK, Schwarz S. Nucleotide sequence and organization of the multiresistance plasmid pSCFS1 from Staphylococcus sciuri. *J Antimicrob Chemother.* 2004;54:936–939.
81. Bannam TL, Crellin PK, Rood JI. Molecular genetics of the chloramphenicol-resistance transposon Tn4451 from Clostridium perfringens, the TnpX site-specific recombinase excises a circular transposon molecule. *Mol Microbiol.* 1995;16:535–551.
82. Villa L, Mammina C, Miriagou V, et al. Multidrug and broad-spectrum cephalosporin resistance among *Salmonella enterica* serotype Enteritidis clinical isolates in Southern Italy. *J Clin Microbiol.* 2002;40:2662–2665.
83. Tosini F, Visca P, Luzzi I, et al. Class 1 integron-borne multiple-antibiotic resistance carried by IncFI and IncL/M plasmids in *Salmonella enterica* serotype Typhimurium. *Antimicrob Agents Chemother.* 1998;42:3053–3058.
84. Yan JJ, Ko WC, Wu JJ. Identification of a plasmid encoding SHV-12, TEM-1, and a variant of IMP-2 metallo-beta-lactamase, IMP-8, from a clinical isolate of *Klebsiella pneumoniae. Antimicrob Agents Chemother.* 2001;45:2368–2371.
85. Hegstad K, Mikalsen T, Coque TM, et al. Mobile genetic elements and their contribution to the emergence of antimicrobial 50 resistant *Enterococcus faecalis* and *Enterococcus faecium. Clinic Microbiol Infect.* 2010;16:541–554.
86. Binda E, Marinelli F, Marcone GL. Old and new Glycopeptide antibiotics: action and resistance. *Antibiotics.* 2014;3:572.
87. Kohler V, Vaishampayan A, Grohmann E. Broad-host-range Inc18 plasmids: occurrence, spread and transfer mechanisms. *Plasmid.* 2018;99:11–21.
88. Marcone GL, Binda E, Berini F, et al. Old and new glycopeptide antibiotics: from product to gene and back in the post-genomic era. *Biotechnol Adv.* 2018;36:534–554.
89. Binda E, Cappelletti P, Marinelli F, et al. Specificity of induction of glycopeptide antibiotic resistance in the producing actinomycetes. *Antibiotics (Basel).* 2018;36.
90. Jiang H, Cheng H, Liang Y, et al. Diverse mobile genetic elements and conjugal transferability of sulfonamide resistance genes (*sul*1, *sul*2, & *sul*3) in *Escherichia coli* isolates from penaeus vannamei and pork from large markets in Zhejiang. *China Front Microbiol.* 2019;10:1787.

91. Johansson MHK, Bortolaia V, Tansirichaiya S, et al. Detection of Mobile genetic elements associated with antibiotic resistance in *salmonella enterica* using a newly developed web tool: mobile element finder. *J Antimicrob Chemother.* 2021;76:101–109.
92. Nohr-Meldgaard K, Struve C, Ingmer H, et al. The tetracycline resistance gene, tet(W) in *Bifidobacterium animalis* subsp. lactis follows phylogeny and differs from tet(W) in other species. *Front Microbiol.* 2021;12(658943).
93. Chaffanel F, Charron-Bourgoin F, Libante V, et al. Resistance genes and genetic elements associated with antibiotic resistance in clinical and commensal isolates of *Streptococcus salivarius. Appl Environ Microbiol.* 2015;81:4155–4163.
94. Chancey ST, Agrawal S, Schroeder MR, et al. (2015) Composite mobile genetic elements disseminating macrolide resistance in *Streptococcus pneumoniae. Front Microbiol.* 2015;6:26.
95. Sugimoto Y, Suzuki S, Nonaka L, et al. The novel mef(C)-mph(G) macrolide resistance genes are conveyed in the environment on various vectors. *J Glob Antimicrob Resist.* 2017;10:47–53.
96. Febler AT, Wang Y, Wu C, et al. Mobile macrolide resistance genes in *staphylococci. Plasmid.* 2018;99:2–10.
97. Robicsek A, Jacoby GA, Hooper DC. The worldwide emergence of plasmid-mediated quinolone resistance. *Lancet Infect Dis.* 2006;6:629–640.
98. Poirel L, Cattoir V, Nordmann P. Is plasmid-mediated quinolone resistance a clinically significant problem? *Clin Microbiol Infect.* 2008;14:295–297.
99. Karah N, Poirel L, Bengtsson S, et al. Plasmid-mediated quinolone resistance determinants qnr and aac(6′)-Ib-cr in *Escherichia coli* and *Klebsiella* spp. from Norway and Sweden. *Diagn Microbiol Infect Dis.* 2010;66:425–431.
100. Rodriguez-Martinez JM, Machuca J, Cano ME, et al. Plasmid-mediated quinolone resistance: two decades on. *Drug Resist Updat.* 2016;29:13–29.
101. Strahilevitz J, Jacoby GA, Hooper DC, et al. Plasmid mediated quinolone resistance: a multifaceted threat. *Clin Microbiol Rev.* 2009;22:664–689.
102. Caniaux I, van Belkum A, Zambardi G, et al. MCR: modern colistin resistance. *Euro J Clin Microbiol Infect Dis.* 2017;36:415–420.
103. Cannatelli A, Principato S, Colavecchio OL, et al. Synergistic activity of colistin in combination with resveratrol against colistin resistant gram-negative pathogens. *Front Microbiol.* 2018;9:1808.
104. Li B, Ke B, Zhao X, et al. Antimicrobial resistance profile of *mcr-1* positive clinical isolates of *Escherichia coli* in China from 2013 to 2016. *Front Microbiol.* 2018;9:2514.
105. Moosavian M, Emam N, Pletzer D, et al. Rough-type and loss of the LPS due to lpx genes deletions are associated with colistin resistance in multidrug-resistant clinical *Escherichia coli* isolates not harbouring *mcr* genes. *PLoS One.* 2020;15:e0233518.
106. Moffatt JH, Harper M, Harrison P, et al. Colistin resistance in *Acinetobacter baumannii* is mediated by complete loss of lipopolysaccharide production. *Antimicrob agents Chemother.* 2010;54:4971–4977.
107. Jiang X, Ellabaan MMH, Charusanti P, et al. Dissemination of antibiotic resistance genes from antibiotic producers to pathogens. *Nat Commun.* 2017;8:15784.
108. Lerner A, Matthias T, Aminov R. Potential effects of horizontal gene exchange in the human gut. *Front Immunol.* 2017;8:1630.
109. Pallecchi LC, Lucchetti A, Bartoloni F, et al. Population structure and resistance genes in antibiotic-resistant bacteria from a remote community with minimal antibiotic exposure. *Antimicrob Agents Chemother.* 2007;51:1179–1184.
110. Davies J, Davies D. Origins and evolution of antibiotic resistance. *Microbiol Mol Biol Rev.* 2010;74:417–433.
111. Gibson MK, Forsberg KJ, Dantas G. Improved annotation of antibiotic resistance determinants reveals microbial resistomes cluster by ecology. *ISME J.* 2015;9:207–216.

112. Hu Y, Yang X, Qin J, et al. Metagenome-wide analysis of antibiotic resistance genes in a large cohort of human gut microbiota. *Nat Commun.* 2013;4:2151.
113. Penders J, Stobberingh EE, Savelkoul PH, et al. The human microbiome as a reservoir of antimicrobial resistance. *Front Microbiol.* 2013;4:87.
114. Geser N, Stephan R, Korczak BM, et al. Molecular identification of extended-spectrum-β-lactamase genes from Enterobacteriaceae isolated from healthy human carriers in Switzerland. *Antimicrob Agents Chemother.* 2012;56:1609–1612.
115. Gijon D, Curiao T, Baquero F, et al. Fecal carriage of carbapenemase-producing Enterobacteriaceae: a hidden reservoir in hospitalized and non-hospitalized patients. *J Clin Microbiol.* 2012;50:1558–1563.
116. Firoozeh F, Zibaei M. *The role of gut microbiota in antimicrobial resistance: a mini-review Anti-Infective Agents.* 2020;18:201–206.
117. Carattoli A. Resistance plasmid families in Enterobacteriaceae. *Antimicrob Agents Chemother.* 2009;53:2227–2238.
118. Schmieder R, Edwards R. Insights into antibiotic resistance through metagenomic approaches. *Future Microbiol.* 2012;7:73–89.

Further reading

119. Wilson BA, Ho M. Pasteurella multocida: from zoonosis to cellular microbiology. *Clin Microbiol Rev.* 2013;26:631–655.
120. McArthur AG, Waglechner N, Nizam F, et al. The comprehensive antibiotic resistance database. *Antimicrob Agents Chemother.* 2013;57:3348–3357.
121. Luo LT, Cunha MPV, Francisco GR, et al. IncX3 plasmid harboring a non-Tn4401 genetic element (NTEKPC) in a hospital-associated clone of KPC-2-producing *Klebsiella pneumoniae* ST340/CG258. *Diagn Microbiol Infect Dis.* 2017;89:164–167.

CHAPTER TWO

Dysbiosis of human microbiome and infectious diseases

Aeshna Gupta[a], Vijai Singh[b], and Indra Mani[c,*]
[a]School of Biology and Environmental Science, University College Dublin, Dublin, Ireland
[b]Department of Biosciences, School of Science, Indrashil University, Rajpur, Mehsana, Gujarat, India
[c]Department of Microbiology, Gargi College, University of Delhi, New Delhi, India
*Corresponding author: e-mail addresses: indramanibhu@gmail.com; indra.mani@gargi.du.ac.in

Contents

Abstract

Since birth, the human body gets colonized by various communities of symbiotic or commensal microorganisms and they persist till the death of an individual. The human microbiome is comprised of the genomes of microorganisms such as viruses, archaea, eukaryotes, protozoa, and, most remarkably, bacteria. The development of "omics" technologies gave way to the Human Microbiome Project (HMP) which aimed at exploring the collection of microbial genes and genomes inhabiting the human body. Eubiosis, i.e., a healthy and balanced composition of such microbes contributes to the metabolic function, protection against pathogens and provides nutrients and energy to the host. Whereas, an imbalance in the diversity of microorganisms, termed dysbiosis, greatly influences the state of health and disease. This chapter summarizes the impact of gut bacteria on the well-being of humans and highlights the protective role played by the human microbiota during bacterial and viral infections. The condition of dysbiosis and how it plays a role in the establishment of various infections and

Progress in Molecular Biology and Translational Science, Volume 192
ISSN 1877-1173
https://doi.org/10.1016/bs.pmbts.2022.06.016

metabolic disorders such as *Clostridioides difficile* infection (CFI), inflammatory bowel disease (IBD), cancer, periodontitis, and obesity are described in detail. Further, treatments such as fecal transplantation, probiotics, prebiotics, phage therapy, and CRISPR/Cas system, which target gut microbiota during digestive diseases are also discussed.

Abbreviations

CDI *Clostridioides difficile* infection
CRC colorectal cancer
CRISPR clustered regularly interspaced short palindromic repeats
FMT faecal microbiota transplantation
GIT gastrointestinal tract
HCC hepatocellular carcinoma
HMP human microbiome project
IBD inflammatory bowel disease
PAMPs pathogen-associated molecular patterns
SBF segmented filamentous bacteria
SCFAs short chain fatty acids
TLRs toll-like receptors

1. Introduction

The human microbiome is constituted of the genomes of microorganisms such as viruses, archaea, eukaryotes, protozoa, and, most notably, bacteria. They reside symbiotically on and inside the human body. Microbial-dominated habitats of our body include the oral cavity, skin, genital organs, respiratory tract and gastrointestinal tract (GIT).[1] The sum of bacterial cells in the intestine (the organ with the greatest number of microorganisms) is 3.8×10^{13} bacteria and this was used to calculate the human microbial count values.[2] The overall count of microbial cells in a 70 kg adult man is approximated to be $\sim 10^{13}–10^{14}$, which is about the same as the total number of cells in the human body.[2] Firmicutes (e.g., *Bacillus, Clostridium*), Bacteroidetes (e.g., *Bacteroides*), Actinobacteria (e.g., *Bifidobacterium*) and Proteobacteria (e.g., *Escherichia, Klebsiella*) are the four main phyla that make up the complex gastrointestinal microbiome.[3] However, *Escherichia coli, Lactobacilli* and *Streptococci* are found in trivial numbers in the gut.[4] Further studies are required to identify and characterize more gastrointestinal-associated microbiota which will help to understand their role in the GIT.

The human body's primary microbial pool is the GIT. As a result, gut bacteria have a major influence on human health. The microbes in the gut and their specific metabolites play an important role in the host's defense

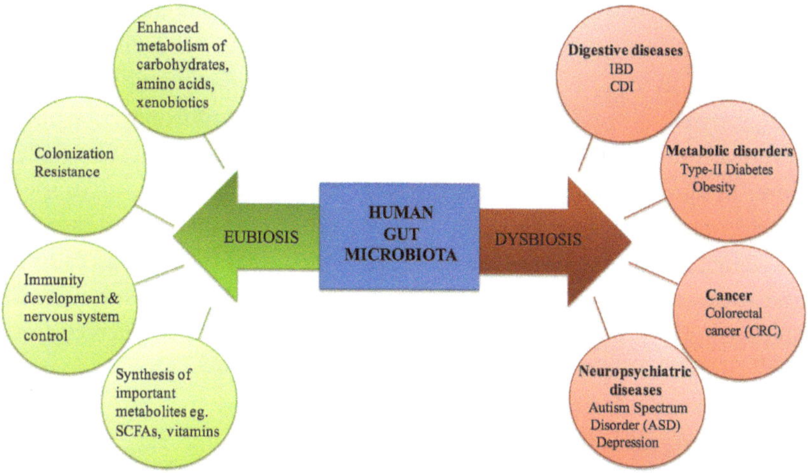

Fig. 1 Overview of the role of human gut microbiota in health and disease. A perfectly balanced microbial composition (eubiosis) of the intestine imparts health benefits such as enhanced metabolic activities, preventing the colonization of pathogenic bacteria and better immunity. On the contrary, a disparity in the gut microbial diversity (dysbiosis) can be the cause of digestive, metabolic and neural disorders and even cancer.

against intruding pathogenic organisms and the regulation of critical host activities such as metabolism and immunity development.[5] Fig. 1 gives an outline of the role of human gut microbiota in health and disease. The gut microbiota enhances polysaccharide, amino acid, xenobiotic, and micronutrient metabolism.[6] They also provide the host with essential amounts of vitamins B5 and B12, which function as coenzymes in the synthesis of acetylcholine and cortisol, both of which are necessary for proper nervous system activity.[7] Another important role of the resident human microbiota is colonization resistance. Dominant non-pathogenic gut microbiota inhabits the niche, suppressing the development and colonization of pathogens (pathobionts, e.g., *Clostridioides difficile*).[8] They also impart protective immunity against lung disorders like tuberculosis.[9] Moreover, a balanced gut microbial composition is associated with proper functioning of the nervous system, imparting good mental health.[10,11]

Given the significance of human microbiota on health, it becomes important to mention that any perturbation in the gut microbiome will undoubtedly give way to complications manifesting as infectious diseases. Variations in the microbial composition of the gut make it prone to pathogenic abuse as the potentially pathogenic bacteria may outgrow the beneficial ones, causing a severe disparity. This disorder in the microbial composition is termed "dysbiosis." which will be the focus of this chapter.[12,13] A multitude of physiological factors like nutrition, physical strain, mental distress, age, use of antibiotics and environmental factors like radiation, humidity and geographical area have been shown to play a role in dysbiosis.[14] For example, the amount of certain *Lactobacillus* species decreases rapidly during flight, which may contribute to anxiety associated with flying; similarly, exposure to metal contamination such as cadmium reduces the amount of *Bacteriodes*, which results in a decrease in metabolites such as short chain fatty acids (SCFAs).[14] An astronaut's microbiome is greatly impacted in space because of ecological factors such as exposure to radiation and microgravity. In an investigation, space travelers were tested before, during and after their spaceflight, and the findings revealed that the overall number of microorganisms in the GIT increased but their diversity declined. In addition, after space travel, the microbial count of pathogenic bacteria *Serratia marcescens* and *Staphylococcus aureus* increased significantly. Moreover, the presence of *S. aureus* among the astronauts suggested that pathogenic bacteria can be transmitted from person to person in a spaceship environment.[15] Despite this, the impact of these changes on the gut microbiome and general wellbeing of space travelers is still uncertain.

2. Diseases associated with dysbiosis

It is highly probable that dysbiosis or a failure to regulate the microbial composition adequately is at the emergence and severity of several diseases, including inflammatory bowel diseases (IBD), irritable bowel syndrome (IBS), diabetes, obesity, and cancer.[16–20] A list of the major human microbiota associated with different human diseases is given in Table 1. Dysbiosis may give rise to diseases in three aspects—(a) gain of function, i.e., pathogens and their mechanisms may be acquired or overgrown on an opportunistic basis (b) loss of function, i.e., health-protective bacteria and their roles may be disrupted or silenced (c) combination of loss and gain of function.[46]

Following are examples of some diseases which result due to dysbiosis: -

(a) ***Clostridioides difficile*** **infection (CDI):** A Firmicutes member in normal gut microbiota, *C. difficile* is a Gram-positive, spore-forming bacterium which grows in obligate anaerobic conditions. It causes a wide range of clinical symptoms, from moderate diarrhea to severe colonic rupture and toxic megacolon.[47,48] Interestingly, administration of broad-spectrum antibiotic was observed to be the leading cause of

Table 1 List of the major human microbiota associated with different human diseases.

S. No	Disease	Species	Reference
1.	Inflammatory Bowel Disease (IBD)	*Faecalibacterium prausnitzii* *Roseburia* sp. *Enterobacteriaceae* sp. *Bacteroides fragilis* *Saccharomyces cerevisiae* *Candida albicans*	21–26
2.	Crohn's disease	*Escherichia coli* *Clostridium leptum* *Clostridium coccoides* *Faecalibacterium prausnitzii* *Lactobacillus coleohominis* *Ruminococcus albus*	27–31
3.	Cancer	*Bacteroides fragilis* *Fusobacterium nucleatum*	32–36
4.	Periodontitis	*Prevotella intermedia* *Prevotella micros* *Fusobacterium nucleatum* *Prevotella gingivalis* *Tannerella forsythia* *Treponema denticola*	37
5.	Obesity	Firmicutes Bacteroidetes	38
6.	Type-II Diabetes	*Lactobacillus* *Clostridium coccoides* *Atopobium* *Prevotella*	39–41
7.	Rheumatoid Arthritis	*Prevotella copri* *Bacteriodes* sp. *Lactobacillus*	42,43
8.	Depression	*Eggerthella* *Holdemania* *Prevotella* *Bacteriodes*	11,44,45

CDI, with diarrhea being the most common side effect.[49,50] Antibiotic administration disrupts gut microbial populations and decreases their diversity, especially dominant microbes that synthesize secondary bile acids which inhibit *C. difficile*'s vegetative growth.[51] This causes

C. difficile to outgrow and colonize, secreting large amounts of toxins that impair intestinal ion absorption, resulting in diarrhea.

(b) Inflammatory Bowel Disease (IBD): IBD is also a gut microbiome-related disease largely defined by gastrointestinal inflammation that is chronic and recurrent. Two common forms of IBD are CD (Crohn's Disease, which involves chronic inflammation affecting any region along the GIT) and UC (Ulcerative Colitis, in which inflammation is limited to the large intestine).[52] Symptoms of both types include fever, discomfort and pain in the abdominal region and diarrhea. In genetically predisposed individuals, antibodies are developed against commensal microbial antigens which lead to dysbiosis, causing an impairment in the gut mucus barrier. The local immunity further gets stimulated because of impaired barrier function contributing to severe gut inflammation.[8] Lowered anti-inflammatory gut Firmicutes such as *Faecalibacterium prausnitzii* and *Roseburia* sp. are examples of dysbiotic characteristics typically seen in IBD patients.[21–23] There has also been observed a surge in pathogenic gut microorganisms like *Enterobacteriaceae* species and *Bacteroides fragilis* which have high levels of endotoxic lipopolysaccharide (LPS).[24,25] Treatment of gastrointestinal inflammation in IBD patients using probiotics and prebiotics to enhance the proliferation and metabolic activity of protective gut bacteria might be an interesting approach.[53,54] Regardless, individual-specific differences in efficacy of therapy which arise due to complex pathogenesis of IBD in each patient offer a significant barrier in developing a treatment that is beneficial for everybody.

(c) Cancer: Dysbiosis has been associated with the establishment and progression of several malignancies, including colorectal cancer (CRC), gastric cancer, pancreatic cancer, hepatocellular carcinoma (HCC), melanoma, and breast cancer.[55] By inducing tumorigenic pathways, causing inflammation, and destroying host DNA, dysbiosis and individual pathogenic bacteria in the gut may trigger or facilitate the development of malignancy.[56] Toll-like receptors (TLRs), a component of innate immune system, get activated via pathogen-associated molecular patterns (PAMPs). This leads to an increase in the synthesis of pro-inflammatory molecules in cells, resulting in accelerated carcinogenesis. In addition to causing inflammation, several bacteria have the potential to cause DNA damage by producing compounds that aid tumorigenesis.[55]

Several bacteria have been studied for their tumor-promoting roles. The enterotoxin produced by *B. fragilis* can lead to inflammation of the gut and can cause DNA damage, both of which play a role in CRC etiology.[32] *Fusobacterium nucleatum*, an oral, Gram-negative commensal bacterium that is more prevalent in colon tumor tissues than in neighboring healthy tissues, has been demonstrated to aid multiplication of tumor cells.[33] Furthermore, it promotes cancer cell autophagy, which increases chemotherapeutic drug resistance and the rate of tumor relapse.[57] Studying the contribution of the host microbiota in cancer can be difficult since it is a life-long phenomenon that unfolds over many years in an individual. Nevertheless, as our understanding of the role of the microbiota in this disease process evolves, prospective therapeutic approaches such as fecal transplantation and probiotics will help us to fight this fatal malignancy.

(d) Periodontitis: Periodontitis, which may result in tooth loss, is an infectious disease involving tissues which support teeth. This inflammatory disease is induced by bacteria and is linked to the oral microbial community of an individual, along with the immune system.[58] The microbial element contributes to the initiation of the disease and immunological element of the host contributes to the progression of the disease.[37] Periodontitis might be caused by dysbiosis in the oral cavity, according to recent studies. Oral dysbiosis is believed to proceed over a lengthy period of time, during which the oral health of the host deteriorates and eventually disease development takes place.[37] At the same time, a set of microbial complexes emerge. The orange complex, which consists of gram-negative anaerobic bacteria such as *Prevotella intermedia*, *P. nigrescens*, *P. micros*, and *F. nucleatum* is the first complex to have been implicated in disease. The microbiota transitions to the red complex as the illness progresses, which includes the periodontal pathogens *Porphyromonas gingivalis*, *Treponema denticola*, and *Tannerella forsythia*.[37] According to the existing research, oral dysbiosis appears to be partly involved in the onset of periodontitis. Therefore, while a microbial change is proven to have a role in the development of periodontitis, healthcare professionals and researchers must also explore genetic, immunological, and environmental elements for a better understanding of the disease.[37] A recent study has suggested that the oral microbiome plays an important role in severe periodontitis and is also indirectly associated between periodontitis and cardiovascular disease.[59]

(e) Obesity: Obesity is a worldwide health concern caused by excess fat deposition as a result of increased energy consumption and reduced energy expenditure. It is correlated with metabolic imbalance in the body, which means obese people are more likely to develop obesity-related disorders such as cardiovascular disease, type-II diabetes, and liver abnormalities.[60,61] The recently discovered gut microbiome, which is substantially implicated in host metabolic control, has been included in studies involving interaction between genetic, behavioral, and environmental variables as a potential driver of obesity. In comparison to lean mice, there is a significant increase in butyrate-producing Firmicutes and a decrease in Bacteroidetes in the gut microbiome of obese patients and genetically obese mice, as revealed by metagenomic studies. This Firmicutes: Bacteroidetes ratio is dependent on body weight and fat deposition, indicating that obese individuals have a more disproportionate ratio of these bacteria.[38] As gut dysbiosis has been linked to obesity, modification of specific microbiota composition through diet changes such as the administration of prebiotics or probiotics, could be a feasible treatment strategy. In animal models, *Lactobacillus rhamnosus* has been shown to reduce plasma cholesterol, triacyl glycerides, and white adipose tissue.[62] More evidence from human models on the other hand, is needed to validate these therapeutic approaches.

3. Protective role of the host microbiota during diseases

Studies of the gut, where the microbial community is the densest, have revealed the beneficial effect of our microbiota in developing the immune system and ensuring homoeostasis.[63,64] In addition to the gut microbiome, the respiratory microbiome also forms a discreet part of the human microbiota.[65] Since 2010, research has linked lung microbiome dysbiosis to a variety of disorders, including chronic obstructive pulmonary disease, cystic fibrosis, and asthma,[66–68] indicating that the microbial community of the lungs affects respiratory health and disease. Furthermore, it has been found that the gut microbiota influences pulmonary immunity via the gut–lung axis.[9,69,70] The following sections emphasize how the host microbiota play a defensive role in respiratory infectious disorders caused by a wide range of airborne pathogens, which can result in acute or chronic infections. In addition, the microbiota protects against viral infections.

3.1 Acute bacterial infections

Researchers have utilized broad-spectrum antibiotic combinations to induce dysbiosis in mice and examine the significance of the bacterial microbiota in response to acute pulmonary diseases such as pneumonia. Antibiotic treatment made mice more vulnerable to pulmonary pathogens such as *Streptococcus pneumoniae* and *Klebsiella pneumoniae*. Particularly, antibiotic-treated mice with *S. pneumoniae* infection had an impairment in lung cytokine synthesis. Infection in mice and the level of cytokines in the lungs were both normalized after fecal transplantation (refer to Section 4.1 for details) with normal gut microbiota, suggesting the importance of gut microbiota in lung immunity.[71] In another investigation, antibiotic-treated mice's tolerance to *S. pneumoniae* and *K. pneumoniae* infection was linked to decreased synthesis of granulocyte macrophage colony-stimulating factor (GMCSF) and interleukin 17A in lungs.[72] Intranasal transfer of normal microbiota from the upper respiratory tract or oral transplantation of fecal matter helped to minimize the infection. This suggested that both proximal and distant microbiota had a positive impact on respiratory immunity.[72] Specific commensals have also been found to have the ability to control host immunity. Segmented filamentous bacteria (SFB), a bacterial lineage found in the gut microbiota, are likely to cause T-helper (T_H-17) cells to develop in the lamina propria in mice. T_H-17 immunity generated by SFB prevents the host from getting infected by gut or respiratory pathogens.[73,74] Further study with utilizing more antimicrobial drugs may provide a better understanding of acute pulmonary or gut infections due to dysbiosis.

3.2 Chronic bacterial infections

Chronic infections by *Mycobacterium tuberculosis*, which manifest as tuberculosis (TB), can be dormant for years before reactivating and causing chronic immunopathological lung tissue damage. The gut microbiota has been examined in people during tuberculosis infections and anti-TB therapy. These investigations indicated that the bacterial population in the guts of tuberculosis patients is disrupted, which may be linked to the disease progression.[75,76] Anti-TB therapy includes antibiotics that target a broad range of bacteria, other than mycobacteria and an analysis found that long-term anti-TB medication causes an imbalance in the gut microbiota of TB patients and that the dysbiotic state continues after treatment is terminated.[77] The first study to explore host microbiota's presumed role in TB immunity reported that replication and spread of *M. tuberculosis* were increased in mice which were given broad-spectrum antibiotics during infection. This was associated with a decrease in tumor necrosis factor (TNF) α-positive and interferon-γ-positive CD4 T-cells and an increase of FoxP3-positive regulatory T-cells (Treg) in the spleen, implying that immune reaction to *M. tuberculosis* infection was greatly impacted by dysbiosis caused after antibiotic treatment.[78] However, a possible concern in such investigations is the use of laboratory mice which have been restricted to very sterile conditions and therefore are not indicative of the microbial diversity in wild mice, ultimately resulting in modified immunological development.[79] This will have to be taken into consideration in future studies.

3.3 Viral infections

Gut microbiome plays a key function in preventing viral infection. Many pathogens, including viruses, enter the body through mucous membranes. Within the gastrointestinal system, secondary lymphoid organs are the first line of defense in protecting the intestinal mucosa and their maturation is supported by gut microbes.[80] Immune resistance in response to infection by influenza virus has been reported.[81] The intestinal microbiota protects against respiratory influenza virus by regulation of immunity through secretion of IgA and appropriate activation of T_H1 cells and CTLs (cytotoxic T-lymphocytes).[82,83]

Bacteria from the *Lactobacillus* genus predominate the vaginal mucosa. In ex vivo cervico-vaginal tissue culture, *Lactobacillus crispatus*, *Lactobacillus gasseri*, and *L. vaginalis* block HIV-1 replication. These actions are mediated via the generation of lactic acid and the acidification of the medium, as well as their adsorption to the virus to limit the number of free virions in the tissue.[84] These findings suggest that commensal bacteria residing in various regions of the mucosa confer protection against pathogenic viruses.

4. Targeting the gut microbiota during digestive diseases

Since 1907, when use of fermented food products harboring *Lactobacillus bulgaricus* was attributed to vitality and healthy life, the prospect of replacing the harmful gut microbiota with more beneficial bacteria has been intensively investigated.[85] Modification of the gut microbiota shows potential as a possible therapy for gut dysbiosis, with the ability to alleviate symptoms of gastrointestinal and systemic disorders while also improving heath.[86] This section describes methods that have been shown to manage and stabilize the intestinal microbiota, as well as to return it to a healthy constitution if it has become dysbiotic due to various digestive ailments.

4.1 Fecal microbiota transplantation (FMT)

Humans have become more prone to infections as a result of the extensive use of antibiotics to cure infectious diseases.[87] Antibiotics such as vancomycin which is used frequently can disturb the normal bowel microbiota.[88] The transfer of a solution of feces from a donor into the digestive tract of a receiver in an attempt to directly enhance the receiver's microbial makeup and impart a health benefit is known as fecal microbiota transplantation (FMT) or fecal bacteriotherapy.[89,90] Usually, the first step is identifying a donor with no family history of autoimmune, metabolic, or malignant ailments, as well as screening for any suspected infections. A filtration step is undertaken after mixing the feces with water or normal saline, to eliminate any particulate matter. The mixture can be administered through various routes via nasogastric tube, nasojejunal tube, colonoscopy, etc. The majority of FMT's clinical experience has resulted from treatment of recurrent CDI.[90] About 90% of CDI patients who undergo microbiota transplantation are cured, compared to 31% of those who receive vancomycin.[91] However, this medication has a number of drawbacks, including an increased risk of transferring additional pathogens.[92] There have been few long-term investigations to evaluate the efficacy of FMT. Researchers are looking into the use of "synthetic stool" products with defined bacterial composition to alleviate these risks.[93] The outcomes of these studies will help to determine if this therapy can be declared both safe and effective.

4.2 Probiotics and prebiotics

World Health Organization (WHO) describes probiotics as "viable microorganisms that provide therapeutic benefits to the host when given in adequate doses"[94]. The pH of gastric juice varies between 3 and 5, and this is the first big challenge that a probiotic microbe must defeat in medical practice to restore gut microbiota dysbiosis. Lactic acid-producing bacteria of the genera *Lactobacillus* and *Bifidobacterium* are the most common probiotic microbes. Other microorganisms have also been studied, such as the yeast *Saccharomyces boulardii* and bioengineered *E. coli* strains named Nissle 1917.[95] Clinical symptoms of inflammatory and necrotizing enterocolitis, irritable bowel syndrome, and acute diarrhea can be prevented or alleviated by the use of probiotics.[96,97] It was discovered that they could control the composition of gut microbiota by physically preventing pathogenic microorganisms from adhering to epithelial cells. This is attributed to the increased formation of a mucosal barrier by goblet epithelial cells.[98]

Prebiotics are non-digestible food ingredients (poorly digestible carbohydrates) which enable beneficial microbes to grow and function in the gut.[99] Intestinal bacteria produce carbohydrate-hydrolyzing enzymes. This, along with the synthesis of hydrogen, methane, carbon dioxide, and SCFAs by fermentation of prebiotics can influence host energy levels and balance of intestinal hormones. Mixtures of probiotic and prebiotic substances can be specifically used to increase the development or functioning of health-promoting bacteria.[100] Although long-term usage of probiotics supports the idea that probiotics are safe, some investigations have found indication of changed metabolic pathways, increased immunological response, gene transfer and gastrointestinal problems. More research is needed to fully determine the frequency and intensity of adverse events associated with probiotics.[101] Indeed, further study will help to understand the role of probiotics comprehensively.

4.3 Phage therapy

Phage therapy is a kind of biological control where bacteriophages are employed to control microbial populations instead of antibacterial agents.[102] Bacteriophages interact with specific receptors on the host membrane and then inject their genetic material into bacteria. Antibiotic-resistant bacteria can be targeted by engineering such phages. With the discovery of specific phage therapies against bacteria like *Pseudomonas aeruginosa*, *Vibrio cholerae*, *Salmonella*, *Shigella dysenteriae*, Vancomycin-resistant *Enterococcus*, methicillin-resistant *S. aureus* (MRSA) strains and antibiotic-resistant *E. coli*, the application of bacteriophages in antibacterial treatment has reached a new frontier.[103,104] Medical findings have revealed some effective broad-ranged, bacteriophage formulations like Enko-phage, SES-bacteriophage, Pyo-bacteriophage, and Intestibacteriophage.[105] In certain clinical situations, Phage-Antibiotic Synergy (PAS), a method which involves the use of bacteriophages and antibiotics in combination, could be beneficial.[106] Metagenomic (culture-independent approach) studies are the primary approach to recognize the increasing number of bacteriophages which reside within the human microbiota. Lytic phages have been found in a variety of habitats, including zoo composting and human and domestic animal waste.[107] They can serve as potential tools for innovation of improved phage-based treatments.

4.4 CRISPR/Cas9 system

Bacteria and Archaea have an intrinsic innate immune function that relies on RNA-mediated DNA interference. Acquisition of small DNA sequences from the invading phage genome helps them to remember that specific phage infection. These virus-acquired sequences are inserted as spacer sequences into the bacterial genome, notably into an array of repeated sequences called Clustered Regularly Interspaced Short Palindromic Repeats or CRISPR, with the aid of the proteins encoded by Cas (CRISPR-associated) family of genes.[108,109] The CRISPR/Cas9 system is a useful tool for a wide range of processes specifically genome editing, due to the relative ease of the method of action and the characteristics of Cas9.[110–117] CRISPR/Cas9 can be used to preferentially eliminate a pathogenic strain or species from a bacterial population.[118] It can be exploited to incorporate selected mutations into important genes which are responsible for virulence or which impart antibiotic resistance. It has earlier been established that introducing a DNA (linear or plasmid) in trans which is homologous to the target sequence allows extremely specific mutations to be introduced into the selected target.[118,119] In the coming years, employing the CRISPR/Cas system to engineer commensal bacteria with enhanced features could be a viable vaccine strategy in healthcare system for disease prevention. Despite significant progress, considerable effort remains to be done to improve target specificity and efficacy of delivery.[120] CRISPR-Cas system has many scopes to be used as a molecular tool to eliminate or kill any harmful microorganisms very precisely.

5. Conclusion and future perspectives

The Human Microbiome Project (HMP), which is an extension of the Human Genome Project (HGP), has uncovered the astounding diversity of microbial populations encountered in different sites on and within the human body. The intestinal microbiota is important for human health in numerous ways, including metabolism, procuring inaccessible nutrients and immune system maturation and activity. It also plays a protective role during various bacterial and viral infections. As highlighted in the chapter, dysbiosis can cause as well as be the consequence of numerous digestive and metabolic diseases. Next generation sequencing (NGS) techniques and high-throughput "omics" technologies such as metagenomics, metatranscriptomics, metaproteomics, metabolomics, etc., offer novel approaches to study the human microbiome. In the foreseeable future, genetic engineering of bacteriophages which target multidrug-resistant bacteria will be achievable and thoroughly investigated. Determining the specifics of the role of gut microbiome in our development and its function in human health and disease has the potential to improve many aspects of our regular lifestyle, from optimal infant formula composition to new strategies in the battle against cancer and obesity pandemics.

References

1. Lloyd-Price J, Abu-Ali G, Huttenhower C. The healthy human microbiome. *Genome Med.* 2016;8:51.
2. Sender R, Fuchs S, Milo R. Revised estimates for the number of human and bacteria cells in the body. *PLoS Biol.* 2016;14(8):e1002533.
3. Tap J, Mondot S, Levenez F, et al. Towards the human intestinal microbiota phylogenetic core. *Environ Microbiol.* 2009;11:2574–2584.
4. Azad M, Sarker M, Li T, Yin J. Probiotic species in the modulation of gut microbiota: an overview. *Biomed Res Int.* 2018;2018:9478630.
5. Kachrimanidou M, Tsintarakis E. Insights into the role of human gut microbiota in *Clostridioides difficile* infection. *Microorganisms.* 2020;8(2):200.
6. Gill SR, Pop M, Deboy RT, et al. Metagenomic analysis of the human distal gut microbiome. *Science.* 2006;312:1355–1359.
7. Andres E, Loukili NH, Noel E, et al. Vitamin B12 (cobalamin) deficiency in elderly patients. *Can Med Assoc J.* 2004;171:251–259.
8. Kho ZY, Lal SK. The human gut microbiome—a potential controller of wellness and disease. *Front Microbiol.* 2018;9:1835.
9. Yang F, Yang Y, Chen L, et al. The gut microbiota mediates protective immunity against tuberculosis via modulation of lncRNA. *Gut Microbes.* 2022;14(1):2029997.
10. Clapp M, Aurora N, Herrera L, Bhatia M, Wilen E, Wakefield S. Gut microbiota's effect on mental health: the gut-brain axis. *Clin Pract.* 2017;7(4):987.
11. Methiwala HN, Vaidya B, Addanki VK, Bishnoi M, Sharma SS, Kondepudi KK. Gut microbiota in mental health and depression: role of pre/pro/synbiotics in their modulation. *Food Funct.* 2021;12(10):4284–4314.
12. Bien J, Palagani V, Bozko P, et al. The intestinal microbiota dysbiosis and *Clostridium difficile* infection: is there a relationship with inflammatory bowel disease? *Therap Adv Gastroenterol.* 2013;6:53–68.
13. Knights D, Lassen K, Xavier R. Advances in inflammatory bowel disease pathogenesis: linking host genetics and the microbiome. *Gut.* 2013;62(10):1505–1510. The BMJ Publishing Group Ltd.
14. Appanna VD. Dysbiosis, probiotics, and prebiotics: in diseases and health. In: *Human Microbes—The Power Within*; 2018:81–122.
15. Voorhies AA, Lorenzi HA. The challenge of maintaining a healthy microbiome during long-duration space missions. *Front Astron Space Sci.* 2016;3:23.

16. Clemente JC, Ursell LK, Parfrey LW, Knight R. The impact of the gut microbiota on human health: an integrative view. *Cell.* 2013;148(6):1258–1270.
17. Brown CT, Sharon I, Thomas BC, Castelle CJ, Morowitz MJ, Banfield JF. Genome resolved analysis of a premature infant gut microbial community reveals a *Varibaculum cambriense* genome and a shift towards fermentation-based metabolism during the third week of life. *Microbiome.* 2013;1(1):30.
18. Brown EM, Sadarangani M, Finlay BB. The role of the immune system in governing host–microbe interactions in the intestine. *Nat Immunol.* 2013;14(2013):660–667.
19. Chang C, Lin H. Dysbiosis in gastrointestinal disorders. *Best Pract Res Clin Gastroenterol.* 2016;30:3–15.
20. Thaiss CA, Zmora N, Levy M, Elinav E. The microbiome and innate immunity. *Nature.* 2016;535:65–74.
21. Sokol H, Pigneur B, Watterlot L, et al. *Faecalibacterium prausnitzii* is an anti-inflammatory commensal bacterium identified by gut microbiota analysis of Crohn disease patients. *Proc Natl Acad Sci U S A.* 2008;105:16731–16736.
22. Willing BP, Dicksved J, Halfvarson J, et al. A pyrosequencing study in twins shows that gastrointestinal microbial profiles vary with inflammatory bowel disease phenotypes. *Gastroenterology.* 2010;139. 1844.e1–1854.e1.
23. Machiels K, Joossens M, Sabino J, et al. A decrease of the butyrate-producing species *Roseburia hominis* and *Faecalibacterium prausnitzii* defines dysbiosis in patients with ulcerative colitis. *Gut.* 2014;63:1275–1283.
24. Ruseler-van Embden JGH, Both-Patoir HC. Anaerobic gram-negative faecal flora in patients with Crohn's disease and healthy subjects. *Antonie Van Leeuwenhoek.* 1983;49:125–132.
25. Darfeuille-Michaud A, Neut C, Barnich N, et al. Presence of adherent *Escherichia coli* strains in ileal mucosa of patients with Crohns disease. *Gastroenterol.* 1998; 115:1405–1413.
26. Aldars-García L, Chaparro M, Gisbert JP. Systematic review: the gut microbiome and its potential clinical application in inflammatory bowel disease. *Microorganisms.* 2021;9 (5):977.
27. Ley RE, Bäckhed F, Turnbaugh P, Lozupone CA, Knight RD, Gordon JI. Obesity alters gut microbial ecology. *Proc Natl Acad Sci U S A.* 2005;102(31):11070–11075.
28. Lozupone C, Knight R. UniFrac: a new phylogenetic method for comparing microbial communities. *Appl Environ Microbiol.* 2005;71(12):8228–8235.
29. Ramasamy D, Mishra AK, Lagier JC, et al. A polyphasic strategy incorporating genomic data for the taxonomic description of novel bacterial species. *Int J Syst Evol Microbiol.* 2014;64(pt 2):384–391.
30. Roediger WE, Macfarlane GT. A role for intestinal mycoplasmas in the aetiology of Crohn's disease? *J Appl Microbiol.* 2002;92(3):377–381.
31. Santana PT, Rosas SLB, Ribeiro BE, Marinho Y, de Souza HSP. Dysbiosis in inflammatory bowel disease: pathogenic role and potential therapeutic targets. *Int J Mol Sci.* 2022;23(7):3464.
32. Boleij A, Hechenbleikner EM, Goodwin AC, et al. The *Bacteroides fragilis* toxin gene is prevalent in the colon mucosa of colorectal cancer patients. *Clin Infect Dis.* 2015;60:208–215.
33. Yang Y, Weng W, Peng J, et al. *Fusobacterium nucleatum* increases proliferation of colorectal cancer cells and tumor development in mice by activating toll-like receptor 4 signaling to nuclear factor-κB, and up-regulating expression of MicroRNA-21. *Gastroenterology.* 2017;152. 851–66.e24.
34. Rubinstein MR, Wang X, Liu W, et al. *Fusobacterium nucleatum* promotes colorectal carcinogenesis by modulating E-cadherin/β-catenin signaling via its FadA adhesin. *Cell Host Microbe.* 2013;14(2):195–206.

35. Abed J, Emgård JE, Zamir G, et al. Fap2 mediates *Fusobacterium nucleatum* colorectal adenocarcinoma enrichment by binding to tumor-expressed gal-GalNAc. *Cell Host Microbe*. 2016;20:215–225.
36. Ranjbar M, Salehi R, Haghjooy Javanmard S, et al. The dysbiosis signature of *Fusobacterium nucleatum* in colorectal cancer-cause or consequences? A systematic review. *Cancer Cell Int*. 2021;21(1):194. https://doi.org/10.1186/s12935-021-01886-z. Erratum in: Cancer Cell Int. 2022; 22(1):134.
37. Nath SG, Raveendran R. Microbial dysbiosis in periodontitis. *J Indian Soc Periodontol*. 2013;17(4):543–545.
38. Ley RE, Turnbaugh PJ, Klein S, Gordon JI. Microbial ecology: human gut microbes associated with obesity. *Nature*. 2006;444:1022–1023.
39. Qin J, Li Y, Cai Z, et al. A metagenome-wide association study of gut microbiota in type 2 diabetes. *Nature*. 2012;490(7418):55–60.
40. Sato J, Kanazawa A, Ikeda F, et al. Gut dysbiosis and detection of "live gut bacteria" in blood of Japanese patients with type 2 diabetes. *Diabetes Care*. 2014;37(8):2343–2350.
41. Ding D, Yong H, You N, et al. Prospective study reveals host microbial determinants of clinical response to fecal microbiota transplant therapy in type 2 diabetes patients. *Front Cell Infect Microbiol*. 2022;12:820367.
42. Liu X, Zou Q, Zeng B, Fang Y, Wei H. Analysis of fecal *Lactobacillus* community structure in patients with early rheumatoid arthritis. *Curr Microbiol*. 2013;67(2):170–176.
43. Scher JU, Sczesnak A, Longman RS, et al. Expansion of intestinal *Prevotella copri* correlates with enhanced susceptibility to arthritis. *Elife*. 2013;2:e01202.
44. Kelly JR, Borre Y, O' Brien C, et al. Transferring the blues: depression-associated gut microbiota induces neurobehavioural changes in the rat. *J Psychiatr Res*. 2016;82: 109–118.
45. Zheng P, Yang J, Li Y, et al. Gut microbial signatures can discriminate unipolar from bipolar depression. *Adv Sci (Weinh)*. 2020;7(7):1902862.
46. Wilkins LJ, Monga M, Miller AW. Defining dysbiosis for a cluster of chronic diseases. *Sci Rep*. 2019;9:12918.
47. Dubberke E. *Clostridium difficile* infection: the scope of the problem. *J Hosp Med*. 2012;7 (suppl S3):S1–S4.
48. Sunenshine RH, McDonald LC. Clostridium difficile-associated disease: new challenges from an established pathogen. *Cleve Clin J Med*. 2006;73:187–197.
49. Pear SM, Williamson TH, Bettin KM, Gerding DN, Galgiani JN. Decrease in nosocomial *Clostridium difficile*–associated diarrhea by restricting clindamycin use. *Ann Intern Med*. 1994;120:272–277.
50. Bartlett JG. Clinical practice. Antibiotic-associated diarrhea. *N Engl J Med*. 2002;46:334–339.
51. Antonopoulos DA, Huse SM, Morrison HG, Schmidt TM, Sogin ML, Young VB. Reproducible community dynamics of the gastrointestinal microbiota following antibiotic perturbation. *Infect Immun*. 2009;77:2367–2375.
52. Lennard-Jones JE. Classification of inflammatory bowel disease. *Scand J Gastroenterol*. 1989;24:2–6.
53. Joossens M, De Preter V, Ballet V, Verbeke K, Rutgeerts P, Vermeire S. Effect of oligofructose-enriched inulin (OF-IN) on bacterial composition and disease activity of patients with Crohn's disease: results from a double-blinded randomised controlled trial. *Gut*. 2012;61(6):958.
54. De Preter V, Joossens M, Ballet V, et al. Metabolic profiling of the impact of oligofructose-enriched inulin in Crohn's disease patients: a double-blinded randomized controlled trial. *Clin Transl Gastroenterol*. 2013;4(1):e30.
55. Chen D, Wu J, Jin D, Wang B, Cao H. Fecal microbiota transplantation in cancer management: current status and perspectives. *Int J Cancer*. 2019;145(8):2021–2031.

56. Lam SY, Yu J, Wong SH, et al. The gastrointestinal microbiota and its role in oncogenesis. *Best Pract Res Clin Gastroenterol.* 2017;31:607–618.
57. Yu T, Guo F, Yu Y, et al. *Fusobacterium nucleatum* promotes chemoresistance to colorectal cancer by modulating autophagy. *Cell.* 2017;170:548–63.e16.
58. Bartold PM, Van Dyke TE. Periodontitis: a hostmediated disruption of microbial homeostasis. UnLearn. Learned concepts. *Periodontol 2000.* 2013;62:203–217.
59. Plachokova AS, Andreu-Sánchez S, Noz MP, Fu J, Riksen NP. Oral microbiome in relation to periodontitis severity and systemic inflammation. *Int J Mol Sci.* 2021;22 (11):5876.
60. Donnelly KL, Smith CI, Schwarzenberg SJ, Jessurun J, Boldt MD, Parks EJ. Sources of fatty acids stored in liver and secreted via lipoproteins in patients with nonalcoholic fatty liver disease. *J Clin Invest.* 2005;115:1343–1351.
61. Cani PD, Amar J, Iglesias MA, et al. Metabolic endotoxemia initiates obesity and insulin resistance. *Diabetes Metab Res Rev.* 2007;56:1761–1772.
62. Lee HY, Park JH, Seok SH, et al. Human originated bacteria, *Lactobacillus rhamnosus* PL60, produce conjugated linoleic acid and show anti-obesity effects in diet-induced obese mice. *Biochim Biophys Acta.* 2006;1761(7):736–744.
63. Brestoff JR, Artis D. Commensal bacteria at the interface of host metabolism and the immune system. *Nat Immunol.* 2013;14:676–684.
64. Lozupone CA, Stombaugh JI, Gordon JI, Jansson JK, Knigh R. Diversity, stability and resilience of the human gut microbiota. *Nature.* 2012;489:220–230.
65. Dumas A, Bernard L, Poquet Y, Lugo-Villarino G, Neyrolles O. The role of the lung microbiota and the gut-lung axis in respiratory infectious diseases. *Cell Microbiol.* 2018;20(12):e12966.
66. Hilty M, Burke C, Pedro H, et al. Disordered microbial communities in asthmatic airways. *PLoS One.* 2010;5(1):e8578.
67. Pragman AA, Kim HB, Reilly CS, Wendt C, Isaacson RE. The lung microbiome in moderate and severe chronic obstructive pulmonary disease. *PLoS One.* 2012;7(10): e47305.
68. Willner D, Haynes MR, Furlan M, et al. Spatial distribution of microbial communities in the cystic fibrosis lung. *ISME J.* 2012;6(2):471–474.
69. Trompette A, Gollwitzer ES, Yadava K, et al. Gut microbiota metabolism of dietary fiber influences allergic airway disease and hematopoiesis. *Nat Med.* 2014;20:159–166.
70. Budden KF, Gellatly SL, Wood DL, et al. Emerging pathogenic links between microbiota and the gut-lung axis. *Nat Rev Microbiol.* 2017;15(1):55–63.
71. Schuijt TJ, Lankelma JM, Scicluna BP. de Sousa e Melo F, Roelofs JJTH, de Boer JD, et al. the gut microbiota plays a protective role in the host defence against pneumococcal pneumonia. *Gut.* 2016;65:575–583.
72. Brown RL, Sequeira RP, Clarke TB. The microbiota protects against respiratory infection via GM-CSF signaling. *Nat Commun.* 2017;8(1):1512.
73. Gauguet S, D'Ortona S, Ahnger-Pier K, et al. Intestinal microbiota of mice influences resistance to *Staphylococcus aureus* pneumonia. *Infect Immun.* 2015;83(10):4003–4014.
74. Ivanov II, Atarashi K, Manel N, et al. Induction of intestinal Th17 cells by segmented filamentous bacteria. *Cell.* 2009;139:485–498.
75. Luo M, Liu Y, Wu P, et al. Alternation of gut microbiota in patients with pulmonary tuberculosis. *Front Physiol.* 2017;8:822.
76. Maji A, Misra R, Dhakan DB, et al. Gut microbiome contributes to impairment of immunity in pulmonary tuberculosis patients by alteration of butyrate and propionate producers. *Environ Microbiol.* 2018;20(1):402–419.
77. Wipperman MF, Fitzgerald DW, Juste MAJ, et al. Antibiotic treatment for tuberculosis induces a profound dysbiosis of the microbiome that persists long after therapy is completed. *Sci Rep.* 2017;7(1):10767.

78. Khan N, Vidyarthi A, Nadeem S, Negi S, Nair G, Agrewala JN. Alteration in the gut microbiota provokes susceptibility to tuberculosis. *Front Immunol.* 2016;7:529.
79. Rosshart SP, Vassallo BG, Angeletti D, et al. Wild mouse gut microbiota promotes host fitness and improves disease resistance. *Cell.* 2017;171(5):1015–1028.e13.
80. Karst SM. The influence of commensal bacteria on infection with enteric viruses. *Nat Rev Microbiol.* 2016;14:197–204.
81. Ichinohe T, Lee HK, Ogura Y, Flavell R, Iwasaki A. Inflammasome recognition of influenza virus is essential for adaptive immune responses. *J Exp Med.* 2009;206:79–87.
82. Ichinohe T, Pang IK, Kumamoto Y, et al. Microbiota regulates immune defense against respiratory tract influenza a virus infection. *Proc Natl Acad Sci U S A.* 2011;108 (13):5354–5359.
83. Wu S, Jiang ZY, Sun YF, et al. Microbiota regulates the TLR7 signaling pathway against respiratory tract influenza a virus infection. *Curr Microbiol.* 2013;67:414–422.
84. Ñahui Palomino RA, Zicari S, Vanpouille C, Vitali B, Margolis L. Vaginal *Lactobacillus* inhibits HIV-1 replication in human tissues ex vivo. *Front Microbiol.* 2017;8:906.
85. Metchnikoff E. *The Prolongation of Life: Optimistic Studies.* Mitchell PC, Trans, New York, NY: GP Putnam's Sons; 1908.
86. Bull MJ, Plummer NT. Part 2: treatments for chronic gastrointestinal disease and gut dysbiosis. *Integr Med (Encinitas).* 2015;14(1):25–33.
87. Forslund K, Sunagawa S, Kultima JR, et al. Country specific antibiotic use practices impact the human gut resistome. *Genome Res.* 2013;23(7):31–39.
88. Andersson DI, Hughes D. Microbiological effects of sublethal levels of antibiotics. *Nat Rev Microbiol.* 2014;12:465–478.
89. Bakken J, Borody T, Brandt L, et al. Treating *Clostridium difficile* infection with fecal microbiota transplantation. *Clin Gastroenterol Hepatol.* 2011;9:1044–1049.
90. Smits L, Bouter K, De Vos W, Borody T, Nieuwdorp M. Therapeutic potential of fecal microbiota transplantation. *Gastroenterology.* 2013;145(5):946–953.
91. Van Nood E, Vrieze A, Nieuwdorp M, et al. Duodenal infusion of donor faeces for recurrent *Clostridium difficile. N Engl J Med.* 2013;368:407–415.
92. Brandt LJ, Reddy SS. Fecal microbiota transplantation for recurrent *Clostridium difficile* infection. *J Clin Gastroenterol.* 2011;45(suppl):S159–S167.
93. Petrof EO, Gloor GB, Vanner SJ, et al. Stool substitute transplant therapy for the eradication of Clostridium difficile infection: 'RePOOPulating' the gut. *Microbiome.* 2013;1(1):3.
94. FAO/WHO. *Guidelines for the Evaluation of Probiotics in Foods*; 2002. Food and Agriculture Organization of the United Nations and World Health Organization Expert Consultation Report. Food and Agricultural Organization of the United Nations and World Health Organization Working Group Report.
95. Ukena SN, Singh A, Dringenberg U, et al. Probiotic *Escherichia coli* Nissle 1917 inhibits leaky gut by enhancing mucosal integrity. *PLoS One.* 2007;2(12):e1308.
96. Gareau MG, Sherman PM, Walker WA. Probiotics and the gut microbiota in intestinal health and disease. *Nat Rev Gastroenterol Hepatol.* 2010;7:503–514.
97. Whelan K, Quigley EM. Probiotics in the management of irritable bowel syndrome and inflammatory bowel disease. *Curr Opin Gastroenterol.* 2013;29:184–189.
98. Etzold S, Kober OI, Mackenzie DA, et al. Structural basis for adaptation of lactobacilli to gastrointestinal mucus. *Environ Microbiol.* 2014;16:888–903.
99. Gibson GR, Roberfroid MB. Dietary modulation of the colonic microbiota: introducing the concept of prebiotics. *J Nutr.* 1995;125:1401–1412.
100. Roberfroid MB. Prebiotics: the concept revisited. *J Nutr.* 2007;137(3):830s–837s.
101. Doron S, Snydman DR. Risk and safety of probiotics. *Clin Infect Dis.* 2015;60(suppl 2): S129–S134.

102. Abedon ST. Phage therapy: eco-physiological pharmacology. *Scientifica*. 2014; 214:581639.
103. Rhoads DD, Wolcott RD, Kuskowski MA, Wolcott BM, Ward LS, Sulakvelidze A. Bacteriophage therapy of venous leg ulcers in humans: results of a phase I safety trial. *J Wound Care*. 2009;18(6):237–244.
104. Kutter EM, Kuhl SJ, Abedon ST. Re-establishing a place for phage therapy in western medicine. *Future Microbiol*. 2015;10(5):685–688.
105. Pelfrene E, Willebrand E, Sanches AC, Sebris Z, Cavaleri M. Bacteriophage therapy: a regulatory perspective. *J Antimicrob Chemother*. 2016;71:2071–2207.
106. Comeau AM, Tetart F, Trojet SN, Prere MF, Krisch HM. Phage-antibiotic synergy (PAS): beta-lactam and quinolone antibiotics stimulate virulent phage growth. *PLoS One*. 2007;2(8):e799.
107. Amgarten D, Martins LF, Lombardi KC, et al. Three novel Pseudomonas phages isolated from composting provide insights into the evolution and diversity of tailed phages. *BMC Genomics*. 2017;18(1):346.
108. Garneau JE, Dupuis ME, Villion M, et al. The CRISPR/Cas bacterial immune system cleaves bacteriophage and plasmid DNA. *Nature*. 2010;468:67–71.
109. Van der Oost J, Westra ER, Jackson RN, Wiedenheft B. Unravelling the structural and mechanistic basis of CRISPR-Cas systems. *Nat Rev Microbiol*. 2014;12:479–492.
110. Ma Y, Zhang L, Huang X. Genome modification by CRISPR/Cas9. *FEBS J*. 2014;281(23):5186–5193.
111. Selle K, Barrangou R. Harnessing CRISPR-Cas systems for bacterial genome editing. *Trends Microbiol*. 2015;23:225–232.
112. Vasdev K, Mani I. CRISPR/Cas-9 system: magnificent tool for genome editing. *Int J Biotech Bioeng*. 2017;3(10):293–297.
113. Bhattacharjee G, Mani I, Gohil N, et al. CRISPR technology for genome editing. In: Faintuch J, Faintuch S, eds. *Precision medicine for investigators, practitioners and providers*. Elsevier; 2019:59–69.
114. Mani I, Arazoe T, Singh V. CRISPR-Cas systems for genome editing of mammalian cells. *Prog Mol Biol Transl Sci*. 2021;181:15–30.
115. Mani I. CRISPR-Cas9 for treating hereditary diseases. *Prog Mol Biol Transl Sci*. 2021;181:165–183.
116. Mani I. Genome editing in cardiovascular diseases. *Prog Mol Biol Transl Sci*. 2021;181:289–308.
117. Bhattacharjee G, Gohil N, Khambhati K, et al. Current approaches in CRISPR-Cas9 mediated gene editing for biomedical and therapeutic applications. *J Control Release*. 2022;343:703–723.
118. Yosef I, Manor M, Kiro R, Qimron U. Temperate and lytic bacteriophages programmed to sensitize and kill antibiotic-resistant bacteria. *Proc Natl Acad Sci U S A*. 2015;112:7267–7272.
119. Jiang W, Bikard D, Cox D, Zhang F, Marraffini LA. RNA-guided editing of bacterial genomes using CRISPR-Cas systems. *Nat Biotechnol*. 2013;31:233–239.
120. Belizário JE, Napolitano M. Human microbiomes and their roles in dysbiosis, common diseases, and novel therapeutic approaches. *Front Microbiol*. 2015;6:1050.

CHAPTER THREE

Gastrointestinal microbiome in the context of *Helicobacter pylori* infection in stomach and gastroduodenal diseases

R.J. Retnakumar[a,b,†], Angitha N. Nath[a,†], G. Balakrish Nair[a], and Santanu Chattopadhyay[a,*]
[a]Rajiv Gandhi Centre for Biotechnology, Trivandrum, Kerala, India
[b]Manipal Academy of Higher Education, Karnataka, India
*Corresponding author: e-mail address: santanu@rgcb.res.in

Contents

† These authors have contributed equally to this work and share first authorship.

Progress in Molecular Biology and Translational Science, Volume 192
ISSN 1877-1173
https://doi.org/10.1016/bs.pmbts.2022.07.001

Abstract

Infectious origins of a set of severe gastroduodenal diseases viz. gastritis, duodenal ulcer, gastric ulcer, gastric adenocarcinoma and gastric mucosa-associated lymphoid tissue (MALT) lymphoma were appreciated only after the discovery of *H. pylori* in 1983. In the past two decades, however, findings from many laboratories suggest that apart from *H. pylori*, several of the trillions of microbes that populate the human gastrointestinal tract and form microbiomes of the respective niches (like oral microbiome, esophageal microbiome, gastric microbiome and intestinal microbiome) may also participate in maintaining the healthy state of stomach and duodenum. Dysbiosis leading to alteration in the relative abundance of the key gastrointestinal microbes is associated with severe gastric diseases. For instance, an increased abundance of genera like *Leptotrichia*, *Prevotella* and *Veillonella* in gastric microbiome and a decreased abundance of *Bifidobacterium* in intestinal microbiome are associated with gastric cancer. *H. pylori* infection, apart from causing direct harm to the gastric epithelium by its virulence proteins like vacuolating cytotoxin A (VacA) and cytotoxin associated gene A (CagA), is also capable of triggering dysbiosis in stomach and intestinal microbiomes. In this chapter, we have discussed the possible roles of bacteria, viruses, fungi, archaea, protozoa and helminths in human gastrointestinal tracts in the context of *H. pylori* infection in stomach and various gastroduodenal diseases.

1. Introduction

Two major gastroduodenal diseases, peptic ulcer disease (PUD) and gastric cancer (GC), are much less prevalent now in the first quarter of the 21st century than they used to be in the 20th century and of course in the 19th century or before.[1,2] But, they still take no less than a million lives every year.[3,4] For centuries, it was believed that these severe gastric diseases are linked to lifestyle, stress, consumption of spicy foods, alcoholic beverages and smoking. While none of these factors are trivial in the development of gastric diseases, it is the discovery of *H. pylori* from human gastric mucosa just four decades ago allowed the clinicians and researchers to appreciate the infectious origin of PUD and GC.[5–7] Now a huge body of literature exists showing various mechanisms employed by a number of *H. pylori* virulence factors that help the bacterium to tolerate the harsh acidity of the gastric juice (e.g., urease), to get attached to the gastric epithelial cells (e.g., the blood group antigen-binding adhesin) and cause pathogenesis (e.g., vacuolating cytotoxin A).[8–10] However, it is poorly understood why some of the *H. pylori* infected individuals suffer from severe gastroduodenal diseases, whereas most of them manage to remain asymptomatic while carrying the most virulent *H. pylori* strains in their stomach for decades or even lifelong. Recent research pointed out that both PUD and GC are complex and multifactorial diseases and apart from *H. pylori* infections, human genetics, geography, climate, food habit, medicine usage and gastrointestinal microbiome together form a "perfect storm" leading to PUD and GC.[11,12] Among these factors, the recognition and attention toward the gastrointestinal microbiome is newest, but now we have begun to appreciate that the contribution of the gastrointestinal microbiome in maintaining the gastroduodenal health is extremely crucial. In this chapter, we discussed how microbes colonized in different regions of the human gastrointestinal tract are either positively or negatively connected to *H. pylori* infection and development of severe gastroduodenal diseases.

2. Gastric diseases

Scientific studies on stomach have been documented as early as 1547, in the book *De Humani Corporis Fabrica*, by Andreas Vesalius. J.B. van Helmont, in 1648 suggested the role of acid in the digestion of food through animal studies.[13] The base of the gastric gland is populated by the chief cells that secrete pepsinogen, while the HCl secreting parietal cells are present in the middle region. The mucus secreting cells are located at the neck region of the gland. The mucosa of the antrum and pylorus contain cells that secrete mucus and bicarbonate.[14] The main gastric diseases are described below.

2.1 Non-ulcer dyspepsia (NUD)

The Non-ulcer dyspepsia (NUD) refers to the "persistent or recurrent discomfort or pain that is centered in and around the upper abdomen,"[15] Although NUD is not a life-threatening condition, it is shown to have association with increased mortality rate.[16] Various causes for the dyspeptic symptoms are hyper secretion or higher sensitivity to gastric acid, motor-disorders of the upper gastrointestinal tract, psychiatric problems like depression and anxiety and most importantly *H. pylori* infection.[17] In Malaysia *H. pylori* infection was found in 31.2% of the NUD patients[18] while Louw et al. observed 63% prevalence of *H. pylori* infection in patients with NUD in South Africa.[19] A population based study in Norway reported that 48% of NUD patients had *H. pylori* infection.[20]

2.2 Gastritis

Gastritis is the inflammation that happens to the gastric mucosa and can be typed into acute and chronic based on the course of disease and functionally into atrophic gastritis and non-atrophic gastritis.[21] The documented history of chronic gastritis started with Broussais, in his publication "*Histoire des phlegmasiesou inflammations chroniques*," where he describes higher incidence of chronic gastritis and development of acute to chronic gastritis cases in Russian soldiers.[22] It is now evident that *H. pylori* infection is the major cause of gastritis and can further progress to PUD and GC.[23,24] Other etiologies of gastritis includes bile acid reflux, autoimmune conditions, infection with bacteria like *Mycobacterium avium-intercellulare* and viruses like *Cytomegalovirus* and *Herpes simplex virus*.[25]

2.3 Peptic ulcer diseases (PUD)

Peptic ulcer disease (PUD) is defined by the disruption of mucosal lining of the stomach [Gastric ulcer (GU)] or proximal intestine [duodenal ulcer (DU)] that can extend through muscularis mucosae and is characterized by epigastric pain which subsides upon intake of food or alkali.[26] The first ever written description of PUD probably dates back to the era of ancient Egyptians or Hippocrates (though it still remains unclear).[27] The factors responsible for the development of PUD can be *H. pylori* infection, alcohol consumption, smoking and extensive use of non-steroidal anti-inflammatory drugs (NSAIDs).[26,28] The incidence of PUD has shown a decrease in the Western countries possibly due to decline in *H. pylori* infection. Studies from Asian countries like Korea shows that *H. pylori* infection in GU shows an increase over time while a decline was seen in DU.[29] It should also be worth mentioning that the primary outcome of *H. pylori* infection in India is DU.[30]

2.4 Gastric cancer (GC)

Gastric Cancer (GC) is one among the major causes of cancer-related mortalities with over 1 million new cases in 2020 taking 769,000 lives. It is the fifth leading cancer and is fourth in cancer-associated deaths globally.[31] Among the pathogens, infection with *H. pylori* and/or Epstein Barr virus are the most important risk factor for developing gastric cancer.[32,33] The incidence of GC is the highest in East Asia (e.g., South Korea, Mongolia and Japan).[34] Mortality rates are very high in Latin America as well as Eastern and Central Asia. However, incidence of GC worldwide has declined over the past 50 years and this might be due to significant reduction in the prevalence of *H. pylori* infection.[35] *H. pylori* colonization can lead to development of distal type gastric adenocarcinoma, which is the common form of gastric adenocarcinoma. The gastric adenocarcinoma, having poor prognosis, usually has a longer history of inflammation and epithelial damage and is diagnosed during late adulthood.[36,37] The incidence of GC is found to be comparatively low in Asia and Africa in spite of higher *H. pylori* prevalence. These phenomena are termed as the Asian and the African enigmas, respectively.[38] Another term, "Indian enigma" represents that most of the *H. pylori* infected Indians do not develop GC.[39]

3. *H. pylori* and gastroduodenal diseases

3.1 *H. pylori* isolation from human stomach and its link to different gastroduodenal diseases

The existence of a spiral bacillus in human stomach was first discovered by Barry J. Marshall and Robin Warren while studying the gastric biopsies. Warren detected a curved, S-shaped bacterium by histology after staining with Warthin Starry silver while Marshall reported isolation of the flagellated spiral microaerophilic bacterium on chocolate agar plate.[5] Soon it was realized that the colonization of the bacterium was consistent in patients with PUDs and gastritis.[6] Since a suitable animal model to prove the pathogenesis of *H. pylori* was lacking, Barry Marshall himself consumed a suspension of *H. pylori* and developed symptoms of gastritis by fifth day.[40] Further in 1994, International Agency for Research on Cancer (IARC) declared *H. pylori* as a Class 1 carcinogen due to high incidence of GC in individuals infected by *H. pylori*.[7] Later, in 2005 Robin Warren and Barry Marshall were awarded with the Nobel Prize in physiology and medicine.[41]

Chronic *H. pylori* colonization leads to a spectrum of disease conditions like non-ulcer dyspepsia, chronic gastritis, gastric adenocarcinoma, peptic ulcer disease (PUD; gastric ulcer and duodenal ulcer) and B-cell mucosa associated lymphoid tissue (MALT) lymphoma. Over half of the world's population (4.4 billion) is currently infected with the bacterium.[42] However, the prevalence of *H. pylori* infection varies across the countries owing to alterations in socio-economic condition, sanitation, population density, ethnicity etc.

3.2 *H. pylori* virulence factors

H. pylori is equipped with array of virulence factors. *H. pylori* has about 5–7 unipolar sheathed flagella, which help the bacterium to enter deep into the mucus layer and remain protected from the luminal gastric acidity.[43] *H. pylori* has constitutively expressing urease, that convert urea into ammonia (Fig. 1). The ammonia produced neutralizes the gastric acidity locally and protects the bacteria.[44] The adherence of *H. pylori* onto the gastric epithelial cells is facilitated by the bacterial outer membrane proteins called the adhesins. The human blood group antigen-binding adhesion (BabA) protein

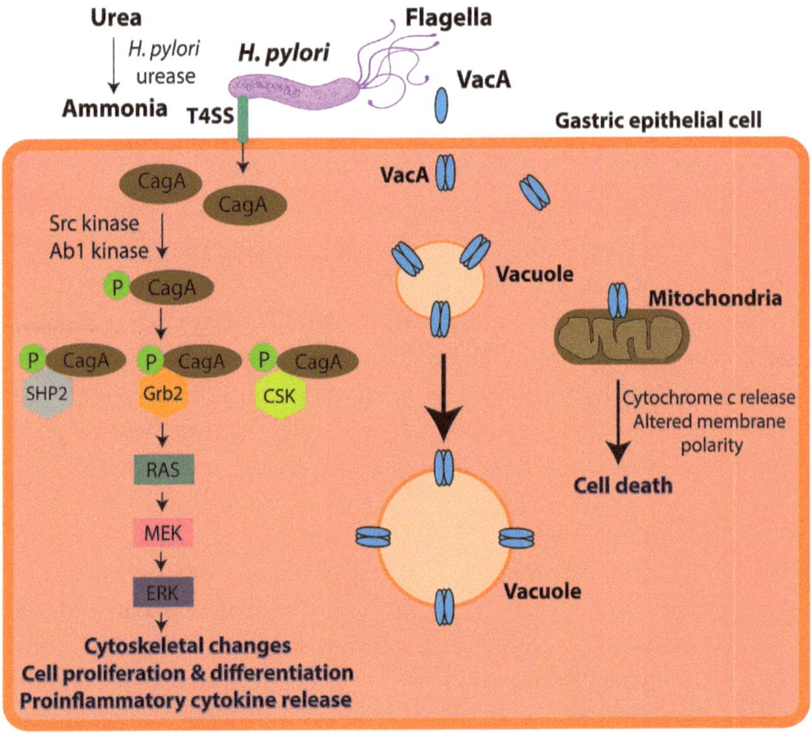

Fig. 1 ***H. pylori* pathogenesis.** *H. pylori* colonization is facilitated by several virulence factors. The initial adhesion to the gastric cells is mediated by surface adhesins like BabA, SabA, HopH and LabA. The enzyme urease helps to neutralize the acidity by producing ammonia. The flagella aid the bacteria to penetrate the mucus layer and protect itself from luminal acidity. The effector proteins like CagA and VacA mediate further pathogenesis. CagA is translocated into the cells by means of a T4SS. In the cell CagA undergoes phosphorylation by Src kinases and Ab1 kinase. The phosphotylated CagA further activated host proteins like SHP2, Grb2 and CSK, which further activated pathways bringing about alteration in cellular morphology, cell proliferation and secretion of proinflammatory cytokines. VacA is a pore forming protein that induces vacuolation, mitochondrial alteration, apoptosis, autophagy and also induce secretion of prostaglandins.

on the bacterial surface recognizes and binds to the difucosylated ABO/ Lewis b (Le^b) antigen, which is present on the epithelial cell surface of the gastric mucosa.[45] Sialic acid-binding adhesin (SabA) or *Helicobacter* outer membrane protein P (HopP) is another bacterial surface protein which binds to sialyl-Lewis a (sLe^a) and sialyl-Lewis X (sLe^x) antigens.[46] Adherence-associated lipoprotein A and B (AlpA/AlpB) can bind to host laminin and mediate bacterial binding to the cell.[47] *H. pylori* outer inflammatory protein A (OipA) or HopH is another surface protein that mediates bacterial adhesion and triggers secretion of IL-8. The *oipA* "on" condition is characterized by enhanced colonization density, IL-8 secretion, neutrophil infiltration and is connected to the increased risk for developing PUD and GC.[48] LacdiNAc-specific adhesin (LabA) is a recently studied adhesin of *H. pylori* that mediates gastric colonization.[47]

Duodenal ulcer promoting gene A (DupA) is involved in the development of DU and gastritis. DupA expression has a negative association with risk of GC. DupA can induce the secretion of proinflammatory cytokines like IL-8, particularly in the gastric antrum and IL-12.[47] One major virulence determinants of *H. pylori* is the *cag*-pathogenicity island (*cag*PAI), which codes the oncoprotein CagA as well as the constituents of Type 4 secretion system (T4SS) which allows the CagA to be translocated inside the gastric cells. The initial event of translocation is mediated by the $\alpha5\beta1$ integrin on the host cell.[49] Once inside the cell, CagA gets attached to the inside surface of the plasma membrane with the Lys-Xn-Arg-X-Arg motif of the CagA N-terminal domain, which binds to the cellular phosphatidyl serine (PS).[50] CagA undergoes tyrosine-phosphorylation at the EPIYA motif by the cellular enzymes like Src family kinases or Abl kinase. Phosphorylated CagA interacts with Src-homology 2 (SH2) domains in proteins like Grb2, SHP2 and CSK and activates downstream pathways resulting in abnormal cell proliferation and differentiation, cytoskeletal alterations and elevated secretion of the pro-inflammatory cytokines such as IL-8 (Fig. 1).[9,51] Depending on flanking region sequence and order of spacers, CagA can be named as A, B and either C or D types. Interestingly, the EPIYA-D motif, which is present in the East-Asian type CagA, binds to SHP2 with higher affinity and activates the effector pathways more intensely than the EPIYA-C motif and results in aggressive diseases in East Asian countries.[52]

Vacuolating cytotoxin A (VacA) is classified as a pore forming toxin that acts on various intracellular locations. The typical cytopathic effect of VacA is the formation of vacuoles in susceptible cell types like parietal cells, gastric epithelial cells, T cells, etc. VacA can induce autophagy and apoptosis. It stimulates the release of cytochrome *c*, decrease mitochondrial transmembrane potential, Bax and Bak activation and damage of the mitochondria. VacA can induce cell permiabilization by forming channels in plasma membrane. VacA can also activate the p38, a MAP kinase, which induces expression of cyclooxygenase-2 (COX-2) resulting in increased prostaglandin E2 secretion (Fig. 1).[53,54]The *vacA* gene encodes the s (signal sequence), i (intermediate), d, m (middle) and c (C-terminal) regions of the VacA protein and exhibits polymorphisms. This results in various allelic types like *s1* (subtypes-*s1a* and *s1b*) and *s2* for *vacA* s region, *i1* (subtypes-*i1a* and *i1b*) and *i2* for *vacA i* region and *m1* (subtypes-*m1a, m1b* and *m1c*) and *m2* for *vacA* m region. The *vacA* allelic combination of *vacAs1i1m1* is almost always linked with a positive *cagA* status and such strains usually exhibit high virulence.[55] Individuals infected with strains having *vacAs1i1m1cagA*+ allelic type shows 4.8-fold increased risk of developing precancerous lesions than the individuals infected with less virulent strains having *vacAs2m2cagA*- type.[56] The *c1* and *d1* allelic types of *vacA* are associated with higher risk of GC. The *d1/c1* allelic type is mostly associated with the *s1m1i1* strains while the *d2/c2* type is common in strains with *s2m2i2* type.[57]

Other virulence determinants of *H. pylori* that help colonization in host include catalase, arginase, superoxide dismutase, phospholipase and lipopolysaccharide. A small non-coding RNA, HPne4160 was found to mediate adaptability in the host. Increased expression of outer membrane proteins and CagA was found associated with decreased expression of HPne4160.[58]

4. Human gastrointestinal microbiome and gastroduodenal diseases

The term microbiome means the total genetic material of all microbial members in our body. The microbes, of course, include bacteria, archaea, viruses, fungi, protozoans and helminths. However, our present knowledge on the bacterial component of the microbiome is much better than our knowledge on other components (Fig. 2). This is because of the fact that the microbiome research mostly depends on metagenomic approaches and metagenomic methods for identifying the bacterial genetic markers or even the bacterial genomes were relatively easier and better developed than the metagenomics methods for the "other" components of the microbiome. Modern microbiome studies typically use metagenomic extraction

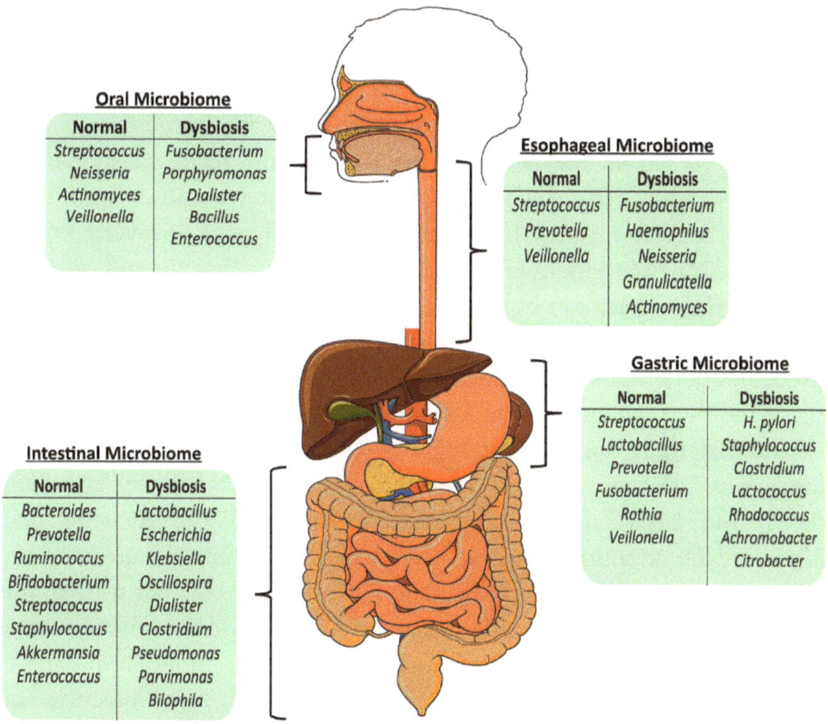

Fig. 2 Human Gastrointestinal Microbiome. Human body harbors millions of different species of microorganism that are either beneficial or detrimental. The composition of human microbiome is also seen to be different across the human body. It is observed that the largest numbers of microbial species are harbored by the intestine. Differences in the composition of microbiome at different locations show how microbiome composition varies in a normal state and in a diseased state.

of the DNA followed by amplification of the V3-V4 regions of the microbial 16S rRNA gene followed by sequence analysis. However, the targeted sequencing, although relatively inexpensive and easier to handle, does not provide several information (e.g., the antibiotic resistance genes and the virulence genes of the bacteria) and therefore, sometimes shotgun metagenomics (also called whole genome metagenome analysis) is necessary for obtaining in depth knowledge of the bacterial strains. Recently, with the use of long read sequencing technique (Oxford Nanopore) the entire bacterial genomes have been assembled from the metagenomics samples.[59]

On the other hand, analysis of virome, the viral component of the microbiome, is relatively challenging since viruses do not carry a common genetic marker like 16S rRNA gene. Therefore, the analysis of virome depends on the shotgun metagenomics, either after enrichment or without enrichment of the total viral particles. The virome analysis has even more complexities since the viral genetic materials could be either DNA or RNA, the viral genomes are comparatively smaller and some of the viral genomes can even be integrated within the host genome. Despite the technical difficulties, the importance of virome is well appreciated in the scientific communities and newer studies are now highlighting the significance of several known viruses and newly discovered viruses in human health and diseases. Although the archaea can be studied by 16S rRNA gene sequencing, their roles in the human body are poorly understood.[60] The fungal, protozoan and helminth component of the microbiome also can be studied by targeted approaches like sequencing the 18S and 28S rRNA genes and sequencing the internal transcribed spacer regions (ITS), which includes the ITS1 (the region between 18S rRNA gene and 5.8S rRNA gene) and ITS2 (the region between 5.8S rRNA gene and 28S rRNA gene).[61–63] However, the attentions toward these "other" microbiomes are new and additional studies are necessary to obtain information on their roles in our body. In this section, we have discussed the bacterial and the other (archaea, viruses, fungi, protozoans and helminths) microbiome in different parts of the healthy gastrointestinal tract and in relation to *H. pylori* infection and different gastroduodenal diseases.

4.1 The "bacterial" component of the gastrointestinal microbiome

4.1.1 Oral microbiome

Oral cavity is the entry point to the interior of the gastrointestinal tract. It is the niche of diverse microorganisms collectively referred as the oral microbiome.[64] *Bacteroides* (now *Porphyromonas*) *gingivalis* was cultivated from periodontal disease lesions in 1970 while *Porphyromonas*, *Peptostreptococcus* (now *Parvimonas*), *Prevotella* were cultivated from decaying root canal and pulp in 1994. The oral microbiome comprises bacteria, fungi, archaea, protozoa and viruses. The Human Oral Microbiome Database (HOMD) is an unique platform launched by the National Institute of Dental and Craniofacial Research in 2010 and provides information about various cultivable and non-cultivable constituents of the oral microbiome along with a repository of their genome sequences.[65]

Culture based and culture independent methods, mainly the 16S rRNA gene sequence-based studies has estimated that the oral microbiome comprises over 700 species of prokaryotes belonging to 185 genera and

12 phyla.[65] The major bacterial phyla include Firmicutes, Bacteroidetes, Fusobacteria, Actinobacteria, Chlamydiae, Proteobacteria, Spirochaetes, Chloroflexi, Absconditabacteria, Gracilibacteria, Saccharibacteria and Synergistetes. The core oral microbiome in most of the healthy individuals comprises genera like *Streptococcus*, *Veillonella*, *Neisseria* and *Actinomyces*.[66] A healthy oral microbiome is maintained by a well-regulated interplay between the host and the resident microbes.

The link between severe gastric diseases and oral microbiome is poorly investigated although it is possible that the pathogenic microbial species from the oral cavity may reach to gastric epithelium through saliva and colonize. Consistent with this, poor oral health (tooth loss, denture associated lesions, gingival bleeding and tooth flossing, but not higher dental plaque or tongue lesions) has been found to be linked with higher risk of GC.[67,68] However, a meta-analysis was able to find a significant association between tooth loss and GC.[69] Interestingly, biomarkers in the tongue coating microbiome have been investigated and higher Firmicutes and lower Bacteroidetes abundance were associated with higher risk of GC.[70,71] The periodontal pathogen burdens as well as the diversity are associated with precancerous lesions of GC.[72,73] Based on the compositions of oral microbiome (e.g., *Veillonella*, *Prevotella*, *Leptotrichia*, *Rothia*, *Capnocytophaga*, *Aggregatibacter*, *Campylobacter*, *Megasphaera*, *Tannerella*, *Granulicatella*) a method for screening GC have been proposed.[74] However, the study suffers from low sample size and high (7.7%) false positive rates. Moreover, infection of *H. pylori* in the stomach was not taken into the account in the study. Taken together, the influence of oral microbiome on gastric diseases seems possible (Table 1), but more studies are required to prove the hypothesis convincingly.

4.1.2 Esophageal microbiome

Esophagus is an 18–25 cm long muscular tube that facilitates transit of food from pharynx to stomach. The sphincters at either ends of the esophagus ensure unidirectional passage of food. It is noticed that at present, GC cases are decreasing while esophageal cancers are increasing and this trend is possibly linked to the changing pattern of microbial composition of the gastrointestinal tract.[75]

Attempts to culture esophageal microflora was carried out from biopsies and aspirates from patients with esophageal carcinoma in early 1980s. The major bacterial species isolated by culture from normal esophageal aspirate were *Streptococcus viridans*, *Klebsiella pneumonia Haemophilus influenzae*,

Table 1 Significant changes in the oral microbiome and its association with different gastroduodenal diseases.

Sl. No	Year	Author	Sample and method	Significance	Association with gastric disease	References
1.	2012	Salazar, Francois, et al.	Biopsy, Oral examination	Observed the association between poor oral hygiene and GC	Gastric cancer	67,68
2.	2018	Ndegwa, Ploner, et al.	Population based cross-sectional survey			
3.	2013	Salazar, Sun, et al.	Biopsy, saliva and plaque sample, Real-time PCR	Observed that periodontal pathogen burdens and the diversity within the oral cavity are associated with increased risk of precancerous lesions	Gastric cancer	72,73
4.	2017	Sun, Zhou, et al.	Whole saliva sample and plaque sample, Real-time PCR			
5.	2018	Wu, Xu, et al.	Midline scrapings of tongue dorsum, 16S rRNA sequencing	Higher abundance of Firmicutes and lower abundance of Bacteroidetes in the tongue coating microbiome is associated with increased risk of GC	Gastric cancer	70,71
6.	2019	Xu, Xiang, et al.	Midline scrapings of tongue dorsum, 16S rRNA sequencing			
7.	2018	Sun, Li, et al.	Subgingival plaque sample, 16S rRNA sequencing	Proposed a method for screening GC that based on the oral microbiome composition such as *Veillonella*, *Prevotella*, *Leptotrichia*, *Rothia*, *Capnocytophaga*, *Aggregatibacter*, *Campylobacter*, *Megasphaera*, *Tannerella*, *Granulicatella*	Gastric cancer	74

Streptococcus group B, *Neisseria catarrhalis* and *Streptococcus faecalis*.[76,77] 16S rRNA sequencing performed from esophageal biopsy DNA from healthy individuals revealed bacteria belonging to 6 phyla namely Firmicutes (70%), Bacteroidetes (20%), Actinobacteria (4%), Proteobacteria (2%), Fusobacteria (2%), and Saccharibacteria (1%).[78] The dominant genera identified were *Streptococcus, Prevotella*, and *Veillonella*.

Altered esophageal microbiome is associated with inflammation and intestinal metaplasia of distal esophagus. Esophageal microbiome is classified based on the dominance of Gram positive taxa, found in normal cases and Gram negative taxa, found in subjects with Barrett's esophagus and GERD into Type I and Type II, respectively.[79] Type I esophageal microbiome was enriched with *Streptococcus*, while the Type II was dominated by *Veillonella, Haemophilus, Prevotella, Neisseria, Granulicatella, Campylobacter, Actinomyces, Porphyromonas, Fusobacterium* and *Rothia*. Moreover, the abundance of *Streptococcus* and *Prevotella* has an inverse relationship.[80,81] Increase in *Campylobacter* was found in individuals with Barrett's esophagus and GERD.[82] Increased expression of IL-18 was observed in Barrett's esophagus and GERD having *Campylobacter* colonization.[83]

Unlike in Barrett's esophagus and GERD, the esophageal microbiome diversity decreases in esophageal adenocarcinoma, with increase in acid-tolerant bacteria like *Lactobacillus fermentum*.[84] *Streptococcus* abundance was lower in high-grade dysplasia and adenocarcinoma when compared to low-grade dysplasia. However, the abundance of Enterobacteriaceae and Verrucomicrobiaceae, particularly *Akkermansia muciniphila was* increased and *Veillonella* was decreased in patients with esophageal adenocarcinoma.[85] Influence of stomach microbiome on esophageal diseases is likely negative, as negative association (although not statistically significant) between *H. pylori* colonization in stomach and gastro esophageal reflex disease (GERD), Barret's esophagus and esophageal adenocarcinoma has been reported.[75,80,86] However, this negative association between infection of *H. pylori* in stomach and diseases of the esophagus remain debatable.[86,87] Overall, the attention toward the esophageal microbiome and its role on esophageal diseases is new. Unfortunately, the significance of esophageal microbiome in the context of gastric diseases is not well documented even though it seems likely that the microbial content of the esophagus should reach to stomach through saliva and esophageal secretion.

4.1.3 Gastric microbiome

Stomach is one of the earliest studied visceral organs. In spite of the constant exposure to microbes via mouth, it has long been thought that microbial colonization is not possible in stomach, because of its unique physiology. The pH of the healthy human stomach is highly acidic, which typically ranges from 2 to 3 and elevates to >5 after food ingestion.[88,89] Further, the stomach is characterized by the reflux of bile, peristaltic movements and the presence of nitric oxide derived from nitrite produced by oral bacteria, which are highly hostile for microbial growth. Additionally, the lack of proper sampling and identification techniques hindered the understanding of the gastric microbes for long. In 1939, Doenges observed spiral organisms in 42% of the gastric autopsy specimens by hematoxylin and eosin staining.[90] Even though several reports on the presence of spiral bacteria were established by histology, attempts to isolate the bacteria were not successful. In 1981, a report in "The Lancet" showed the presence of acid resistant bacteria in stomach by culture from the gastric juice.[91] Now it is well appreciated that stomach, despite having a harsh milieu, has a unique microbiome, although the bacterial count is as low as 10^2–10^4 CFU/mL.[92]

Conventional culture-based methods to study the gastric microflora were inaccurate and biased as 80% of the bacteria are non-cultivable, although attempts have been made.[93] By 16S rRNA gene cloning and library preparation, Bik et al. in 2006, studied the microbiome of the gastric mucosa from 23 healthy individuals. The study revealed 1056 non-*H. pylori* clones and 127 phylotypes. *Streptococcus, Prevotella, Rothia, Fusobacterium* and *Veillonella* were identified to be the dominant genera. Later Li et al. in 2009 analyzed the microbiome from 10 healthy individuals by cloning and 16S rRNA gene sequencing and detected 1223 non-*H. pylori* clones, 133 phylotypes, with *Streptococcus, Prevotella, Neisseriae, Haemophilus* and *Porphyromonas* as the dominant genera. Pyrosequencing studies has identified that *Prevotella, Streptococcus, Veillonella, Rothia* and Pasturellaceaea are the dominant members in the stomach and also added that the composition did not vary much between antrum and body.[94,95] It was also observed that Firmicutes, Bacteroidetes and Actinobacteria were the dominant phyla present in the gastric fluid, while Firmicutes and Proteobacteria were predominant in the gastric mucosal samples.[95]

4.1.3.1 Healthy gastric microbiome

Normally, human stomach is dominated by five phyla namely Firmicutes, Bacteroidetes, Actinobacteria, Fusobacteria and Proteobacteria. Studies have also found that the normal healthy gastric microbiome are more diverse than the ones observed in patients with severe gastric diseases.[96] Factors that affect the composition of human gastric microbiome includes age, diet, medication, overuse of antibiotic and *H. pylori* colonization.[95,97]Proton pump inhibitors (PPI) alter the gastric microbiome composition and increases the presence of oropharyngeal-like and fecal-like bacteria in the stomach.[95] PPIs are one of the most commonly prescribed drugs for gastric-acid related disorder which work by suppressing the release of H^+ ion. Gastric acid normally keeps the number of many gastric microbiome under check, reduction of gastric acid leads to an overall shift in the composition of gastric microbiome with increased diversity.[98] Bacterial overgrowths have been detected in the stomach at a pH 3.8 and above.[95]

Gastric microbiome dysbiosis is connected to various gastric diseases and conversely, the composition of gastric microbiome may contribute to gastric diseases development (Table 2). Importantly, *H. pylori* infection also alters normal gastric microbiome. A decrease in the gastric microbial diversity is observed in *H. pylori* infected individuals as compared to individuals without infection.[111,112] Consistent to this observation, an increase in microbial diversity was also seen following successful *H. pylori* eradication.[113] A higher relative abundance of Proteobacteria, Spirochaetes and Acidobacteria and a lower relative abundance of Bacteroidetes, Firmicutes and Actinobacteria were observed in *H. pylori* positive cases than the *H. pylori* negative cases.[114] A different study reported enrichment of the phyla like Proteobacteria, Firmicutes and Actinobacteria in the *H. pylori* positive stomachs, while phyla like Firmicutes, Bacteroidetes and Actinobacteria tends to be abundant in stomach without the *H. pylori* infection.[115] The C57BL/6 N mice grown by two different vendors, Taconic Farms and Charles River Lab showed significant difference in the degree of inflammation, gastritis and metaplasia upon infection by a strain of *H. pylori*. It was found that the mice from Charles River Lab exhibited higher inflammatory response when compared to that from Taconic Farms. Interestingly, the Taconic Farms mice had significantly higher load of *Lactobacillus* species ASF360, while the Charles River Lab mice had more *Lactobacillus* species ASF361. This suggests that colonization status of different *Lactobacillus* strains in the two groups of animals play crucial roles in deciding the manifestation of *H. pylori* infection.[116]

4.1.3.2 Gastric microbiome in gastritis

Gastritis is often considered as the primary stage that eventually leads to GC. Thus, it is important to further understand the alterations in gastric microbiome in association with gastritis. It is identified that *H. pylori* induced chronic gastritis indeed alters the microbial flora of the stomach.

Table 2 Significant changes in the gastric microbiome and its association with different gastroduodenal diseases.

Sl. No	Year	Author	Sample and method	Significance	Association with Gastric Diseases	References
1.	1985	Sjöstedt, Heimdahl, et al.	Biopsy and gastric juice, culture technique	Observed increased abundance of *Lactobacillus, Streptococcus, Prevotella, Veillonella* as well as *Bifidobacterium* in the gastric mucosa of patients with GC	Gastric Cancer	99
2.	1988	Sjöstedt, Kager, et al.	Biopsy and gastric juice, culture technique			100
3.	2009	Dicksved, Lindberg, et al.	Biopsy, 16S rRNA sequencing	Observed increased abundance of *Streptococcus*, *Lactobacillus*, *Veillonella* and *Prevotellain* Swedish patients with GC	Gastric cancer	101
4.	2012	Hu, He, et al.	Biopsy, MALDI-TOF MS	Low abundance of non *H. pylori* bacteria in patients with GU when compared with non-ulcer dyspepsia	Gastric ulcer	102
5.	2014	Eun, Kim, et al.	Biopsy, 16S rRNA sequencing	Relative abundance of *H. pylori* in gastric environment in GC was lower when compared with chronic gastritis	Gastric cancer	103
6.	2014	Eun, Kim, et al.	Biopsy, 16S rRNA sequencing	The relative abundance of *H. pylori* is higher in chronic gastritis than that observed in GC. Also observed the increased abundance of Bradyhizobiaceae, Caulobacteraceae, Lactobacillaceae, and Burkholderiaceae in patients with chronic gastritis not associated with *H. pylori* when compared to patients with *H. pylori* associated chronic gastritis	Chronic gastritis	103

7.	2016	Wang, Zhou, et al.	Biopsy, 16S rRNA sequencing	Observed higher abundance of Proteobacteria, Firmicutes, Bacteroidetes, Fusobacteria and Actinobacteria in the stomach of GC patients in China	Gastric cancer	104
8.	2016	Yang, Woltemate, et al.	Biopsy, 16S rRNA sequencing	Observed difference in gastric microbiome of two set of population in Colombia, where people with high risk for GC showed increased abundance of *Lepotrichiawadie* and *Veillonella* while people with low risk had higher abundance of *Staphylococcus*, *Neisseria flavescens* and *Flavobacterium* in their gastric environment	Gastric cancer	105
9.	2017	Parsons, Ijaz, et al.	Biopsy, 16S rRNA sequencing	Observed increased abundance as well as diversity of microbes with *Streptococcus* dominating the gastric environment in patients with autoimmune gastritis when compared to normal gastric environment	Autoimmune gastritis	96
10.	2018	Ferreira, Pereira-Marques, et al.	Biopsy, 16S rRNA sequencing	Observed increased abundance of Actinobacteria and Firmicutes as well as decreased abundance of Bacteroidetes and Fusobacteria in patients with GC	Gastric cancer	106
11.	2018	Liu, Xue, et al.	Biopsy, MALDI-TOF MS	Observed significant abundance of *Streptococcus mitis*, *Neisseria flavencens*, and *Neisseria perlava* in patients with chronic gastritis	Chronic gastritis	107

Continued

Table 2 Significant changes in the gastric microbiome and its association with different gastroduodenal diseases.—cont'd

Sl. No	Year	Author	Sample and method	Significance	Association with Gastric Diseases	References
12.	2018	Hsieh, Tung, et al.	Biopsy, 16S rRNA sequencing	Observed that the most abundant genera in GC are *Clostridium*, *Fusobacterium* and *Lactobacillus* along *Veillonella* and *Lactococcus*	Gastric cancer	11,108
13.	2019	Liu, Shao, et al.	Biopsy, 16S rRNA sequencing	Gastric tumor environment is dominated by *Prevotella melaninogenica, Streptococcus anginosus* and *Propionibacterium acnes* while *H. pylori, Prevotella copri* and *Bacteroides uniformis are* significantly reduced	Gastric cancer	109
14.	2020	Wang, Xin, et al.	Biopsy, 16S rRNA sequencing	Increased abundance of Fusobacteria, Bacteroidetes, Patescibacteria are observed in signet-ring cell carcinoma while in Caucasian population with adenocarcinoma have increased abundance of Proteobacteria and Acidobacteria. *Novosphingobium*, *Ralstonia, Ochrobactrum*, *Anoxybacillus*, and *Pseudoxanthomonas* are seen abundant in the early stages of GC while *Burkholderia, Tsukamurella, Uruburuella,* and *Salinivibrio* are abundant in the advanced stages	Gastric cancer	110

Using pyrosequencing, a study based on Korean population observed that the relative abundance of *H. pylori* is higher in gastritis patients than GC patients.[103] This is confirmed by another study, which has also found increase in relative abundance of *H. pylori* in chronic gastritis patients than in GC patients. *Streptococcus, Neisseria* and *Prevotella* were found to be abundant in both *H. pylori* induced chronic gastritis and in chronic gastritis without *H. pylori* infection.[106] Another study showed that there were variations in microbial communities in stomachs of the patients with *H. pylori* associated chronic gastritis and for the patients with chronic gastritis, but no *H. pylori* infection. It is also interesting to note that the relative abundance of several bacterial families such as Lactobacillaceae, Bradyhizobiaceae, Burkholderiaceae and Caulobacteraceae were higher in the stomach of the chronic gastritis patients without *H. pylori* infection as compared to chronic gastritis patients having *H. pylori* infection.[103] Higher abundance of Firmicutes was observed in gastric environment of antral gastritis patients, while Proteobacteria was abundant in normal individuals. Other bacterial species that are found abundant during chronic gastritis (excluding *H. pylori*) are *Streptococcus mitis, Neisseria perlava and Neisseria flavencens.*[117]

In atrophic gastritis, gastric glands are replaced by connective tissue or intestinal-type tissue. *H. pylori* is the well documented causative agent for atrophic gastritis and early detection of atrophic gastritis is extremely important as it has been proved to be a precancerous condition.[89,97] In a gastric microbiome profiling study in gastritis vs normal cases, it was observed that *Streptococcus* genus was dominant in the stomach of antral gastritis patients.[118] Also, a higher *Streptococcus* abundance and a lower *Prevotella* abundance were seen in atrophic gastritis when compared to normal cases. Taxa like Helicobacteraceae, Streptococcaceae, Fusobacteriaceae, and Prevotellaceae were found to be higher in atrophic gastritis. Moreover, the level of *Tannerella, Prevotella and Treponema* were found to be lower in atrophic gastritis. A study based on matrix-assisted laser desorption ionization time-of-flight (MALDI-TOF) mass spectrometry analysis found that the abundance of non-*H. pylori* bacteria were higher in NUD cases when compared with GU cases. *Streptococcus, Neisseria, Staphylococcus* and *Rothia* were the genus dominantly seen in the stomach of the *H. pylori* positive individuals. Chronic dyspepsia is characterized by the abundance of *Staphylococcus* and *Lactobacillus*, while *Streptococcus, Pseudomonas mosselii, Escherichia coli* and *Klebsiella pneumoniae* dominated the stomach of the non-dyspeptic patients.[119] A study done in UK population showed abundance of taxa like Helicobacteraceae, Streptococcaceae, Fusobacteriaceae and Prevotellaceae in chronic atrophic gastritis.[96]

Autoimmune gastritis is a different form of gastritis which is characterized by the chronic inflammation of mucosa that may lead to atrophic gastritis. Though there are only few studies on understanding the connection between gastric microbiome and autoimmune atrophic gastritis, it is found that the patients with autoimmune atrophic gastritis had 38% Streptococcaceae as the predominant group in their stomach followed by 9% Prevotellaceae, 7% Flavobacteriaceae, 7% Campylobacteraceae, 5% Enterobacteriaceae and 5% Pasteurellaceae.[96,97] In these patients, rather than a simple change in the bacterial population, a drastic gain or loss of certain bacteria were seen, like the colonization of *Gemella* and *Bosea.*[96] Firmicutes were found to be dominant in the gastric microbiome of autoimmune atrophic gastritis patients as compared to the patients with chronic atrophic gastritis. The autoimmune atrophic gastritis is also associated with a higher microbial diversity than the *H. pylori* associated atrophic gastritis.[97]

4.1.3.3 Gastric microbiome in gastric cancer (GC)

GC is still one of the major health problem globally with over a million cases being reported every year.[120] The incidence of GC varies from one population to another with the rates soaring highest in East-Asia, South and Central America and Eastern Europe.[121] The main cause of GC is *H. pylori* colonization in stomach (causing approximately 90% of cases globally) and it is the first bacterium to be described as a carcinogen. Several studies have found the positive correlation between *H. pylori* infection and progression of GC and it happens via a cascade of inflammation, gastric atrophy, intestinal metaplasia and dysplasia.[89] The associations of *H. pylori* with GC vary with different populations. It was seen that there is neutral or negative correlation of *H. pylori* with GC in western countries while in Eastern population have strong evidence of a positive correlation.[121]

In a study on the diversity of microbiome in GC patients from China, it was found that the abundance of Proteobacteria, Firmicutes, Fusobacteria, Actinobacteria and Bacteroidetes were higher in the stomach of patients with GC.[104] The dominant genera that are associated with GC are *Clostridium, Fusobacterium* and *Lactobacillus* along with enrichment of *Veillonella* and *Lactococcus.*[11,108] A recent study on Portuguese population suggests a lower abundance of *H. pylori* and a decreased diversity in gastric microbiome in stomach of the patients with GC. Also the relative abundance of bacteria like *Lactobacillus, Citrobacter, Achromobacter, Rhodococcus* and *Clostridium* were higher in GC. These bacteria are normally seen as intestinal commensals and can act as opportunistic pathogens. Contradicting to the existing understanding on

H. pylori inducing GC, in a susceptible mouse model, it was shown that *Citrobacter rodentium* have the potential to induce epithelial proliferation and facilitate carcinogenesis.[106] These studies indicate the significance of non-*H. pylori* bacteria in GC. A study where researchers looked into the gastric microbiome of two different populations in Columbia showed significant variation from each other. It was found that people with high risk for developing GC, had increased abundance of *Leptotrichia wadie* and *Veillonella* while *Staphylococcus, Neisseria flavescens, Flavobacterium* and *Rothia* spp. were abundant in patients with lower risk of GC development.[117]

Similarly, another study showed an increased relative abundance of *Lactobacillus, Streptococcus, Prevotella and Veillonella in the stomach of* Swedish patients with GC.[101] A decrease in *H. pylori, Bacteroides uniformis* and *Prevotella copri* was associated with GC in China.[109] Furthermore, a culture dependent study revealed that the abundance of anaerobic bacteria like *Clostridium* and *Bacteroides* were higher for the patients with GC.[95,101] Gastric microbiome also varies with the stage of GC. In early stages of GC, the abundance of *Novosphingobium, Ochrobactrum, Ralstonia, Pseudoxanthomonas* and *Anoxybacillus* are higher, while *Burkholderia, Uruburuella, Salinivibrio* and *Tsukamurella* are higher in the advanced stages.[110] *Prevotella melaninogenica, Streptococcus anginosus* and *Propionibacterium acnes* are found to be higher in tumor tissues compared to normal and pre-tumor tissues.[109] In signet-ring cell carcinoma, Fusobacteria, Bacteroidetes, Patescibacteria were found to be more abundant while in adenocarcinoma, Proteobacteria and Acidobacteria are enriched in Caucasian population.[122]

4.1.3.4 Gastric microbiome in peptic ulcer disease (PUD)

H. pylori colonization accounts for about 95% of duodenal ulcer (DU) and 70% of gastric ulcers (GU).[89] Apart from *H. pylori*, a culture based study suggests the association of *Streptococcus* with GU.[123] A study from China demonstrated higher abundance of acid susceptible genus such as *Staphylococcus, Rothia* and *Neisseria* in PUD when compared to NUD cases[102].

4.1.4 Intestinal microbiome

Human intestine is the most densely colonized part of our body. Human gut microbiome consist over 35,000 species.[124] Analysis of human intestinal microbiome evolved to be an extensively studied area in the past decade in the context of many communicable and non-communicable diseases. These studies, apart from appreciating the microbial link with many diseases, have also suggested beneficial roles of several intestinal microbes. The intestinal microbiome is heavily altered during severe gastric diseases like PUD and GC and the relative abundance of the beneficial microbes like *Bifidobacterium* is decreased.

Most studies related to intestinal microbiome actually analyze fecal microbiome, which do not require any invasive procedures. However, it is also possible to study mucosal biopsy and intestinal fluid for the identification of the colonized microbes. In a study published in 1969, Drasar et al. described different types of tubing systems for obtaining the contents from small intestine for the isolation of bacteria. In this culture-dependent study, the identified bacteria from the upper small intestine (jejunum) were Enterobacteria, *Enterococci, Streptococcus, Neisseria, Staphylococci, Lactobacilli, Bacteroides, Bacteroides melaninogenicus,* Fusobacteria, Bifidobacteria and Clostridia. The bacterial load was found to be increased toward the terminal ileum and the composition, like the large intestine, was found to be dominated by *Bacteroides* and *Bifidobacterium.*[125]

The large intestine is the niche of ~70% of all the microbes in our body.[124] Each gram of human feces contains about 10^{12} bacteria.[126] Early culture-dependent studies showed that 60–80% of the total fecal microbiome belong to phylum Bacteroidetes and Firmicutes.[126] In a metagenomic study of 98 samples, phylum Proteobacteria, Bacteroidetes, Spirochaetes, Firmicutes and Actinobacteria were seen in all individuals, but the relative abundance of each phylum varied. For instance, the abundance of Bacteroidetes varied from 0.37% to 98.82% among the study group. *Bacteroides fragilis* was identified in all the samples studied. *Bifidobacterium bifidum* BGN4 strain was identified in 97.9% of the samples. According to the Gut Feeling Knowledge Base database, the most abundant class under the phylum Firmicutes is Clostridia (20.3%), which constitute members of order Clostridiales. The other dominant classes are Bacteroidia (18.5%), Bifidobacteriales (16.6%), Enterobacterales (14%) and Lactobacillales (14%). *Bifidobacterium longum* is found to be the abundant species under Bifidobacteriaceae family and its abundance varied from 0.003% to 10.30% among the study group. The most abundantly seen family was Bacteroidaceae, followed by Lachnospiraceae, Ruminococcaceae, Odoribacteraceae, Rikenellaceae, Bifidobacteriaceae, Enterobacteriaceae and Tannerellaceae. *Bacteroides* is the most dominant genus found with *Bacteroides dorei* being the most abundant species.[127] Increased abundance of *Bacteroides, Prevotella* and *Ruminococcus* with decreased abundance of Proteobacteria, indicates a healthy gut.[128] Microbial diversity varies between the lumen and mucosa. *Bifidobacterium, Bacteroides, Streptococcus, Clostridium,*

Lactobacillus, Ruminococcus and Enterococcus are the genus dominating the luminal microbiome while *Lactobacillus, Clostridium, Akkermansia* and *Enterococcus* are mostly associated with the mucosa.[124]

As compared to gastric microbiome, the intestinal microbiome is less studied with respect to *H. pylori* infection and related gastrointestinal diseases (Table 3). However, recent literature clearly showed that *H. pylori* colonization in stomach not only alters the gastric microbiome but it also has the potential to disrupt the normal intestinal microbiome. The C57Bl/6 mice colonized with *H. pylori* strain PMSS1 (10^5–10^6 CFU/g of stomach) showed altered intestinal microbiome.[129] The significantly altered genera were *Lactobacillus, Turicibacter* and *Allobaculum* in ileum and *Turicibacter* and *Anaeroplasma* in cecum.[129] In human, a study from Germany showed that the higher load of *H. pylori* in stomach is linked to higher diversity and increased abundance of pathogenic bacteria in the intestinal microbiome.[136] A study in Sharjah, United Arab Emirates showed that the intestinal microbiome of *H. pylori* positive individuals was abundant in Succinivibrio, Coriobacteriaceae, Enterococcaceae and Rikenellaceae.[131] *H. pylori* induced gastritis was also associated with dysbiosis in intestine with noticeable decrease in the Firmicutes/Bacteroidetes ratio in Chinese children.[132] Metagenomic analyses of bacterial V3-V4 regions of 16S rRNA gene, using Illumina MiSeq revealed that the intestinal microbiome composition varies between the *H. pylori* positive and negative subjects in Chinese and Indian populations.[130,137] For both populations, *H. pylori* infections were linked to increased intestinal microbial diversities.[130,137] However, the alteration in gut microbiome composition in the presence or absence of *H. pylori* varies among the populations. A higher relative abundance of Firmicutes and Proteobacteria and a lower relative abundance of Bacteroidetes were found in the Chinese population with past *H. pylori* infections.[130] On the other hand, *H. pylori* positive individuals from India had lower Actinobacteria abundance, but higher abundance of TM7 (Saccharibacteria) in their intestine as compared to *H. pylori* negative individuals.[137] Further analyses revealed that the relative abundance of the genus *Bifidobacterium* was lower and the relative abundance of the genera *Dialister* and *Prevotella* was higher in *H. pylori* infected group than *H. pylori* negative group.[137] For the Indian population, specifically in the intestine of the *H. pylori* positive individuals with severe gastric diseases like GU or GC, the relative abundance of several *Bifidobacterium* species (e.g., *B. adolescentis, B. longum,* and *B. bividum*) was found to be significantly low.[137] Similarly, a lower abundance of *Bifidobacterium* along with few other genera like *Lachnoclostridium, Parabacteroides* and *Barnesiella* was observed in the

Table 3 Significant changes within the intestinal microbiome and its association with different gastroduodenal diseases.

Sl. No	Year	Author	Sample and method	Significance	Association with gastric disease	References
1.	2016	Kienesberger, Cox, et al.	Stomach autopsy samples, 16S rRNA sequencing	Studies in mice model showed that *H. pylori* colonization alters intestinal microbiome and the altered genera in ileum were *Lactobacillus*, *Turicibacter* and *Allobaculum* while in cecum, *Turicibacter* and *Anaeroplasma* were altered	*H. pylori* infection	129
2.	2018	Gao, Zhang, et al.	Fecal sample, 16S rRNA sequencing	Higher abundance of Firmicutes and Proteobacteria while decreased abundance of Bacteroidetes were observed in Chinese patients with past *H. pylori* infections	Gastric cancer	130
3.	2019	Dash, Khoder, et al.	Fecal sample, 16S rRNA sequencing	Study conducted in Sharjah, showed that *H. pylori* infected patients have higher abundance of *Succinivibrio*, *Coriobacteriaceae*, *Enterococcaceae* and *Rikenellaceae* in their intestinal microbiome	*H. pylori* infection	131
4.	2019	Yang, Zhang, et al.	Fecal sample, 16S rRNA sequencing	Study conducted in Chinese children showed increased abundance of Bacteroidaceae, Bacteroides, Fusobacteriaceae, Megasphaera, Enterobacteriaceae and Porphyromonadaceae in children with non *H. pylori* gastritis while in children with *H. pylori* induced gastritis have significantly low Firmicutes/Bacteroidetes ratio	Gastritis	132

5.	2019	Qi, Sun, et al.	Fecal sample, 16S rRNA sequencing	Observed increased abundance of *Lactobacillus*, *Veillonella* and *Streptococcus* while *Lachnospira*, and *Tyzzerella_3* were found to be decreased. A strong positive correlation between *Veillonella*, *Lactobacillus* and *Streptococcus* were also identified	Gastric cancer	133
6.	2020	Guo, Zhang, et al.	Fecal sample, 16S rRNA sequencing	Observed increase in the abundance of *Bifidobacterium* in intestines of patients who underwent anti-*H. pylori* therapy	Following *H. pylori* eradication	134
7.	2021	Sarhadi, Mathew, et al.	Fecal sample, 16S rRNA sequencing	Observed lower abundance of Bifidobacteriaceae family in patients with a subtype of GC. Also observed significantly lower abundance of *Oscillibactor*, *Lachnoclostridium*, *Bifidobacterium*, *Parabacteroides* and *Barnesiella*	Gastric Cancer	135
8.	2021	Devi, Devadas, et al.	Fecal sample, 16S rRNA sequencing	*H. pylori* positive individuals in India have lower abundance of *Bifidobacterium* (*B. adolescentis*, *B. longum, and B. bividum*) and higher abundance of *Dialister* and *Prevotella* when compared with *H. pylori* negative individuals	Gastric cancer and gastric ulcer	46
9.	2021	Sarhadi, Mathew, et al.	Fecal sample, 16S rRNA sequencing	Observed increase in the abundance of Lactobacillaceae in patients with gastrointestinal stromal tumors (GIST) and gastric adenocarcinoma	Gastric cancer and GIST	135

intestine of Finnish patients with gastric adenocarcinoma.[135] How the lower abundance of a known beneficial bacterium like *Bifidobacterium* in the lower gut is linked to aggressive diseases is not understood at present, but it is worth mentioning that previous studies have reported anti-tumor and anti-ulcer properties of *Bifidobacterium*.[138,139] Furthermore, a higher relative abundance of *Bifidobacterium* in the intestine was also noticed in patients who underwent anti-*H. pylori* therapy and showed successful eradication.[134] A lower abundance of Lactobacillaceae was also seen in the gut of the patients with gastric adenocarcinoma and gastrointestinal stromal tumors (GIST).[135]

Several taxa were also found in relatively higher abundance in the individuals with severe gastric diseases. For example, higher abundance of *Oscillospira* in the gut of *H. pylori* positive patients from India having severe gastric diseases and higher abundance of Enterobacteriaceae in the gut of the Finnish patients with GC were observed.[135,137] The increased abundance of Enterobacteriaceae is associated with elevated inflammatory response and has strong correlation with the progression from premalignant GC and metaplasia. For the Finnish population, the microbial diversity was lower in diffused subset of gastric adenocarcinoma, followed by intestinal subtype and GIST.[135] However, in a Chinese population, intestinal microbiome richness was found to be higher in GC patients with an increase in *Lactobacillus*, *Escherichia* and *Klebsiella*.[133]

5. The "other" gastrointestinal microbiomes and their relationships with *H. pylori* infection and gastroduodenal diseases

5.1 Virome

Human virome refers to the collection of viruses in and on the body. Human body typically harbors huge number of viruses and phages form majority of the viruses. Studies using fluorescent microscopy shows that saliva contain around 10^8 virus-like particles (VLPs) per ml, with Caudovirales being the most abundant among the phages.[140] *Anelloviridae, Redondoviridae, Herpesviridae* and *Papillomaviridae* are the most common eukaryotic viruses in the oral cavity.[141,142] However, the prevalence of *Redondoviridae* can vary from 2% to 15% in different populations.[141,143,144] Recent studies showed that the phages can traverse the gut epithelium by transcytosis and reach blood circulation.[144] There are at least 10^9 VLPs that are present in each gram of human feces.[145] Metagenomic studies revealed that *Caudovirales*, which consists of the tailed phages and *Microviridae* dominate the human gut virome. CrAssphage (cross-assembly phage), particularly Podoviruses that infect the Bacteroidetes, constitute about 90% of the gut virome.[146] Eukaryotic viruses are seen in lower abundance in healthy human gut and include DNA viruses like *Geminiviridae, Anelloviridae, Nanoviridae, Papillomaviridae, Parvoviridae, Adenoviridae Polyomaviridae, Circoviridae* and *Herpesviridae.*[147] The common RNA viruses seen are *Caliciviridae, Picornaviridae* and *Reoviridae.*[144,148]

5.1.1 *Epstein-Barr virus (EBV) or human herpesvirus 4 (HHV-4)*

EBV is a dsDNA virus belonging Herpesviridae that primarily infects B-cells and epithelial cells and is associated with Burkett's lymphoma, infectious mononucleosis, nasopharyngeal carcinoma, GC and autoimmune diseases.[33] Like *H. pylori*, EBV is also classified as a Class 1 carcinogen by IARC.[149] The virus may enter a latent phase, where it maintains minimum gene expression. EBV is associated with about 10% of the total gastric carcinoma cases.[150] Ephrin receptor A2, integrins and non-muscle myosin heavy chain IIA (NMHCIIA) facilitate the entry of virus into the epithelial cells.[151] Inside the cell, the naked linear genomic DNA gets assembled into circular mini-chromosome in the host nucleus.[152] The assembly provides protection from DNA damage and also helps to regulate gene expression. Methylation of the CpG motifs facilitates the establishment of latency.[153] Viral latent membrane protein 2A (LMP2A) cause STAT3 phosphorylation and induce activation of DNMT1 and methylation of *PTEN* promoter.[154] This in turn facilitates the progression and maintenance of cancer.

It is estimated that 45% of the world population are co-infected with *H. pylori* and EBV.[155] The risk of GC has a significant association with the coexistence of EBV and *H. pylori* in the human stomach.[156] Further, the load of EBV DNA and anti-EBV IgG titer were found to be higher for the *H. pylori* positive individuals, suggesting *H. pylori*'s role in the lytic phase of the virus.[157,158] CagA protein of *H. pylori* is a strong activator of PLCγ pathway which facilitates the reactivation of latent EBV. Moreover, it was observed that the occurrence of EBV-*H. pylori* coinfection was higher in patients with DU and GC as compared to the patients with GERD.[155,158,159]

5.2 Mycobiome

Mycobiome, the fungal components of the human microbiome, is a much overlooked area. The human mycobiome forms not more than 0.1% of the total gut microbiome.[160] An elevation in Basidiomycota/Ascomycota ratio was observed in the fecal microbiome of the patients having inflammatory bowel disease. Further increased abundance of *Candida tropicalis* was associated with Corn's disease and correlated positively with the abundance of *Serratia marcescens* and *Escherichia coli*. Formation of biofilm was observed involving *Candida tropicalis, Serratia marcescens* and *Escherichia coli*, using electron microscope. It was also found that the pili of *Serratia marcescens* mediate the binding between *Candida tropicalis* and *Escherichia coli*. [160]

Pyrosequencing of oral rinse samples revealed that, the oral cavity of a healthy individual is mostly inhabited by *Candida, Cladosporium,* and *Aspergillus* with smaller proportion of *Fusarium, Penicillium, Glomus, Alternaria, Saccharomycetales, Zygosaccharomyces, Cryptococcus, Ophiosoma, Phoma* and *Schizosaccharomyces*.[161] The most dominant fungal phyla present in gut is Ascomycota, followed by Zygomycota and Basidiomycota.[162] Culture-dependent studies have identified member of the fecal mycobiome as, *Candida albicans, Candida deformans, Yarrowia lipolytica, Cryptococcus neoformans, Cryptococcus saitoi, Aspergillus glaucus, Lichtheimia ramosa, Pleurostomophora richardsiae, Candida glabrata, Rhodotorula mucilaginosa, Mucor circinelloides and Trichosporon asahii*. On the other hand, *Malassezia, Saccharomyces, Candida, Cyberlindnera, Cladosporium, Debaryomyces, Aspergillus, Pichia, Clavispora, Gloeotinia/Paecilomyces, Galactomyces* and *Penicillium* were found dominant in culture-independent studies including ITS2 and 18S rRNA sequencing.[163,164] A pyrosequencing study revealed that the composition of healthy fecal mycobiome primarily includes *Wallemia*, Trichomaceae, *Rhodotorula*, Saccharomycetaceae, Pleosporaceae, Agaricaceae, Cystofilobasidiaceae, Metschnikowiaceae, Ascomycota and Amphisphaeriaceae.[165] 18S rDNA sequencing using colonic biopsy tissues revealed the presence of *R. mucilaginosa, Cryptococcus carnescens, C. albicans, Galactomyces geotrichum, Septoria epambrosiae, Bulleracrocea, Candida dubliensis, Raciborskiomyces longisetosum, Penicillium italicum and Cladosporium cladospoirioides*.[166]

A study compared the fungal diversity between in the cancer lesion and the adjacent non-cancerous area in patients with gastric cancer using the ITS2 sequence. An elevation in the relative abundances of *Alternaria* and *Candida* and a decrease in *Saitozyma* and *Thermomyces* were observed in GC cases. Abundance of species like *Arcopilus aureus, Fusicolla acetilerea, Fusicolla aquaeductuum* and particularly *C. albicans* were increased while *Aspergillus montevidensis, Candida glabrata, Penicillium arenicola* and *Saitozyma podzolica* were decreased in patients with GCs. The study also suggests increased abundance of *C. albicans* as a significant biomarker for GC.[167]

Candida is one of the dominant components of gastrointestinal mycobiome and gastric candidiasis is mostly seen in immunocompromised individuals. However, *Candida* associated GU is seen in both healthy and immunocompromized individuals.[168] But certain conditions like administration of antibiotics, steroids, immunosuppressant and anti-cancer drug can alter the microenvironment and results in overgrowth of the fungus.[169] Usually *Candida* associated ulcer have slow healing rate.[168] GU with fungal infections appeared to be malignant and larger than the normal ulcer.[170] There needs to further understanding regarding the mycobiome and its association with disease conditions.

5.3 Protozoa and helminths

Specific pattern in the intestinal microbiome composition has been associated with various parasitic infections. A study from Bangladesh showed that infants with diarrheagenic *Entamoeba histolitica* infection, the abundance of *Prevotella copri* was found to be higher.[171] The abundance of *Prevotella copri* and *Prevotella stercorea* were found to be lower in individuals with asymptomatic *Entamoeba* infection. Higher abundance of Proteobacteria along with lower abundance of Verrucomicrobia and Bacteroidetes are associated with protection against *Cryptosporedium* infection.[172,173] On the other hand, *Cryptosporedium* infected group shows higher relative abundance of *Bacteroides pyogenes, Bacteroides fragilis, Akkermansia muciniphila* and *Prevotella bryantii*. An increase in Clostridia levels with a decrease in Enterobacteriaceae was observed in *Blastocystis* infection.[173]

The association between parasitic components of the human microbiome and the development of gastroduodenal diseases is not well studied. J. A. G. Fibiger in 1926, demonstrated the development of GC in mice model infected with nematode *Spiroptera*.[174] However, the authenticity of this study was questioned later by many researchers.[175] *Schistosoma japonicum*, an intestinal trematode, is mostly associated with colorectal cancer but several studies from Japan and China suggest its association with the cancers of stomach and esophagus.[176–178] In a study done on Chinese population, it was found that the coinfection of *Schistosoma japonicum* with *H. pylori* is linked to alteration in IgG response to *H. pylori* and diminished gastric atrophy.[179] Patients having co-infection exhibited lower *H. pylori* IgG titer and CagA seropositivity, indicating an antagonistic association.[179]

5.4 Archaea

The domain Archaea is considered as a distinct group and is different from bacteria based on the rRNA gene sequence.[180] The members of Archaea exist in the microbiome of human skin, oral cavity and gastrointestinal tract.[181] The most abundant Archaeal group in the human gut are the Methanogens (methane producing), which belong to Methanobacteriales and Methanomassiliicoccales.[182] The other dominant orders present in normal individuals include Methanosarcinales, Halobacteriales and Haloferacales.[183] In a study from Korea, where the Archaeal population in fecal samples were studied using 16S rRNA sequencing, predominance of phyla Euryarchaeota and Crenarchaeota and *Methanobrevibacter* and *Methanosphaera* were observed.[184]

Present literatures does not have evidence showing dysbiosis in the archaeal component of the gastrointestinal microbiome is linked to severe gastroduodenal diseases, although the possibility cannot be ignored.

6. Factors affecting the gastrointestinal microbiome

The human gastrointestinal microbiome is in a state of dynamic equilibrium and is under the influence of various environmental and intrinsic host factors. Among 160 different bacterial species that inhabit human gastrointestinal tract 75% are similar in 50% of the population. A comparative study of intestinal microbiome shows that *Prevotella* was the dominant among native Tibetans while *Bacteroides* was enriched in the Hans.[185] In a study from India, the gut microbiome of rural high altitude population of Leh was characterized by an increase in the abundance of *Bacteroidetes* and a decrease in the abundance of Proteobacteria. On the other hand, population from Ballabhgarh, which is at sea level showed enrichment of Firmicutes and Proteobacteria.[186]

Diet is another key variable determining the microbiome composition. The fiber content of the food can significantly affect the gut microbiome. Increased abundances of *Bifidobacterium* and *Lactobacillus* were associated with intake of prebiotics like fructans and galacto-oligosaccharides.[187] A Brassica rich diet was found to be associated with reduced abundance of sulfur-producers and also the bacteria belonging to Rikenellaceae, Ruminococcaceae, Mogibacteriaceae, and Clostridiales in the intestinal microbiome when compared to low Brassica diet.[188]

Diet rich in fruits was associated with an increase in *Akkermansia muciniphila, Faecalibacterium prausnitzii*, Ruminococcaceae, Clostridiales and *Acidaminococcus* in the fecal microbiome.[189] Further a decreased abundance of *Fusobacterium* was associated with high intake of fruits.[189] An increase in the ratio of Firmicutes to Bacteroidetes as well as the abundance of Lachnospiraceae and *Ruminococcus,* and a decrease in abundance of

Lactobacillus in the gut microbiome were found to be associated with high salt intake.[190] The growth of beneficial gut bacteria like *Bifidobacterium* and *Lactobacillus* is enhanced by the fermentation of prebiotic carbohydrates like fructo-oligosaccharides and inulin.[191]

Personal lifestyle habits also affect the individuals' microbiome composition. An increase in genera like *Actinomyces*, *Leptotrichia*, *Cardiobacterium* and *Neisseria* was observed in the oral microbiome of individuals with higher alcohol consumption.[192] An increase in Proteobacteria and Firmicutes and a decrease in Bacteroidetes and Verrucomicrobia in the intestinal microbiome was observed in alcoholics.[193] In a study in pediatric population, a significant reduction in alpha diversity and a lower abundance of *Bacteroides* in the gut microbiome was observed following antibiotic treatment.[194] A 4 days intervention with meropenem, gentamicin and vancomycin resulted in an increase in Enterobacteriaceae in the gut microbiome. An increase in *Enterococcus faecalis* and *Fusobacterium nucleatum* and a decrease in *Bifidobacterium* was also observed following the intake of the antibiotic combinations.[195] Treatment with broad spectrum antibiotics can alter the gut microbiome and can decrease the colonization resistance to opportunistic pathogens in the gut like *Clostridium difficile*. *Clostridium difficile* is the common etiology for antibiotics associated diarrhea, especially following clindamycin treatment.[196,197] Colonization of *H. pylori* in human stomach leads to a decrease in the abundance of *Bifidobacterium* in the fecal microbiome.[137] Eradication of *H. pylori* from the stomach is linked with higher abundance of *Bifidobacterium* in the intestine.[134]

7. Conclusion and future perspectives

Studies on stomach had begun as early as 1547 and diseases like PUD as well as GC were known from ancient times. Even now, gastric diseases like PUD and GC accounts for over a million deaths every year. Several claims regarding the existence of microbes in stomach was ignored by scientific communities for nearly a century until the famous and serendipitous discovery of *H. pylori* from human gastric tissues was reported by Warren and Marshall.[5] Soon it was realized that chronic *H. pylori* colonization in the gastric mucosa leads to a spectrum of diseases such as NUD, gastritis, PUD and GC, though not every *H. pylori* infected individual develops such diseases. The major virulence factors are CagA and VacA, along with DupA, BabA, SabA and OipA, which contribute to the development of the gastroduodenal diseases.

In recent past, several serious efforts have been made by metagenomic approach to understand the connection between the gastrointestinal microbiome and gastroduodenal diseases. A higher Firmicutes abundance and lower Bacteroidetes abundance in oral microbiome is found to be associated with GC.[70] A study also suggested using oral microbiome composition as marker of GC.[74] An increase in Enterobacteriaceae and Verrucomicrobiaceae, particularly *Akkermansia muciniphila* and a decrease in *Veillonella* in the esophageal microbiome was observed in patients with esophageal adenocarcinoma.[85] However, the link between severe gastric disease and esophageal microbiome is poorly understood and needs further research.

The core gastric microbiome of a healthy individual consists of major five phyla namely Firmicutes, Bacteroidetes, Actinobacteria, Fusobacteria and Proteobacteria.[96] *Streptococcus, Neisseria* and *Prevotella* were found to be abundant in chronic gastritis cases, irrespective of the *H. pylori* status.[106] While in GC, increased abundance of Proteobacteria, Fusobacteria, Firmicutes, Bacteroidetes, and Actinobacteria were observed.[104] Human intestine is one of the densely colonized parts of the human body with around 35,000 different species inhabiting it. A healthy gut consists of higher abundance of *Bacteroides, Prevotella* and *Ruminococcus* and a lower abundance of *Proteobacteria.*[128] A lower abundance of *Bifidobacterium* is observed in patients with severe gastric disease such as GC or GU in India and in Finland.[135,137] The *Bifidobacterium* abundance increases upon successful *H. pylori* eradication.[134] The 'other' components of the human gastrointestinal microbiome, archaea, viruses, fungi, protozoans and helminthes are also likely to have an impact on the gastroduodenal health, but fewer studies have been conducted on these. For virome, it is known that a healthy gut consists of *Geminiviridae, Herpesviridae, Anelloviridae, Parvoviridae, Polyomaviridae, Papillomaviridae, Adenoviridae, Circoviridae* and *Nanoviridae.*[147] Epstein-Barr Virus, which primarily infects B-cell and epithelial cells, is found to be associated with at least 10% of the total GC cases. A healthy human mycobiome consists of *Candida, Cladosporium,* and *Aspergillus,* with smaller proportion of *Glomus, Penicillium, Fusarium, Alternaria, Saccharomycetales, Phoma, Ophiosoma, Cryptococcus, Schizosaccharomyces* and *Zygosaccharomyceras.*[161] An elevation in the relative abundances of *Alternaria* and *Candida* and a decrease in *Saitozyma* and *Thermomyces* were observed in GC cases. On the other hand, lower abundance of *Aspergillus montevidensis, Saitozyma podzolica, Penicillium arenicola* and *Candida glabrata* was associated with GC.[167] Candida may also be involved in delaying the healing of GU and DU. The association between gastrointestinal diseases and the archaeal, protozoal and helminth components of the human microbiome is poorly understood. However, it is noticed that *Schistosoma japonicum*, an intestinal trematode, which is observed to be associated with colorectal cancer, is also associated with the cancers of stomach and esophagus.[175]

Several factors determine the microbiome composition. Diet plays an important role in shaping one's gastrointestinal microbiome. It is observed that *Bacteroides* increase with the dietary fiber intake. High fat diet fed mice had decreased abundance of Firmicutes and Bacteroidetes as well as *Bifidobacterium* spp.[198] Geography and host genetics also plays an important factor in determining the microbiome composition. Other factors such as smoking and alcohol consumption are also known to alter the microbiome composition. However, presently it is not precisely understood how the dysbiosis in the gastrointestinal tract is linked to *H. pylori* infection in stomach and gastroduodenal diseases.

Human gastrointestinal tract harbors wide variety of microorganisms (bacteria, archaea, viruses, fungi, protozoans and helminths), which is collectively called as microbiome. With the advent of newer metagenomics techniques, it is now possible to study those microbes that were long thought to be uncultivable. The microbiome composition contributes in maintaining a healthy gastrointestinal tract. Recent studies have exposed the interplay between the gastrointestinal microbiome, *H. pylori* infection and the development of gastroduodenal diseases. Further studies addressing the dysbiosis in each niche of the gastrointestinal tract is necessary to understand the contributions from different factors influencing the gastroduodenal diseases.

Acknowledgments

We thank Prof. Chandrabhas Narayana, Director, RGCB for generous support. This work was supported by a grant (ECR/2016/000171) from the Science and Engineering Research Board (SERB) of Department of Science and Technology (DST), Government of India and a grant (MED/2017/46) from the Department of Biotechnology (DBT), Government of India to SC as well as by the institutional support from the Rajiv Gandhi Centre for Biotechnology, an autonomous institute under the Department of Biotechnology, Government of India.

References

1. Peter WT, DPK P, Tepper JE. Cancers of the stomach. In: DeVita Jr VT, Lawrence TS, Rosenberg SA, eds. *Cancer: Principles & Practice of Oncology*. Vol. 1. 8th ed. Lippincott Williams & Wilkins; 2008.
2. Xie X, Ren K, Zhou Z, Dang C, Zhang H. The global, regional and national burden of peptic ulcer disease from 1990 to 2019: A population-based study. *BMC Gastroenterol.* 2022;22(1):1–13.

3. GBD 2016 Causes of Death Collaborators. Global, regional, and national age-sex specific mortality for 264 causes of death, 1980-2016: A systematic analysis for the global burden of disease study 2016. *Lancet.* 2017;390(10100):1151–1210.
4. Bray F, Ferlay J, Soerjomataram I, Siegel RL, Torre LA, Jemal A. Global cancer statistics 2018: GLOBOCAN estimates of incidence and mortality worldwide for 36 cancers in 185 countries. *CA Cancer J Clin.* 2018;68(6):394–424.
5. Warren JR, Marshall B. Unidentified curved bacilli on gastric epithelium in active chronic gastritis. *Lancet.* 1983;1(8336):1273–1275.
6. Marshall BJ, Warren JR. Unidentified curved bacilli in the stomach of patients with gastritis and peptic ulceration. *Lancet.* 1984;1(8390):1311–1315.
7. No authors listed. Schistosomes, liver flukes and helicobacter pylori. IARC working group on the evaluation of carcinogenic risks to humans. Lyon, 7-14 June 1994. *IARC Monogr Eval Carcinog Risks Hum.* 1994;61:1–241.
8. Yamaoka Y. Mechanisms of disease: Helicobacter pylori virulence factors. *Nat Rev Gastroenterol Hepatol.* 2010;7(11):629–641.
9. Hatakeyama M. Helicobacter pylori CagA and gastric cancer: A paradigm for hit-and-run carcinogenesis. *Cell Host Microbe.* 2014;15(3):306–316.
10. Isomoto H, Moss J, Hirayama T. Pleiotropic actions of helicobacter pylori vacuolating cytotoxin. *VacA Tohoku J Exp Med.* 2010;220(1):3–14.
11. Alexander SM, Retnakumar RJ, Chouhan D, et al. Helicobacter pylori in human stomach: The inconsistencies in clinical outcomes and the probable causes. *Front Microbiol.* 2021;12:713955.
12. Correa P, Piazuelo MB. Helicobacter pylori infection and gastric adenocarcinoma. *US Gastroenterol Hepatol Rev.* 2011;7(1):59–64.
13. Rosenfeld L. William Prout: Early 19th century physician-chemist. *Clin Chem.* 2003;49(4):699–705.
14. Soybel DI. Anatomy and physiology of the stomach. *Surg Clin North Am.* 2005;85 (5):875–894.
15. Talley N. Functional dyspepsia: A classification with guidelines for diagnosis and management. *Gastroenterol Int.* 1991;4:145–160.
16. Mahadeva S, Goh K-L. Epidemiology of functional dyspepsia: A global perspective. *World J Gastroenterol: WJG.* 2006;12(17):2661.
17. Fisher RS, Parkman HP. Management of nonulcer dyspepsia. *N Engl J Med.* 1998;339 (19):1376–1381.
18. Goh KL. Prevalence of and risk factors for helicobacter pylori infection in a multi-racial dyspeptic Malaysian population undergoing endoscopy. *J Gastroenterol Hepatol.* 1997;12 (6):S29–S35.
19. Louw JA, Jaskiewicz K, Girdwood A, et al. Helicobacter pylori prevalence in non-ulcer dyspepsia ethnic and socio-economic differences. *S Afr Med J.* 1993;83(3):169–171.
20. Br B, Johnsen R, Bostad L, Br S, Sommer A-I, Burhol PG. Is helicobacter pylori the cause of dyspepsia? *Br Med J.* 1992;304(6837):1276–1279.
21. Rugge M, Meggio A, Pennelli G, et al. Gastritis staging in clinical practice: The OLGA staging system. *Gut.* 2007;56(5):631–636.
22. Cheli R, Perasso A, Giacosa A. *Definition of Gastritis.* Gastritis: Springer; 1987:3–11.
23. Sipponen P. Natural history of gastritis and its relationship to peptic ulcer disease. *Digestion.* 1992;51(Suppl. 1):70–75.
24. Kuipers EJ, Pérez-Pérez GI, Meuwissen SG, Blaser MJ. Helicobacter pylori and atrophic gastritis: Importance of the cagA status. *J Natl Cancer Inst.* 1995;87 (23):1777–1780.
25. Azer SA, Akhondi H. *Gastritis*; 2019.
26. Graham DY. History of helicobacter pylori, duodenal ulcer, gastric ulcer and gastric cancer. *World J Gastroenterol: WJG.* 2014;20(18):5191.

27. Goldstein H. Ulcer and cancer of the stomach in the middle ages. *J Int Coll Surg.* 1943;6:482–489.
28. Kuna L, Jakab J, Smolic R, Raguz-Lucic N, Vcev A, Smolic M. Peptic ulcer disease: A brief review of conventional therapy and herbal treatment options. *J Clin Med.* 2019;8(2):179.
29. Sung J, Kuipers E, El-Serag H. Systematic review: The global incidence and prevalence of peptic ulcer disease. *Aliment Pharmacol Ther.* 2009;29(9):938–946.
30. Tovey F. Peptic ulcer in India and Bangladesh. *Gut.* 1979;20(4):329.
31. Sung H, Ferlay J, Siegel RL, et al. Global cancer statistics 2020: GLOBOCAN estimates of incidence and mortality worldwide for 36 cancers in 185 countries. *CA Cancer J Clin.* 2021;71(3):209–249.
32. Kuipers E. Helicobacter pylori and the risk and management of associated diseases: Gastritis, ulcer disease, atrophic gastritis and gastric cancer. *Aliment Pharmacol Ther.* 1997;11(S1):71–88.
33. Hatton OL, Harris-Arnold A, Schaffert S, Krams SM, Martinez OM. The interplay between Epstein–Barr virus and B lymphocytes: Implications for infection, immunity, and disease. *Immunol Res.* 2014;58(2):268–276.
34. Gantuya B, El-Serag HB, Matsumoto T, et al. Gastric microbiota in helicobacter pylori-negative and-positive gastritis among high incidence of gastric cancer area. *Cancer.* 2019;11(4):504.
35. Rawla P, Barsouk A. Epidemiology of gastric cancer: Global trends, risk factors and prevention. *Gastroenterol.* 2019;14(1):26.
36. Atherton JC. The pathogenesis of helicobacter pylori–induced gastro-duodenal diseases. *Annu Rev Pathol Mech Dis.* 2006;1:63–96.
37. Blaser MJ. Helicobacter pylori: Its role in disease. *Clin Infect Dis.* 1992;386–391.
38. Singh K, Ghoshal UC. Causal role of helicobacter pylori infection in gastric cancer: An Asian enigma. *World J Gastroenterol: WJG.* 2006;12(9):1346.
39. Graham DY, Lu H, Yamaoka Y. African, Asian or Indian enigma, the east Asian helicobacter pylori: Facts or medical myths. *J Dig Dis.* 2009;10(2):77–84.
40. Marshall BJ, Armstrong JA, McGechie DB, Glancy RJ. Attempt to fulfil Koch's postulates for pyloric campylobacter. *Med J Aust.* 1985;142:436–439.
41. Institutet TNAaK. *The Nobel Prize in Physiology or Medicine 2005*; 2005. https://www.nobelprize.org/prizes/medicine/2005/summary/.
42. Hooi JK, Lai WY, Ng WK, et al. Global prevalence of helicobacter pylori infection: Systematic review and meta-analysis. *Gastroenterology.* 2017;153(2):420–429.
43. Ottemann KM, Lowenthal AC. Helicobacter pylori uses motility for initial colonization and to attain robust infection. *Infect Immun.* 2002;70(4):1984–1990.
44. Mobley H. The role of helicobacter pylori urease in the pathogenesis of gastritis and peptic ulceration. *Aliment Pharmacol Ther.* 1996;10(Sup1):57–64.
45. Ilver D, Arnqvist A, Ögren J, et al. Helicobacter pylori adhesin binding fucosylated histo-blood group antigens revealed by retagging. *Science.* 1998;279(5349):373–377.
46. Matsuo Y, Kido Y, Yamaoka Y. Helicobacter pylori outer membrane protein-related pathogenesis. *Toxins.* 2017;9(3):101.
47. Baj J, Forma A, Sitarz M, et al. Helicobacter pylori virulence factors—Mechanisms of bacterial pathogenicity in the gastric microenvironment. *Cell.* 2021;10(1):27.
48. Liu J, He C, Chen M, Wang Z, Xing C, Yuan Y. Association of presence/absence and on/off patterns of helicobacter pylori oipA gene with peptic ulcer disease and gastric cancer risks: A meta-analysis. *BMC Infect Dis.* 2013;13(1):1–10.
49. Kaplan-Türköz B, Jiménez-Soto LF, Dian C, et al. Structural insights into helicobacter pylori oncoprotein CagA interaction with β1 integrin. *Proc Natl Acad Sci.* 2012;109(36):14640–14645.

50. Takahashi-Kanemitsu A, Knight CT, Hatakeyama M. Molecular anatomy and pathogenic actions of helicobacter pylori CagA that underpin gastric carcinogenesis. *Cell Mol Immunol.* 2020;17(1):50–63.
51. Papadakos KS, Sougleri IS, Mentis AF, Hatziloukas E, Sgouras DN. Presence of terminal EPIYA phosphorylation motifs in helicobacter pylori CagA contributes to IL-8 secretion, irrespective of the number of repeats. *PLoS One.* 2013;8(2):e56291.
52. Hayashi T, Senda M, Suzuki N, et al. Differential mechanisms for SHP2 binding and activation are exploited by geographically distinct helicobacter pylori CagA oncoproteins. *Cell Rep.* 2017;20(12):2876–2890.
53. Foegeding NJ, Caston RR, McClain MS, Ohi MD, Cover TL. An overview of helicobacter pylori VacA toxin biology. *Toxins.* 2016;8(6):173.
54. Chauhan N, Tay ACY, Marshall BJ, Jain U. Helicobacter pylori VacA, a distinct toxin exerts diverse functionalities in numerous cells: An overview. *Helicobacter.* 2019;24(1): e12544.
55. Sugimoto M, Zali M, Yamaoka Y. The association of vacA genotypes and helicobacter pylori-related gastroduodenal diseases in the Middle East. *Eur J Clin Microbiol Infect Dis.* 2009;28(10):1227–1236.
56. Chang W-L, Yeh Y-C, Sheu B-S. The impacts of H. pylori virulence factors on the development of gastroduodenal diseases. *J Biomed Sci.* 2018;25(1):1–9.
57. Thi Huyen Trang T, Thanh Binh T, Yamaoka Y. Relationship between vacA types and development of gastroduodenal diseases. *Toxins.* 2016;8(6):182.
58. Kinoshita-Daitoku R, Kiga K, Miyakoshi M, et al. A bacterial small RNA regulates the adaptation of helicobacter pylori to the host environment. *Nat Commun.* 2021;12(1):1–12.
59. Moss EL, Maghini DG, Bhatt AS. Complete, closed bacterial genomes from microbiomes using nanopore sequencing. *Nat Biotechnol.* 2020;38(6):701–707.
60. Yarza P, Yilmaz P, Pruesse E, et al. Uniting the classification of cultured and uncultured bacteria and archaea using 16S rRNA gene sequences. *Nat Rev Microbiol.* 2014;12 (9):635–645.
61. Ishaq SL, Wright AD. Design and validation of four new primers for next-generation sequencing to target the 18S rRNA genes of gastrointestinal ciliate protozoa. *Appl Environ Microbiol.* 2014;80(17):5515–5521.
62. Kounosu A, Murase K, Yoshida A, Maruyama H, Kikuchi T. Improved 18S and 28S rDNA primer sets for NGS-based parasite detection. *Sci Rep.* 2019;9(1):15789.
63. Seed PC. The human mycobiome. *Cold Spring Harb Perspect Med.* 2014;5(5), a019810.
64. Gao L, Xu T, Huang G, Jiang S, Gu Y, Chen F. Oral microbiomes: More and more importance in oral cavity and whole body. *Protein Cell.* 2018;9(5):488–500.
65. Deo PN, Deshmukh R. Oral microbiome: Unveiling the fundamentals. *J Oral Maxillofac Pathol.* 2019;23(1):122.
66. Perera M, Al-Hebshi NN, Speicher DJ, Perera I, Johnson NW. Emerging role of bacteria in oral carcinogenesis: A review with special reference to perio-pathogenic bacteria. *J Oral Microbiol.* 2016;8(1):32762.
67. Ndegwa N, Ploner A, Liu Z, Roosaar A, Axell T, Ye W. Association between poor oral health and gastric cancer: A prospective cohort study. *Int J Cancer.* 2018;143 (9):2281–2288.
68. Salazar CR, Francois F, Li Y, et al. Association between oral health and gastric precancerous lesions. *Carcinogenesis.* 2012;33(2):399–403.
69. Yin XH, Wang YD, Luo H, et al. Association between tooth loss and gastric cancer: A meta-analysis of observational studies. *PLoS One.* 2016;11(3):e0149653.
70. Wu J, Xu S, Xiang C, et al. Tongue coating microbiota community and risk effect on gastric cancer. *J Cancer.* 2018;9(21):4039–4048.

71. Xu J, Xiang C, Zhang C, et al. Microbial biomarkers of common tongue coatings in patients with gastric cancer. *Microb Pathog.* 2019;127:97–105.
72. Sun J, Zhou M, Salazar CR, et al. Chronic periodontal disease, periodontal pathogen colonization, and increased risk of precancerous gastric lesions. *J Periodontol.* 2017;88 (11):1124–1134.
73. Salazar CR, Sun J, Li Y, et al. Association between selected oral pathogens and gastric precancerous lesions. *PLoS One.* 2013;8(1):e51604.
74. Sun JH, Li XL, Yin J, Li YH, Hou BX, Zhang Z. A screening method for gastric cancer by oral microbiome detection. *Oncol Rep.* 2018;39(5):2217–2224.
75. Blaser MJ. Disappearing microbiota: Helicobacter pylori protection against esophageal adenocarcinoma. *Cancer Prev Res (Phila).* 2008;1(5):308–311.
76. Mannell A, Plant M, Frolich J. The microflora of the oesophagus. *Ann R Coll Surg Engl.* 1983;65(3):152.
77. Lau W, Wong J, Lam KH, Ong GB. Oesophageal microbial flora in carcinoma of the oesophagus. *ANZ J Surg.* 1981;51(1):52–55.
78. Pei Z, Bini EJ, Yang L, Zhou M, Francois F, Blaser MJ. Bacterial biota in the human distal esophagus. *Proc Natl Acad Sci.* 2004;101(12):4250–4255.
79. Yang L, Lu X, Nossa CW, Francois F, Peek RM, Pei Z. Inflammation and intestinal metaplasia of the distal esophagus are associated with alterations in the microbiome. *Gastroenterology.* 2009;137(2):588–597.
80. Gall A, Fero J, McCoy C, et al. Bacterial composition of the human upper gastrointestinal tract microbiome is dynamic and associated with genomic instability in a Barrett's esophagus cohort. *PLoS One.* 2015;10(6):e0129055.
81. Park CH, Lee SK. Exploring esophageal microbiomes in esophageal diseases: A systematic review. *J Neurogastroenterol Motil.* 2020;26(2):171.
82. Blackett K, Siddhi S, Cleary S, et al. Oesophageal bacterial biofilm changes in gastro-oesophageal reflux disease, Barrett's and oesophageal carcinoma: Association or causality? *Aliment Pharmacol Ther.* 2013;37(11):1084–1092.
83. Liu F, Ma R, Wang Y, Zhang L. The clinical importance of campylobacter concisus and other human hosted campylobacter species. *Front Cell Infect Microbiol.* 2018;8:243.
84. Elliott DRF, Walker AW, O'Donovan M, Parkhill J, Fitzgerald RC. A non-endoscopic device to sample the oesophageal microbiota: A case-control study. *Lancet Gastroenterol Hepatol.* 2017;2(1):32–42.
85. Snider EJ, Compres G, Freedberg DE, et al. Alterations to the esophageal microbiome associated with progression from Barrett's esophagus to esophageal adenocarcinoma. *Cancer Epidemiol Biomark Prev.* 2019;28(10):1687–1693.
86. Malfertheiner P, Link A, Selgrad M. Helicobacter pylori: Perspectives and time trends. *Nat Rev Gastroenterol Hepatol.* 2014;11(10):628–638.
87. Malfertheiner P, Megraud F, O'Morain CA, et al. Management of Helicobacter pylori infection- -the Maastricht IV/ Florence consensus report. *Gut.* 2012;61(5):646–664.
88. Lu P-J, Hsu P-I, Chen C-H, et al. Gastric juice acidity in upper gastrointestinal diseases. *World J Gastroenterol: WJG.* 2010;16(43):5496.
89. Petra CV, Rus A, Dumitraşcu DL. Gastric microbiota: Tracing the culprit. *Clujul Med.* 2017;90(4):369.
90. Blaser MJ. Gastric campylobacter-like organisms, gastritis, and peptic ulcer disease. *Gastroenterology.* 1987;93(2):371–383.
91. Reed P, Haines K, Smith P, House F, Walters C. Gastric juice N-nitrosamines in health and gastroduodenal disease. *The Lancet.* 1981;318(8246):550–552.
92. Delgado S, Cabrera-Rubio R, Mira A, Suárez A, Mayo B. Microbiological survey of the human gastric ecosystem using culturing and pyrosequencing methods. *Microb Ecol.* 2013;65(3):763–772.

93. Pereira V, Abraham P, Nallapeta S, Shetty A. Gastric bacterial Flora in patients Harbouring helicobacter pylori with or without chronic dyspepsia: Analysis with matrix-assisted laser desorption ionization time-of-flight mass spectroscopy. *BMC Gastroenterol.* 2018;18(1):1–8.
94. Sheh A, Fox JG. The role of the gastrointestinal microbiome in helicobacter pylori pathogenesis. *Gut Microbes.* 2013;4(6):505–531.
95. Nardone G, Compare D. The human gastric microbiota: Is it time to rethink the pathogenesis of stomach diseases? *United European Gastroenterol J.* 2015;3(3):255–260.
96. Parsons BN, Ijaz UZ, D'Amore R, et al. Comparison of the human gastric microbiota in hypochlorhydric states arising as a result of helicobacter pylori-induced atrophic gastritis, autoimmune atrophic gastritis and proton pump inhibitor use. *PLoS Pathog.* 2017;13(11):e1006653.
97. Zhang S, Shi D, Li M, Li Y, Wang X, Li W. The relationship between gastric microbiota and gastric disease. *Scand J Gastroenterol.* 2019;54(4):391–396.
98. Ohno H, Satoh-Takayama N. Stomach microbiota, helicobacter pylori, and group 2 innate lymphoid cells. *Exp Mol Med.* 2020;52(9):1377–1382.
99. Sjöstedt S, Heimdahl A, Kager L, Nord C. Microbial colonization of the oropharynx, esophagus and stomach in patients with gastric diseases. *Eur J Clin Microbiol.* 1985;4(1):49–51.
100. Sjöstedt S, Kager L, Heimdahl A, Nord CE. Microbial colonization of tumors in relation to the upper gastrointestinal tract in patients with gastric carcinoma. *Ann Surg.* 1988;207(3):341.
101. Dicksved J, Lindberg M, Rosenquist M, Enroth H, Jansson JK, Engstrand L. Molecular characterization of the stomach microbiota in patients with gastric cancer and in controls. *J Med Microbiol.* 2009;58(4):509–516.
102. Hu Y, He L-H, Di Xiao G-DL, Gu Y-X, Tao X-X, Zhang J-Z. Bacterial flora concurrent with helicobacter pylori in the stomach of patients with upper gastrointestinal diseases. *World J Gastroenterol: WJG.* 2012;18(11):1257.
103. Eun CS, Kim BK, Han DS, et al. Differences in gastric mucosal microbiota profiling in patients with chronic gastritis, intestinal metaplasia, and gastric cancer using pyrosequencing methods. *Helicobacter.* 2014;19(6):407–416.
104. Wang L, Zhou J, Xin Y, et al. Bacterial overgrowth and diversification of microbiota in gastric cancer. *Eur J Gastroenterol Hepatol.* 2016;28(3):261.
105. Yang I, Woltemate S, Piazuelo MB, et al. Different gastric microbiota compositions in two human populations with high and low gastric cancer risk in Colombia. *Sci Rep.* 2016;6(1):1–10.
106. Ferreira RM, Pereira-Marques J, Pinto-Ribeiro I, et al. Gastric microbial community profiling reveals a dysbiotic cancer-associated microbiota. *Gut.* 2018;67(2):226–236.
107. Liu J, Xue Y, Zhou L. Detection of gastritis-associated pathogens by culturing of gastric juice and mucosa. *Int J Clin Exp Pathol.* 2018;11(4):2214.
108. Hsieh Y-Y, Tung S-Y, Pan H-Y, et al. Increased abundance of clostridium and fusobacterium in gastric microbiota of patients with gastric cancer in Taiwan. *Sci Rep.* 2018;8(1):1–11.
109. Liu X, Shao L, Liu X, et al. Alterations of gastric mucosal microbiota across different stomach microhabitats in a cohort of 276 patients with gastric cancer. *EBioMedicine.* 2019;40:336–348.
110. Wang L, Xin Y, Zhou J, et al. Gastric mucosa-associated microbial signatures of early gastric cancer. *Front Microbiol.* 2020;11:1548.
111. Andersson AF, Lindberg M, Jakobsson H, Bäckhed F, Nyrén P, Engstrand L. Comparative analysis of human gut microbiota by barcoded pyrosequencing. *PLoS One.* 2008;3(7):e2836.

112. Das A, Pereira V, Saxena S, et al. Gastric microbiome of Indian patients with helicobacter pylori infection, and their interaction networks. *Sci Rep*. 2017;7(1):1–9.
113. Li TH, Qin Y, Sham PC, Lau K, Chu K-M, Leung WK. Alterations in gastric microbiota after H. pylori eradication and in different histological stages of gastric carcinogenesis. *Sci Rep*. 2017;7(1):1–8.
114. Maldonado-Contreras A, Goldfarb KC, Godoy-Vitorino F, et al. Structure of the human gastric bacterial community in relation to helicobacter pylori status. *ISME J*. 2011;5(4):574–579.
115. Iliev HI, Kovacheva-Slavova MD, Angelov TA, Valkov HY, Bedran A, Vladimirov BG. Gastric microbiota: Between health and disease. In: *Gastrointestinal Stomas*. IntechOpen; 2019.
116. Rolig AS, Cech C, Ahler E, Carter JE, Ottemann KM. The degree of helicobacter pylori-triggered inflammation is manipulated by preinfection host microbiota. *Infect Immun*. 2013;81(5):1382–1389.
117. Ozbey G, Sproston E, Hanafiah A. Helicobacter pylori infection and gastric microbiota. *Euroasian J Hepatogastroenterol*. 2020;10(1):36.
118. Li X-X, Wong GL-H, To K-F, et al. Bacterial microbiota profiling in gastritis without helicobacter pylori infection or non-steroidal anti-inflammatory drug use. *PLoS One*. 2009;4(11):e7985.
119. Shiotani A, Matsumoto H, Fukushima S, Katsumata R, Kawano M, Saito M. H. Pylori and human gut microbiota. *J Bacteriol Mycol*. 2018;5(8):1–5.
120. Thrift AP, El-Serag HB. Burden of gastric cancer. *Clin Gastroenterol Hepatol*. 2020;18(3):534–542.
121. Yu G, Hu N, Wang L, et al. Gastric microbiota features associated with cancer risk factors and clinical outcomes: A pilot study in gastric cardia cancer patients from Shanxi, China. *Int J Cancer*. 2017;141(1):45–51.
122. Yang J, Zhou X, Liu X, Ling Z, Ji F. Role of the gastric microbiome in gastric cancer: From carcinogenesis to treatment. *Front Microbiol*. 2021;12.
123. Khosravi Y, Dieye Y, Poh BH, et al. Culturable bacterial microbiota of the stomach of helicobacter pylori positive and negative gastric disease patients. *Sci World J*. 2014;2014.
124. Jandhyala SM, Talukdar R, Subramanyam C, Vuyyuru H, Sasikala M, Reddy DN. Role of the normal gut microbiota. *World J Gastroenterol: WJG*. 2015;21(29):8787.
125. Drasar BS, Shiner M, McLeod G. Studies on the intestinal flora: I. the bacterial flora of the gastrointestinal tract in healthy and achlorhydric persons. *Gastroenterology*. 1969;56(1):71–79.
126. Marchesi J, Shanahan F. The normal intestinal microbiota. *Curr Opin Infect Dis*. 2007;20(5):508–513.
127. King CH, Desai H, Sylvetsky AC, et al. Baseline human gut microbiota profile in healthy people and standard reporting template. *PLoS One*. 2019;14(9):e0206484.
128. Hollister EB, Gao C, Versalovic J. Compositional and functional features of the gastrointestinal microbiome and their effects on human health. *Gastroenterology*. 2014;146(6):1449–1458.
129. Kienesberger S, Cox LM, Livanos A, et al. Gastric helicobacter pylori infection affects local and distant microbial populations and host responses. *Cell Rep*. 2016;14(6):1395–1407.
130. Gao JJ, Zhang Y, Gerhard M, et al. Association between gut microbiota and helicobacter pylori-related gastric lesions in a high-risk population of gastric cancer. *Front Cell Infect Microbiol*. 2018;8:202.
131. Dash NR, Khoder G, Nada AM, Al Bataineh MT. Exploring the impact of helicobacter pylori on gut microbiome composition. *PLoS One*. 2019;14(6):e0218274.

132. Yang L, Zhang J, Xu J, et al. Helicobacter pylori infection aggravates dysbiosis of gut microbiome in children with gastritis. *Front Cell Infect Microbiol.* 2019;9:375.
133. Qi YF, Sun JN, Ren LF, et al. Intestinal microbiota is altered in patients with gastric cancer from Shanxi Province, China. *Dig Dis Sci.* 2019;64(5):1193–1203.
134. Guo Y, Zhang Y, Gerhard M, et al. Effect of helicobacter pylori on gastrointestinal microbiota: A population-based study in Linqu, a high-risk area of gastric cancer. *Gut.* 2020;69(9):1598–1607.
135. Sarhadi V, Mathew B, Kokkola A, et al. Gut microbiota of patients with different subtypes of gastric cancer and gastrointestinal stromal tumors. *Gut Pathog.* 2021;13(1):11.
136. Frost F, Kacprowski T, Ruhlemann M, et al. Helicobacter pylori infection associates with fecal microbiota composition and diversity. *Sci Rep.* 2019;9(1):20100.
137. Devi TB, Devadas K, George M, et al. Low Bifidobacterium abundance in the lower gut microbiota is associated with helicobacter pylori-related gastric ulcer and gastric cancer. *Front Microbiol.* 2021;12:631140.
138. Nagaoka M, Hashimoto S, Watanabe T, Yokokura T, Mori Y. Anti-ulcer effects of lactic acid bacteria and their cell wall polysaccharides. *Biol Pharm Bull.* 1994;17(8): 1012–1017.
139. Sivan A, Corrales L, Hubert N, et al. Commensal Bifidobacterium promotes antitumor immunity and facilitates anti-PD-L1 efficacy. *Science.* 2015;350(6264):1084–1089.
140. Pride DT, Salzman J, Haynes M, et al. Evidence of a robust resident bacteriophage population revealed through analysis of the human salivary virome. *ISME J.* 2012;6(5): 915–926.
141. Abbas AA, Taylor LJ, Dothard MI, et al. Redondoviridae, a family of small, circular DNA viruses of the human oro-respiratory tract associated with periodontitis and critical illness. *Cell Host Microbe.* 2019;25(5):719–729.
142. Baker JL, Bor B, Agnello M, Shi W, He X. Ecology of the oral microbiome: Beyond bacteria. *Trends Microbiol.* 2017;25(5):362–374.
143. Spezia PG, Macera L, Mazzetti P, et al. Redondovirus DNA in human respiratory samples. *J Clin Virol.* 2020;131:104586.
144. Liang G, Bushman FD. The human virome: Assembly, composition and host interactions. *Nat Rev Microbiol.* 2021;1–14.
145. Shkoporov AN, Hill C. Bacteriophages of the human gut: The "known unknown" of the microbiome. *Cell Host Microbe.* 2019;25(2):195–209.
146. Dutilh BE, Cassman N, McNair K, et al. A highly abundant bacteriophage discovered in the unknown sequences of human faecal metagenomes. *Nat Commun.* 2014; 5(1):1–11.
147. Rascovan N, Duraisamy R, Desnues C. Metagenomics and the human virome in asymptomatic individuals. *Annu Rev Microbiol.* 2016;70:125–141.
148. Zhang T, Breitbart M, Lee WH, et al. RNA viral community in human feces: Prevalence of plant pathogenic viruses. *PLoS Biol.* 2006;4(1):e3.
149. Parkin DM. The global health burden of infection-associated cancers in the year 2002. *Int J Cancer.* 2006;118(12):3030–3044.
150. Takada K. Epstein-Barr virus and gastric carcinoma. *Mol Pathol.* 2000;53(5):255.
151. Nanbo A, Kachi K, Yoshiyama H, Ohba Y. Epstein–Barr virus exploits host endocytic machinery for cell-to-cell viral transmission rather than a virological synapse. *J Gen Virol.* 2016;97(11):2989–3006.
152. Lieberman PM. Keeping it quiet: Chromatin control of gammaherpesvirus latency. *Nat Rev Microbiol.* 2013;11(12):863–875.
153. Salamon D, Takacs M, Ujvari D, et al. Protein-DNA binding and CpG methylation at nucleotide resolution of latency-associated promoters Qp, cp, and LMP1p of Epstein-Barr virus. *J Virol.* 2001;75(6):2584–2596.

154. Hino R, Uozaki H, Murakami N, et al. Activation of DNA methyltransferase 1 by EBV latent membrane protein 2A leads to promoter hypermethylation of PTEN gene in gastric carcinoma. *Cancer Res*. 2009;69(7):2766–2774.
155. Singh S, Jha HC. Status of Epstein-Barr virus coinfection with helicobacter pylori in gastric cancer. *J Oncol*. 2017;2017.
156. Ferrasi AC, Pinheiro NA, Rabenhorst SHB, et al. Helicobacter pylori and EBV in gastric carcinomas: Methylation status and microsatellite instability. *World J Gastroenterol: WJG*. 2010;16(3):312.
157. Shukla SK, Prasad K, Tripathi A, et al. Epstein-Barr virus DNA load and its association with helicobacter pylori infection in gastroduodenal diseases. *Braz J Infect Dis*. 2011;15:583–590.
158. Buzás GM, Konderák J. Co-infection with helicobacter pylori and Epstein–Barr virus in benign upper digestive diseases: An endoscopic and serologic pilot study. *Eur Gastroenterol J*. 2016;4(3):388–394.
159. Teresa F, Serra N, Capra G, et al. Helicobacter pylori and epstein–barr virus infection in gastric diseases: Correlation with IL-10 and IL1RN polymorphism. *J Oncol*. 2019;2019.
160. Gu Y, Zhou G, Qin X, Huang S, Wang B, Cao H. The potential role of gut mycobiome in irritable bowel syndrome. *Front Microbiol*. 2019;10:1894.
161. Ghannoum MA, Jurevic RJ, Mukherjee PK, et al. Characterization of the oral fungal microbiome (mycobiome) in healthy individuals. *PLoS Pathog*. 2010;6(1):e1000713.
162. Chin VK, Yong VC, Chong PP, Amin Nordin S, Basir R, Abdullah M. Mycobiome in the gut: A multiperspective review. *Mediators Inflamm*. 2020;2020.
163. Chen Y, Chen Z, Guo R, et al. Correlation between gastrointestinal fungi and varying degrees of chronic hepatitis B virus infection. *Diagn Microbiol Infect Dis*. 2011;70 (4):492–498.
164. Zhang L, Zhan H, Xu W, Yan S, Ng SC. The role of gut mycobiome in health and diseases. *Therap Adv Gastroenterol*. 2021;14 [17562848211047130].
165. Dollive S, Peterfreund GL, Sherrill-Mix S, et al. A tool kit for quantifying eukaryotic rRNA gene sequences from human microbiome samples. *Genome Biol*. 2012;13 (7):1–13.
166. Ott SJ, Kühbacher T, Musfeldt M, et al. Fungi and inflammatory bowel diseases: Alterations of composition and diversity. *Scand J Gastroenterol*. 2008;43(7):831–841.
167. Zhong M, Xiong Y, Zhao J, et al. Candida albicans disorder is associated with gastric carcinogenesis. *Theranostics*. 2021;11(10):4945.
168. Sasaki K. Candida-associated gastric ulcer relapsing in a different position with a different appearance. *World J Gastroenterol: WJG*. 2012;18(32):4450.
169. Goyal P, Bansal S, Kaur P, Goyal O. Candida associated giant non-healing gastric ulcer in an immunocompetent host. *J Gastroenterol Liver Dis*. 2016;1(1):1003.
170. Jung MK, Jeon SW, Cho CM, et al. Treatment of gastric candidiasis in patients with gastric ulcer disease: Are antifungal agents necessary? *Gut Liver*. 2009;3(1):31.
171. Gilchrist CA, Petri SE, Schneider BN, et al. Role of the gut microbiota of children in diarrhea due to the protozoan parasite Entamoeba histolytica. *J Infect Dis*. 2016;213 (10):1579–1585.
172. Chappell CL, Darkoh C, Shimmin L, et al. Fecal indole as a biomarker of susceptibility to cryptosporidium infection. *Infect Immun*. 2016;84(8):2299–2306.
173. Burgess SL, Gilchrist CA, Lynn TC, Petri Jr WA. Parasitic protozoa and interactions with the host intestinal microbiota. *Infect Immun*. 2017;85(8):e00101–e00117.
174. Fibiger J. Investigations on Spiroptera carcinoma and the experimental induction of cancer. *Nobel Lecture*. 1927;122–150.
175. Peterson MR, Weidner N. Gastrointestinal neoplasia associated with bowel parasitosis: Real or imaginary? *J Trop Med*. 2011;2011.

176. Cancer IAfRo. *Schistosomes, Liver Flukes and Helicobacter pylori*. Vol 61. IARC Lyon; 1994.
177. Guo Z, Ni Y, Wu J. Epidemiological study on relationship between schistosomiasis and colorectal cancer. *Jiangsu Med J*. 1984;4:209.
178. Liu B, Rong Z, Sun X, Wu Y, Gao R. Geographical correlation between colorectal cancer and schistosomiasis in China. *Zhongguo yi xue ke xue Yuan xue bao. Acta Acad Med Sin*. 1983;5(3):173–177.
179. Du Y, Agnew A, Ye X-p, Robinson PA, Forman D, Crabtree JE. Helicobacter pylori and Schistosoma japonicum co-infection in a Chinese population: Helminth infection alters humoral responses to H. pylori and serum pepsinogen I/II ratio. *Microbes Infect*. 2006;8(1):52–60.
180. Woese CR, Kandler O, Wheelis ML. Towards a natural system of organisms: Proposal for the domains archaea, bacteria, and Eucarya. *Proc Natl Acad Sci*. 1990;87(12): 4576–4579.
181. Pausan MR, Csorba C, Singer G, et al. Exploring the archaeome: Detection of archaeal signatures in the human body. *Front Microbiol*. 2019;10:2796.
182. Gaci N, Borrel G, Tottey W, O'Toole PW, Brugère J-F. Archaea and the human gut: New beginning of an old story. *World J Gastroenterol: WJG*. 2014;20(43):16062.
183. Coker OO, Wu WKK, Wong SH, Sung JJ, Yu J. Altered gut archaea composition and interaction with bacteria are associated with colorectal cancer. *Gastroenterology*. 2020;159(4):1459–1470.
184. Kim JY, Whon TW, Lim MY, et al. The human gut archaeome: Identification of diverse haloarchaea in Korean subjects. *Microbiome*. 2020;8(1):1–17.
185. Li K, Dan Z, Gesang L, et al. Comparative analysis of gut microbiota of native Tibetan and Han populations living at different altitudes. *PLoS One*. 2016;11(5):e0155863.
186. Das B, Ghosh TS, Kedia S, et al. Analysis of the gut microbiome of rural and urban healthy Indians living in sea level and high altitude areas. *Sci Rep*. 2018;8(1):1–15.
187. So D, Whelan K, Rossi M, et al. Dietary fiber intervention on gut microbiota composition in healthy adults: A systematic review and meta-analysis. *Am J Clin Nutr*. 2018;107(6):965–983.
188. Kellingray L, Tapp HS, Saha S, Doleman JF, Narbad A, Mithen RF. Consumption of a diet rich in brassica vegetables is associated with a reduced abundance of sulphate-reducing bacteria: A randomised crossover study. *Mol Nutr Food Res*. 2017;61(9) [1600992].
189. Jiang Z, Sun T-y, He Y, et al. Dietary fruit and vegetable intake, gut microbiota, and type 2 diabetes: Results from two large human cohort studies. *BMC Med*. 2020; 18(1):1–11.
190. Wang C, Huang Z, Yu K, et al. High-salt diet has a certain impact on protein digestion and gut microbiota: A sequencing and proteome combined study. *Front Microbiol*. 2017;8:1838.
191. Hemarajata P, Versalovic J. Effects of probiotics on gut microbiota: Mechanisms of intestinal immunomodulation and neuromodulation. *Therap Adv Gastroenterol*. 2013;6(1):39–51.
192. Fan X, Peters BA, Jacobs EJ, et al. Drinking alcohol is associated with variation in the human oral microbiome in a large study of American adults. *Microbiome*. 2018; 6(1):1–15.
193. Mutlu EA, Gillevet PM, Rangwala H, et al. Colonic microbiome is altered in alcoholism. *American journal of physiology-gastrointestinal and liver. Phys Ther*. 2012;302(9): G966–G978.
194. Yassour M, Vatanen T, Siljander H, et al. Natural history of the infant gut microbiome and impact of antibiotic treatment on bacterial strain diversity and stability. *Sci Transl Med*. 2016;8(343):[343ra381-343ra381].

195. Palleja A, Mikkelsen KH, Forslund SK, et al. Recovery of gut microbiota of healthy adults following antibiotic exposure. *Nat Microbiol.* 2018;3(11):1255–1265.
196. Theriot CM, Young VB. Interactions between the gastrointestinal microbiome and Clostridium difficile. *Annu Rev Microbiol.* 2015;69:445–461.
197. Rashid M-U, Zaura E, Buijs MJ, et al. Determining the long-term effect of antibiotic administration on the human normal intestinal microbiota using culture and pyrosequencing methods. *Clin Infect Dis.* 2015;60(suppl_2):S77–S84.
198. Cani PD, Neyrinck AM, Fava F, et al. Selective increases of bifidobacteria in gut microflora improve high-fat-diet-induced diabetes in mice through a mechanism associated with endotoxaemia. *Diabetologia.* 2007;50(11):2374–2383.

CHAPTER FOUR

Respiratory tract microbiome and pneumonia

Lekshmi Narendrakumar[a] and Animesh Ray[b,*]
[a]Molecular Genetics Laboratory, Centre for Human Microbial Ecology, Translational Health Science and Technology Institute, Faridabad, India
[b]Department of Medicine, All India Institute of Medical Sciences, New Delhi, India
*Corresponding author: e-mail address: doctoranimeshray@gmail.com

Contents

Abstract

The respiratory system, like the gut, harbors a vast variety of microorganisms which include bacteria, viruses and fungi. The advent of next generation sequencing and multi-omic approaches has revealed the diversity and functional significances of microorganisms in the respiratory health. It has been identified that there has been a co-evolution of indigenous respiratory microbiota and the human immune system. However, an immune response is usually generated when the homeostasis of the microbiota is disturbed. The respiratory microbiome has been identified to be important in shaping the respiratory immunity. Gut microbiota and oral microbiota are also known to be pivotal in shaping the immune system of the respiratory tract and influence its microbial dynamics. *Proteobacteria*, *Firmicutes*, and *Bacteroidetes* have been identified

Progress in Molecular Biology and Translational Science, Volume 192
ISSN 1877-1173
https://doi.org/10.1016/bs.pmbts.2022.07.002

to be predominant in the respiratory system. While, *Streptococcus, Prevotella, Fusobacteria,* and *Veillonella* forms the major part, potential pathogens, such as *Haemophilus* and *Neisseria*, also form a small fraction of the healthy lung microbiome. Dysbiosis of respiratory microbiome can lead to increased colonization of opportunistic pathogens that can lead to respiratory infections such as pneumonia. This chapter describes the microbial diversity of respiratory system and the role of respiratory microbiome during respiratory infections like pneumonia. The chapter also discusses few strategies that have been proved effective in preventing pneumonia.

1. Introduction

The respiratory system, specifically the lung, was considered to be sterile as evidenced by early experiments of Hildebrandt in 1888 as excised nasal and tracheal mucosa of rabbits were identified to grow only minimal bacteria by culture-dependent methods.[1] However, this argument was contested by various researchers as they believed that the air inhaled was unsterile and the upper respiratory tract and lung are in constant exposure to microorganisms that can colonize these surfaces.[2,3] Even though presence of microorganisms in both upper and lower respiratory tract was evidenced by several studies over the years, there are still disputes over the knowledge of indigenous microorganisms and acquired microorganisms in the respiratory tract and lungs.[4] Further, there is ecological adjacency of the lung and the upper respiratory tract which causes overlap of microorganisms isolated from these regions causing misperception of specific microbiome of these niches.[5] Additionally, the bacterial community interaction in the respiratory tract and lung has been identified to play major role in chronic respiratory diseases such as pneumonia, chronic obstructive pulmonary, chronic rhinosinusitis, asthma, otitis media, cystic fibrosis and non-CF bronchiectasis.[6] This chapter expounds the diversity of microbes in the upper and lower respiratory tract, the immunoecology of these microbes, respiratory tract microbiome changes during an infection such as pneumonia and ways to modulate microbiome to prevent respiratory infections.

2. Respiratory system and respiratory tract microbiome

The respiratory system consists of organs and related body parts that are involved in breathing. The human respiratory system consists of nose and nasal cavity, sinuses, adenoids, tonsils, mouth, pharynx, larynx, trachea, lungs, bronchi, bronchioles, alveoli and the diaphragm. Broadly, the respiratory system is divided as upper respiratory tract (URT) and the lower respiratory tract (LRT). The URT consist the nose and nasal cavity, sinuses, adenoids, tonsils, mouth, pharynx and the larynx. Trachea connects the URT to the LRT which consists of bronchi, bronchioles, alveoli which makes up the lung. The nasal opening in the nose allows air to pass through into the respiratory tract and the nasal cavity consists of conchae lined by mucous membranes. The function of conchae is to greatly increase the surface area of the mucus membrane so that the air is sufficiently filtered. Further, the mucus membrane contains goblet cells that secrete mucus and additionally possess extensive network of blood vessels that maintains warmth and moisture.[7] The pharynx is the common pathway to air and food that connects oral and nasal cavities to the larynx and esophagus. The trachea extends from the larynx and divides to form primary bronchi which further branches to form secondary and tertiary bronchi and finally to the bronchioles. The bronchioles terminate into several sacs like alveoli which are surrounded by capillaries that facilitate the air exchange between the lung and the blood stream.

The respiratory microbiome is known to be dynamic which relies on several host and environmental factors. The microbiome comprises of bacterial, viral and fungal assemblages which prevent the growth and dissemination of pathogenic microorganisms to some extent.[8] Further, the respiratory microbiome also provides signals to the host immune system that are important for immune training, organogenesis and immune tolerance homeostasis.[9,10] Specific sites in the respiratory tract are colonized by different microbial communities though there are some overlaps. Anatomy of human respiratory system and the sites of microbial colonization have been depicted in Fig. 1. With the advent of next generation sequencing technologies, our understanding of the microbial diversity and their possible functions has been greatly increased. The different microbial communities and their possible role at the specific sites are summarized in the following sections.

2.1 Upper respiratory tract (URT) microbiome

The URT is known to harbor a higher microbial density than the LRT. Recent studies have evidenced that many microbial communities present in the URT play a major role in preventing colonization and spread of respiratory pathogens into the LRT. This inhibition of pathogen colonization, also known as the "colonization resistance" is of utmost importance to respiratory health. The initial acquisition of microorganisms initiates the

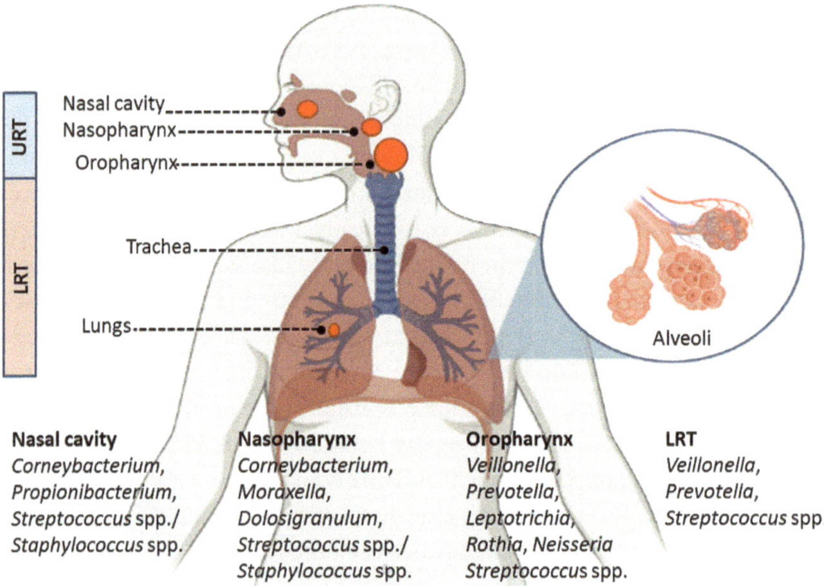

Fig. 1 Anatomy of human respiratory system and the sites of microbial colonization (red bubble). The size of the red bubble corresponds to the diversity of microbial community. Important bacterial colonizers in each respiratory niche are depicted in the figure. The figure was created with BioRender software (BioRender.com).

establishment of URT microbiome. In contrast to the long standing hypothesis that the air we breathe after birth serves as the source of microbes in the respiratory tract, recent studies suggested that respiratory microbiome development starts *in uetro*.[11] Studies by Dominguez-Bello and Bosch showed that a wide variety of microorganisms were identified from the URT of healthy term neonates even during the initial hours after birth.[12,13] At the initial phases of life, the URT is mostly known to be colonized by *Staphylococcus* spp., presumed to be of maternal origin. Later, the *Staphylococcus* spp. is predominated by *Corynebacterium* spp. and *Dolosigranulum* spp., and at further stages by *Moraxella* spp.[12] However, a decreased abundance of *Corynebacterium* spp. and *Dolosigranulum* spp., were identified in children who were not born vaginally and whose mothers consumed antibiotics during pregnancy or immediately post-delivery.[14,15] Thus, birth mode and feeding type were identified to be major factors determining URT microbiome during early development.[16] Also antibiotic regime during or post pregnancy is identified as an important perturbing factor which determines the microbial homeostasis.[17] Other factors that affect URT microbiome include genetics of the person, varying seasons, hygiene and smoking habit.[18,19] Charlson et al. demonstrated that cigarette smoking select the growth and dominance of gram negative bacteria such as *Pseudomonas aeruginosa* and *Klebsiella* spp. In an earlier study, *Prevotella* and *Peptostreptococcus* spp. were identified to be abundant in the URT of nonsmokers while *Streptococcus pneumoniae*, *Haemophilus influenzae*, and *Moraxella catarrhalis* were abundant in smokers.[20]

The nostril microbiome consists of Actinobacteria such as *Corynebacterium* and *Propionibacterium* spp. and Firmicutes such as *Streptococcus* spp. in children and *Staphylococcus* spp. in adults. Also there is a small amount of Bacteroidetes which are anaerobes. The nasopharynx predominantly contains *Dolosigranulum, Moraxella, Corynebacterium, Staphylococcus* or *Streptococcus* spp. However, the oropharynx is identified to harbor greater diversity of microorganism than the nasopharynx which includes *Streptococcal* spp., *Neisseria* spp., *Rothia* spp. and anaerobes, which includes *Veillonella* spp., *Prevotella* spp. and *Leptotrichia* spp.[21] Apart from the bacterial commensals, the URT is also known to be the home for a wide variety of virus and fungi such as human rhinovirus (HRV), human bocavirus, polyomaviruses, human adenovirus, human coronavirus and *Aspergillus* spp., *Penicillium* spp., *Candida* spp. and *Alternaria* spp.[22,23] Apart from the normal commensal flora, the URT is also prone to many pathogenic bacteria and viruses. The incidence of *S. pneumonia*, a potential bacterial pathogen colonizing the URT is common which subsequently spread into the LRT and cause infection.[24] However, commensal microbiota in the URT such as the *Dolosigranulum* spp. and *Corynebacterium* spp., confer colonization resistance to pathogenic bacteria such as pneumococci thus maintaining health.[25] Furthermore *S. epidermidis* is known to exclude *S. aureus* nasal colonization.[26] Thus, the URT microbiota and microbiome plays a significant role in preventing pathogenic microorganisms from reaching the LRT.

2.2 Lower respiratory tract (LRT) microbiome

Though traditionally thought sterile, with the advent of next generation sequencing facilities the understanding of LRT microbiome has changed. However, there have been disputes whether these microbiota are indigenous or acquired by external contamination. LRT of young children were identified to be dominated by *Moraxella* spp., *Haemophilus* spp., *Staphylococcus* spp. and *Streptococcus* spp. while that of adults is dominated by Firmicutes such as *Streptococcus* spp., *Veillonella* spp. and Bacteroidetes such as *Prevotella* spp.[27,28] Though significant overlap was seen in the microbiome of URT and LRT, few bacterial species such as the *Tropheryma whipplei* was exclusively found to be enriched in the LRT.[27] Apart from the bacterial microbiome, the LRT has been identified to house a repertoire of viral and fungal microbiome. The virome include members of the *Anelloviridae* while the mycobiome include *Eremothecium*, *Systenostrema* and *Malassezia* genera, and the Davidiellaceae family.[29,30] Also, the load of bacteriophages in the LRT was identified to be higher than that in the URT. Though an extended diversity of microorganisms has been identified in the LRT, the microbiome is known to be transient and primarily relies on the balance between microbial immigration and elimination. Moreover the exact function of this LRT microbiome is not clear unlike the URT microbiome that plays an eminent role in prevention of colonization and spread of pathogens into the LRT.

3. Immunoecology of microbes in lungs

The respiratory tract microbiome is essential for programming of respiratory immunity. Immunoecology deals with the participation of the microbes at a specific location of the host body in regulating the host immunity. It explains the major microbial players that regulate the host immunity in a way that the pathogens and cleared and the beneficial microbes are sustained.[31] It has been demonstrated in mice model that microbial colonization in the URT is essential for the expression of pulmonary dendritic cells that play an important role in identifying allergens in the later stages of life.[32] The immunoecology of microbiota in the respiratory tract especially the lungs primarily depends on three important factors which are (i) microbial migration into the airway and lungs, (ii) elimination of microbes in the airway and lungs by the host cells and the (iii) elimination of microorganism by other microorganisms. Microbial elimination is primarily established by mucociliary clearance and host innate and adaptive immunity. It is understood that there is a significant interaction between commensal microbes of the respiratory system and the innate immunity.[33] Further there have been reports of mutualism between the commensal microbiome and the adaptive immune system. However, the exact mechanism by which the commensal microbiota of the respiratory system evades the host immunity is relatively poorly understood. It has been identified that secretory IgA antibodies produced by B cells maintain homeostasis and balance of respiratory microbiome. Infectious agents such as bacteria are known to cause epithelial damage and trigger host immune system producing cytokines such as IL-25, and IL-33 which in turn activates IL-5 and IL-13 stimulating eosinophils and goblet cells. Goblet cell expansion increases the mucus secretion that helps to clear pathogenic bacteria to some extent, but further accumulation of mucus also leads to the colonization of bacterial pathogens.[34]

4. Pneumonia

Pneumonia has been the common cause of morbidity and mortality worldwide particularly in the developing and underdeveloped nations. Globally pneumonia is one of the major causes of death associated with infectious diseases.[35] The infection affects the lungs particularly the air sacs causing it to get inflamed and filled with purulent material which makes breathing difficult. Pneumonia can be caused by a number of different infectious agents such as bacteria, virus or fungi. According to the World Health Organization (WHO), *S. pneumoniae* and *H. influenzae* type b is the most common cause of bacterial pneumonia in children while respiratory syncytial virus is the most common cause of viral pneumonia.[36] While bacterial pneumonia is most common in infants, viral pneumonia has been identified to be more common in older children. Neonates has been known to acquire group B streptococci, *Klebsiella, Escherichia coli, and Listeria monocytogenes* that cause early onset pneumonia and *S. pneumoniae, Streptococcus pyogenes,* and *S. aureus* that cause late onset pneumonia.[37] Fungal pneumonia is most common when the host is immunocompromised. *Pneumocystis jirovecii* is the most common fungal species identified to cause pneumonia in immunocompromised children and HIV patients. The infection is acquired by the inhalation of infectious agents via air borne droplets and the most predominant symptoms of viral/ bacterial pneumonia includes cough and/or difficult breathing. *H. influenza, Mycoplasma pneumoniae, Chlamydia pneumonia, Legionella,* gram-negative bacilli and *S. aureus* are the common pathogens associated with pneumonia in adults. Further, pneumonia is broadly classified as community-acquired pneumonia (CAP) or nosocomial (hospital-acquired) pneumonia (HAP) based on how and from where it is acquired. Community acquired pneumonia is termed when pneumonia is acquired by a patient or population who is/are not hospitalized whereas nosocomial or hospital acquired pneumonia usually develop in patients who are hospitalized for at least 48–72 h or on mechanical ventilation for 48 h or longer. Details regarding the differential bacterial diversity associated with CAP and HAP are presented in the following sections of the chapter.

4.1 Community acquired pneumonia (CAP)

Community acquired pneumonia is the main cause of reason for hospital admission of people in the developing countries. The incidence of CAP has been increasing over the past years especially in the aged people and there have been several factors associated with the infection. One of the major reasons for CAP is the reduced lung immunity due to age, comorbidities or lifestyle. The most common causative agent of CAP has been identified as *S. pneumoniae* while other pathogens such as *H. influenzae, S. aureus, M. catarrhalis* and *Legionella pneumophila* have also been identified to cause CAP.[38] CAP can be acquired from various sources and the pathogens that are acquired from these sources also differ. The major source of infection by Influenza A, other respiratory viruses, *Chlamidophila pneumoniae,* and *Mycoplasma pneumonia* is known to be by the inhalation of infected droplets while pathogens such as the *S. pneumoniae, H. influenza, S. aureus, M. catarrhalis* and other gram negative enteric pathogens are acquired by the aspiration of nasopharyngeal commensals. Pathogens such as *L. pneumophila* and *P. aeruginosa* that cause CAP are known to be acquired from the environment. The currently ongoing pandemic due to the SARS-CoV-19 infection can be also considered as an example of CAP.

Clinical presentation of CAP at early stages is similar to that of a common cold. Mild pneumonia is characterized by cough, fever, sputum production, mild difficulty in breathing and chest pain while severe pneumonia can even lead to respiratory failure, sepsis and death. Further, the clinical presentation of CAP differs on the infectious organisms. For example, CAP due to *Legionella* is presented by dry cough or elevated lactate dehydrogenase levels whereas CAP caused by *Mycoplasma* is presented by acute psychosis or encephalitis and pneumococcal pneumonia by cough, dyspnoea, and pleuritic pain.[39,40] Though the cases of methicillin resistant *S. aureus* (MRSA) associated CAP is less common, it is one of the most difficult forms of pneumonia to be treated. Another major concern is the increasing rate of CAP caused by antibiotic resistant *P. aeruginosa.*[41] Apart from these, the major problem with CAP is the development of sepsis, respiratory failure, and acute respiratory distress syndrome.[42]

The major diagnosis of CAP is by performing microbiological tests of respiratory specimens (most commonly sputum).[43] Further, analysis of urinary antigen is used to detect *S. pneumoniae* and *L. pneumophila* and are today included in CAP algorithms. However, despite a high specificity, there is a low sensitivity of urinary antigen test, as it is evident only during severe pneumonia.[44] Multiplex PCR that can detect various pathogens at a time and real time PCR are fast and reliable methods to diagnose CAP.[45] When the causative pathogen is identified, antibiotics are started targeting specifically the microorganism. However, empirical antibiotic treatment is also usual in CAP which most commonly includes amoxicillin, doxycycline, a respiratory fluoroquinolone or a macrolide antibiotic.[45] Moreover, depending on severity of pneumonia, combination therapies are also used to treat CAP which includes a combination of a β-lactam and macrolide antibiotics or amoxicillin–clavulanic/cephalosporin and a macrolide/doxycycline.[45]

4.2 Hospital acquired pneumonia (HAP) and ventilator acquired pneumonia (VAP)

Hospital acquired pneumonia (HAP) usually occur in patients who are hospitalized for 48 h or more. Development of HAP depends on factors like age of the patient, underlying comorbidities and immune status. Mortality due to HAP is estimated to be between 5% and 33%.[46] Ventilator-associated pneumonia (VAP) is a major subset of HAP which develops 48 h or more after mechanical ventilation. The incidence of HAP and VAP has significantly increased in the recent time. The ongoing COVID-19 pandemic has increased the number of patients getting admitted to intensive care units (ICU), and the use of invasive mechanical ventilation (IMV) which exposes patients to the risk of VAP. Studies have revealed that VAP among ventilated COVID-19 patients is 40–86% and mortality rate due to VAP in COVID-19 patients is 60%.[47,48] The major predisposing factors for VAP is the use of proton-pump inhibitors and histamine-2-receptor blocker therapy, cleanliness of the hospital environment, patient's history of any other respiratory illness, age and immune status of the host. Other major hospital associated risk factors that cause VAP includes increased mechanical ventilation time, longer duration of hospital stay, disorders of consciousness, burn wounds, invasive operations, other comorbidities and gene polymorphisms.[49] Common pathogens of both HAP and VAP include gram-negative bacteria such as *P. aeruginosa*, *E. coli*, *Klebsiella pneumoniae*, *Enterobacter* spp., *Acinetobacter* spp. and gram-positive bacteria including *S. aureus* and *Streptococcus* spp.[50] However, the bacterial profile has been identified to be different based on different geographical locations. The major concern in HAP and VAP is the emergence of multidrug resistant pathogens that makes treatment options limited. Extremely high rates of MDR pathogens such as MRSA, enterobacteria, *P. aeruginosa*, *Acinetobacter baumanii*, and *Stenotrophomonas maltophilia* have been detected in various HAP and VAP studies from Singapore while *H. influenzae*,

S. pneumoniae, *P. aeruginosa* and anaerobes have been reported from Japan. The rate of incidence of MRSA and drug resistant *P. aeruginosa* causing HAP is more than CAP. Further, the mortality of HAP/VAP patients is higher as compared to CAP patients.[50]

Signs and symptoms of HAP include fever, tachypnea, higher purulent secretions, rhonchi, difficulty in breathing and bronchospasm. Diagnosis of HAP and VAP is challenging owing to the limited tests that have a greater turnaround time, are fast and reliable and also the fact that the patient may be already on antibiotic treatment. Blood culture is a common diagnosis method as in more than 20% cases, HAP and VAP develops bacteremia.[51] Sputum culture is the recommended diagnostic method to detect HAP. However both these diagnosis methods are reliable when the patient has not started antibiotics. Apart from culturing techniques, molecular techniques such as multiplex PCR that detect *Enterobacteriaceae* spp., *S. aureus* and their antibiotic resistance genes are commonly used to diagnose HAP and VAP. GenXpert (Cepheid) that utilizes an automated microfluidic system that depends on real-time PCR and BD GeneOhm (Becton-Dickinson) that utilizes a real-time PCR with a fluorogenic target hybridization probe have been recently used to detect MRSA causing HAP or VAP.[52] However, today, real-time PCR, hybridization and mass spectrometry-based platforms are the most widely used diagnostic methods for HAP and VAP though these techniques are less predominantly used in developing countries.[53] Despite few limitations, the molecular diagnostic techniques have a great potential to identify the causative organism in HAP and also CAP. However, like in that of CAP, empirical treatment is usually started in VAP patients before the identification of the causative microorganisms. A study by Alvarez-Lerma et al. reported that patients who received empirical antibiotic treatment had lower compilations such as septic shock, hemorrhage and mortality compared to patients who were not administered antibiotics at an early stage of the infection.[54] The Infectious Disease Society of America (IDSA) and the American Thoracic Society (ATS) have jointly prepared guidelines for the management of HAP and VAP. Imipenem and ciprofloxacin are among the common antibiotics used for the empirical treatment of HAP and VAP. However, as the cases of HAP and VAP caused by MDR pathogens are increasing, combination therapy of a broad-spectrum β-lactam with an aminoglycoside or β-lactam with fluoroquinolone is commonly preferred. Apart from the natural microbiome changes occurring during pneumonia infection, the use of antibiotics during the infection also cause stress and dysbiosis of the indigenous microorganisms of the respiratory tract.

5. Respiratory microbiome changes during pneumonia

Pneumonia, both CAP and VAP is known to disturb the normal respiratory tract microbial homeostasis. Microorganisms that are otherwise normal flora of the URT and co-exist with other microorganisms can become pathogenic if the host immunity is compromised and the normal equilibrium is disturbed. This can lead to the over growth of the pathogenic organism and dissemination of the organism to the LRT leading to pneumonia. Compared to CAP, more studies have focused on microbial dysbiosis during VAP primarily because of the difficulty in getting samples prior to the development of CAP. It is today a well-known fact that orotracheal intubation impairs the natural lung immune system. In a recent study by Zakharkina and team revealed that there is a significant decrease in the alpha diversity of microbes detected in the endotracheal aspirate (ETA) in VAP patients compared to that of patients who have no VAP but are intubated.[55] Similarly, in another study, Woo et al. identified that the predominant microbial species in pneumonia patients (*Pseudomonas, Corynebacterium* and *Rothia*) differed from that of people who have no pneumonia (*Streptococcus* and *Prevotella*).[56] Further, it was identified that *Gammaproteobacteria* were predominant in sample collected on the third day of VAP patients as compared to *Streptococcus, Enterococcus, Lactobacillus* and *Staphylococcus* predominant in patients who were admitted to the ICU but not intubated.[57] Interestingly, Sommerstein et al. noticed that *Enterobacteriaceae* in the oropharynx during the first 5 days of mechanical ventilation was later associated with *Enterobacteriaceae* VAP.[58] Overall it has been identified that during mechanical ventilation, there occurs a predominance of single bacterial species and the overall bacterial diversity is decreased.

Additionally, there have been various studies on the shift of orotracheal microbiome to various biofilm communities. Biofilms are bacteria present in a self-assembled structure generally comprising of bacterial extracellular matrix and these biofilms provide resistance to antibiotics and host immune system to the bacteria within these biofilms. In a recent study it has been noticed that the bacterial biofilms on endotracheal tube (ETT) is mostly composed of nosocomial pathogens rather than indigenous bacteria.[59] Moreover, Hotterbeekx et al. identified that 89% of ETT phyla belonged to the family *Proteobateriaceae*, 86% was *Phylobacteriaceae* and 77% was *Enterobacteriaceae*. Relative abundance of *Pseudomonadaceae* and *Staphylococcaceae* was 4.6% and 70.8%, respectively. The study also revealed that *Actinomyces*,

Corynebacterium or *Bifidobacterium adolescentis* were predominantly found on ETT of patients who did not develop VAP.[60] Additionally various studies have pointed out that oral and oropharyngeal microbiome play a major role in VAP as these bacteria have a higher chance of getting translocated to LRT during mechanical ventilation. A study by Kalanuria et al. revealed that oropharyngeal microbial dysbiosis was common in mechanically ventilated patients and a wide variety of microbes associated with dental plaque have been isolated from LRT.[61] There has been also a report that the recent SARS CoV-2 infection that has caused the COVID-19 pandemic creates a hypoxic condition in the lungs favoring the growth of facultative anaerobes and anaerobes of oral origin causing secondary infections such as pneumonia. *Capnocytophaga*, *Veillonella* are two important oral opportunistic pathogens that can cause pneumonia.[62] Although respiratory virome and mycobiome have not been well described, *Candida albicans* and *Candida glabrata* are the most common fungi isolated from ETTs of mechanically ventilated patients.[60] Further, significantly high relative abundance of *Streptococcus (pseudo)pneumonia and Rothia* were identified in elderly CAP patients as compared to normal elderly controls. The study also revealed lower abundance of *Gemellales*, *Prevotella melaninogenica*, *Veillonella dispar*, *Parascardovia* and *Leptotrichia* in pneumonia patients.[63] On the whole, most of the previous studies revealed that reduction of beneficial bacterial diversity is a major cause for the increase in pathogenic microorganisms that leads to pneumonia and other respiratory tract infections.

6. Oral microbiome relation to pulmonary microbiome

Microbes that have migrated from the oral cavity are known to be an important source of microbiota in the lungs under normal conditions and are estimated to be higher than that of the microbes migrated from the nasal cavity. It has been noticed that many microbiota present in the saliva is also present in the lungs. Several studies have suggested that the LRT microbiome is more similar to the oropharynx microbiome than the nasopharynx microbiome.[5,64] Two common ways by which the oral microorganisms can reach the LRT is by hematogenous spread and aspiration. Few of the common bacteria that are present in the oropharynx as well as LRT are *Prevotella*, *Veillonella*, and *Streptococcus* and also *Fusobacterium* and *Haemophilus*. These bacteria in the LRT are associated with subclinical inflammation by triggering neutrophils and lymphocytes. Endotyping of the LRT has identified that these microbes are associated with Th17 phenotype, increased $CD4^+$ $IL\text{-}17^+$ lymphocytes, and IL-1α.[65] Further it has been identified that like the URT microbiome playing a crucial role in maintaining LRT health, the oral microbiome also play a key role in preventing colonization of pathogenic bacteria and their dispersal into the lungs. The saliva in mouth also plays a prominent role in respiratory health as the salivary components such as the lysozyme, secretory immunoglobulins and lactoperoxidases prevent infectious agents to surpass the oral cavity.[66] However, during an event of oral dysbiosis, there are alterations in salivary components that can cause the colonization of respiratory pathogens. Oral dysbiosis can be caused due to various factors such as lifestyle (smoking, use of tobacco), environment, any specific disease, poor oral health, or the use of instruments to provide mechanical ventilation such as endotracheal tubes.

7. Pulmonary-gut microbiome cross talk

There have been emerging evidences that there is significant cross talk between gut and pulmonary microbiome. Gut microbiota plays a major role in maintaining pulmonary immunity via a gut-lung cross talk also known as the gut-lung axis. This bidirectional axis is known to exchange vital signals in the form of microbial metabolites, cytokines, endotoxins and hormones. Studies have indicated that at the phylum level there is significant similarity in the microbial predominance in the gut and respiratory system. While *Bacteroidetes* and *Firmicutes* predominate in the gut, *Bacteroidetes*, *Firmicutes*, and *Proteobacteria* predominates the respiratory system.[67] It has been identified by several studies that gut microbial dysbiosis play an important role in pulmonary diseases such as pneumonia, cystic fibrosis, lung inflammatory disease, chronic obstructive pulmonary disease (COPD), asthma and even lung cancer.[68–71] The immune response modulation has been identified by inflammatory cytokines and microbial metabolites. It has been identified that infants with low gut microbial diversity due to antibiotic exposure is known to have a higher probability of developing asthma at the age of 5–10 years as the early gut microbiome plays an important role in avoiding airway inflammation by balancing T cell subsets Th1/Th2.[72,73] Demirci et al. had showed that decreased abundance of *Akkermansia muciniphila* and *Faecalibacterium prausnitzii* were observed in the guts of children having asthma compared to the healthy group.[74] In another study conducted by Charlson and coworkers (2011), *Enterobacter cloacae*, *Citrobacter*, *Eggerthella*, *Pseudomonas*, *Anaerococcus*, *Proteus*, *Clostridium difficile*, and *Salmonella* belonging to the proteobacteria group were identified to be increased in patients with COPD compared to the healthy control.[75]

Further, there have been evidences of relation between respiratory infections and gut microbiome. A variation in gut microbiome particularly the *Prevotella* and *Lachnospira* were identified to be decreased in tuberculosis patients.[76] An association was delineated between gut microbiome and lung macrophage inducible C-type lectin (mincle) expression which is important for the survival of *Mycobacterium tuberculosis* in the lung.[77] Additionally, a decrease in Firmicutes abundance that regulates immune responses such as IFN-γ, IL-6 and CCL2 against viral infections was identified to cause lung inflammation.[78] Gut microbiota dysbiosis by antibiotics was known to significantly increase the *S. aureus* associated pneumonia as a result of impaired Toll-like receptor 4 (TLR4) function. Further, mast cell migration from gut to lung is an important factor to keep *S. aureus* pneumonia at bay. However, when the gut microbiota is disturbed, mast cell migration gets hampered thus increasing the chance of *S. aureus* pneumonia. Also a reduced caspase-3 and Bax/Bcl-2 and NF-κB signaling in the gut is observed during *S. aureus* pneumonia.[79] Similar results regarding significant decrease in immunomodulation of gut microbes particularly *Bacteroides* species were observed in cystic fibrosis patients. The inflammation associated with cystic fibrosis was identified to be the result of increased IL-8 which was regulated by *Bacteroides* species in the gut.[68] Apart from *Bacteroides* species, a significant reduction in Firmicutes, *F. prausnitzii, Bifidobacterium adolescentis* and *Eubacterium rectale*, along with increased *Streptococcus, Staphylococcus, V. dispar, C. difficile, P. aeruginosa*, and *E. coli* were observed in children having cystic fibrosis.[80] Fecal proteome analysis of chronic cystic fibrosis patients revealed a major reduction in butyrate reducers *F. prausnitzii*. Short chain fatty acids are understood to contribute to cystic fibrosis-specific alterations during airway infection and inflammation.[81] In the same way, gut microbiome is also known to play a major role in the pathology of lung cancer. Decreased level of *Bifidobacterium* sp., *Enterococcus* sp. and *Actinobacteria* sp. were identified to have a positive correlation with lung cancer. Given the vital role of gut microbiota in shaping the respiratory immunity and regulating the progression of several respiratory diseases and infections, manipulation of gut microbiota has been considered a major prophylactic and therapeutic option for preventing these diseases.

8. Strategies to prevent pneumonia by respiratory and gut microbiome modulation

Pneumonia remains to be a major public health concern globally owing to its high mortality rate among children and the economic cost to the society. Though we rely on antibiotics to treat pneumonia, the overuse of antibiotics is known to be a major driver for antimicrobial resistance (AMR) and the emergence of multidrug resistant pathogens. Bacteria harboring AMR to new generation antibiotics can possibly cause an "antibiotic apocalypse" where such bacterial infections would no longer be treatable. Thus, serious efforts have to be taken to prevent pneumonia rather than to treat the infection. One major method by which this can be achieved is by community vaccination. *S. pneumoniae* and influenza are two major pathogens known to cause CAP and there are safe and effective vaccines for both the pathogens.[82] Vaccination is paramount for people who are greater risk of acquiring pneumonia such as children, elderly and immune compromised so as to prevent pneumonia. Additionally, passive serum therapy with antibodies developed in pneumonia patients has been identified to be a promising strategy to prevent pneumonia caused by Gram-negative bacteria. The strategy has been successful in animal models and is being translated to human.[83] Many strategies have been followed to prevent HAP especially when the patient is on mechanical ventilation. The prime method is to reduce the exposure of the patient to mechanical ventilation. Other methods that have been identified to be effective in reducing VAP incidence include, (i) elevation of the head of the bed to 30–45 °C and (ii) daily "sedation vacation" and daily assessment of readiness to extubate. These methods are known as "VAP Bundles" that are a group of best practices that when used have been found to be effective.[84] Another major method is to provide good oral hygiene to patients who are on mechanical ventilation. It has been identified that oral care with chlorhexidine could potentially lower the incidence of VAP in mechanically ventilated patients.[85] Additionally, performing subglottic secretion suctioning to aspirate secretions that accumulate around the endotracheal tube at constant intervals has been identified to prevent the development of VAP by about 50%.[86] Proper positioning of the patient on mechanical ventilator to reduce reflux has been also identified to be important to prevent VAP.

Another alternative strategy for the prevention of pneumonia is by targeting the indigenous microbiome. There has been growing evidence suggesting that alterations of intestinal microbiome during pneumonia and antibiotic therapy play a major role in determining the course of the infection. Oral bacteriotherapy has grown popular for the treatment of pneumonia during the recent times where beneficial bacteria are administered orally to strengthen the gut microbiome. The bacteria thus administered orally could be a probiotic strain that has health benefitting property or that can boost the host immunity. Certain strains of lactobacilli and bifidobacteria are known to enhance immunity and exert antibacterial activity.[87] A meta-analysis by Batra et al. had revealed that use of probiotics reduces the incidence of VAP and decrease the duration of mechanical ventilation.[88] Additionally, studies by Cheng et al. and Su et al. revealed that use of probiotics on mechanically ventilated patients decreased the incidence of VAP and improved their total health.[89,90] Strengthening of the intestinal microbiome by probiotic is also known to have an effect on intestinal immunity as it regulates the pro-inflammatory cytokine production.[91] Further, probiotic lactic acid bacteria were administered directly in the respiratory tract to prevent viral pneumonia.[92] A recent study by De Boeck et al. revealed that Lactobacilli isolated from the nasal cavity of healthy people portrayed probiotic activity and could be used as nasal spray to prevent VAP.[93] The isolate prevented colonization of pathogenic bacteria in the respiratory tract as well as stimulated the respiratory mucosal immunity. Further, administration of probiotic strains is known to increase the levels of type I interferons, and stimulate NK cells and T cells thus increasing the levels of systemic and mucosal antibodies in the lungs. However, it is not always necessary that probiotic bacteria have to be administered in the respiratory mucosa to have an effect on respiratory diseases. Probiotics administered orally elicit indirect enhancement of respiratory immunity via the Peyer's Patches in the gut. Thus, probiotics play an important role in modulating both the gut and the respiratory microbiome thus reducing incidence of pneumonia and other respiratory infections. Few major cohort studies that have portrayed reduction in respiratory infections by probiotic therapy have been listed in Table 1.

Table 1 List of major cohort studies portraying reduction in respiratory infections by probiotic therapy.

Sl No	Probiotic intervention	Outcome	Duration of probiotic supplementation	Study group	References
1.	*L. acidophilus, L. casei*	Reduced incidence of respiratory tract infections	3-months	Children	94
2.	*L. gasseri* PA 16/8, *B. longum* SP 07/3, *B. bifidum* MF 20/5	Shortened common cold incidence and reduced the symptoms	7-months	Adults	95
3.	*L. casei*	Reduced duration of respiratory tract infections	5-months	Children	96

Table 1 List of major cohort studies portraying reduction in respiratory infections by probiotic therapy.—cont'd

Sl No	Probiotic intervention	Outcome	Duration of probiotic supplementation	Study group	References
4.	*B. longum* BL999	Reduced the incidence of respiratory tract infections	7-months	Children	97
5.	*L. rhamnosus* GG, *L. acidophilus* LA-5, and *Bifidobacterium* Bb-12	Reduced common respiratory infections	7-months	Children	98
6.	*L. plantarum*, *L. rhamnosus*, *B. lactis*	Reduced incidence of cold episode	3-months	Adults	99
7.	*Saccharomyces cerevisiae*	Reduced incidence of cold episode	3-months	Adults	100
8.	*L. acidophilus* NCFM, *B. animalis* subsp. *lactis* Bi-07	Reduce fever, rhinorrhea, and cough incidence	6-months	Children	101
9.	*L. rhamnosus* GG, *B. lactis* Bb-12	Reduced incidence of early and recurrent upper respiratory infections	12-months	Children	102
10.	*L. helveticus* R0052, *B. infantis* R0033, *B. bifidum* R0071	Reduced common winter diseases	3-month	Children	103
11.	*L. rhamnosus* GG	Reduced common respiratory infections	3-months	Children	104
12.	*L. delbrueckii* ssp. *bulgaricus* OLL1073R-1, *S. themophilus* OLS3059	Reduced incidence of cold episode	2-months	Adults	105
13.	*L. rhamnosus* HN001	Reduced acute respiratory infection	3-months	Children	106
14.	*L. casei* DN-114001	Reduced common respiratory infections	3-months; 1-month follow-up	Adults	107
15.	*L. casei*, *S. thermophiles*, *L. bulgaricus*	Reduced incidence of common	3-months	Children	108

Continued

Table 1 List of major cohort studies portraying reduction in respiratory infections by probiotic therapy.—cont'd

Sl No	Probiotic intervention	Outcome	Duration of probiotic supplementation	Study group	References
		respiratory infections			
16.	*L. plantarum*, *L. paracasei*	Incidence of common cold episodes	3-months	Adults	109
17.	*L. rhamnosus* GG	Reduced the occurrence of respiratory illness	7-months	Children	110
18.	*L. fermentum* CECT5716	Reduced upper respiratory infections	6-month	Children	111
19.	*L. acidophilus*, *B. bifidum*	Reduced common cold	3-month	Children	112
20.	*L. casei* strain *Shirota*	Reduced incidence of acute URI	5-months	Children	113
21.	*L. rhamnosus* GG, *B. animalis* ssp. *lactis* BB-12	Reduced duration of upper respiratory tract infections	3-months	Adults	114
22.	*L. acidophilus* DDS-1 (NCIMB 30333), *B. lactis* UABLA-12 (NCIMB 30334)	Shortened acute respiratory infection	2 weeks	Children	115

Note: Common respiratory infections include any one or more of the following: tonsillitis, pharyngitis, laryngitis, sinusitis, otitis media, certain types of influenza, and the common cold. LRTIs include bronchitis and pneumonia.

9. Future directions and way forward

The infection of the LRT or pneumonia is caused by a variety of microbes which are ubiquitous. Few strategies that require further research for decreasing the global pneumonia burden is depicted in Fig. 2. (i) One of the major areas that require attention is the AMR development in microorganism that causes pneumonia. A heightened amount of resistance has been observed in pathogens isolated from patients having pneumonia especially VAP. The emergence and spread of antibiotic resistant microbes in the environment and hospital settings has to be closely monitored. The efforts to

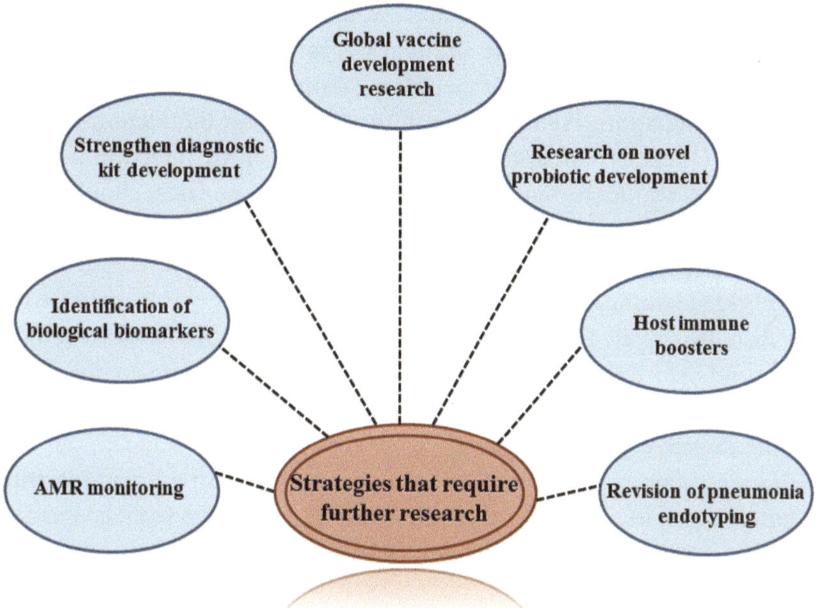

Fig. 2 Few strategies that require further research to be adopted for decreasing the global pneumonia burden. The figure was created using powerpoint.

identify and report the emergence of AMR has to be streamlined at the national and international level. Source tracking of antibiotic resistant pathogens isolated from pneumonia patients has to be performed. (ii) Apart from the microbial pathobionts, the host immune system plays a pivotal role in the pathogenesis of the infection. Though a plethora of research is available on the causative organisms of pneumonia, no in depth research is available portraying the host immunity. Further research on the host-related factors is highly warranted. This should include research on host susceptibility before infection, host response during the infection and host immunity and memory after the infection. Use of multi-omics approaches to predict pneumonia susceptibility in the host and identification of specific biological biomarkers that indicate levels of pneumonia susceptibility is of utmost importance to develop pneumonia prevention strategies. Understandings on the immune resistance of host to various bacterial viral and fungal infections are still unclear. Further, microbial changes during the infection and the function of each microbial population in the progress of the infection are at infancy. As of today, five diagnostic biomarkers (Procalcitonin, C reactive protein, CD163, High Mobility Group Box-1 and Soluble triggering receptor expressed on myeloid cells-1) have been linked to pneumonia prognosis. However, there is considerable ambiguity regarding the efficiency of using these biomarkers as they are not specific to pneumonia and are also produced during inflammation. Hence, research that focuses on the identification of pneumonia specific biomarker for early prognosis of the infection is necessary. (iii) It is also important that advanced reliable and fast technologies to identify the pathogens associated with pneumonia has to be developed. Development and use of rapid diagnostic tests is highly warranted for fast and reliable identification and characterization of the causative microorganism. Also, diagnostic kits that can give information about microbiome shifts during and before the development of VAP can serve to implement proper preventive measures. The major advantage of using rapid diagnostic kits is that the causative pathogen can be detected at early stages and this can help in choosing appropriate antibiotic therapy rather than using broad spectrum empirical therapies. Dipsticks that can detect the causative pathogen and its antibiotic resistant genes (ARGs) would be a breakthrough as it can significantly reduce the time taken for conventional bacterial identification and antibiotic susceptibility profiles. (iv) Additionally, research should focus on identification and development of natural compounds that can strengthen the normal gut and respiratory microbiome that has a major role in prevention of pathogen colonization. Traditional and complementary medicine (T&CM) has been well recommended as immune boosters for promoting pulmonary health. Research that can evidence the improvement in mucosal immunity by steam inhalation with medicinal herbs, oil pulling, nasal irrigation, aromatherapy and by inclusion of herbal medicines as part of the diet should be encouraged. (v) Other alternative strategies such as the use of bacteriophages as bio-control agents and use of probiotic bacteria to replenish and strengthen gut and respiratory microbiome can be investigated. In one of the largest probiotic clinical trials performed on new born babies by administering synbiotics (*Lactobacillus plantarum* and fructo-oligosaccharides) reduced the rate of respiratory tract infections by 40%.[116] Thus, development of prophylactic probiotic therapy that can boost host immunity and thereby prevent infection will be very useful as it would also indirectly reduce the use of antibiotics. (vi) It has been now understood that for all diseases, host directed therapies has a better outcome and thus, the same has to be followed in pneumonia. The current pneumonia typing is focused on the source from where pneumonia is acquired (CAP or VAP), the causative organisms (bacterial, viral, fungal) or the severity of the infection. Host focused endotyping for pneumonia is a major shortcoming. Immunoparalysis has been identified to be a possible endotype of pneumonia which utilizes blood transcriptome differences to identify decreased antigen presentation or dysregulations of innate immune pathways. Improved understanding on mechanism of lung defense during the infection in a recovering patient and lung response in patients that have severe pneumonia could help in formulating host directed therapies understanding immune resistance and tissue resilience against pneumonia. Host focused endotyping methods are at large in its infancy and more elaborate research on the same would help in efficient treatment and cure of pneumonia.

10. Conclusion

In summary, next generation sequencing has shed light on various microbiome niches in the respiratory tract that have role in disease cause and prevention. Microbial dynamics in the respiratory tract is linked to many respiratory infections like pneumonia, asthma, ARDS and COPD. Gut microbiota and oral microbiota are also known to play an important role in shaping the immune system of the respiratory tract and influence its microbial dynamics. Microbes in the oropharynx act as a microbial reservoir for the LRT and oral hygiene is considered to be a major protective factor against VAP. Overall, a better understanding about the indigenous microbial communities within the respiratory system is important and will likely lead to a more targeted approach to develop personalized therapy for each individual patient. The use of probiotics and prebiotics to modulate the respiratory and gut microbiome is gaining importance. Other therapies attempting to modify the respiratory microbiota include bacteriophage therapy and the use of anti-bacterial conjugate vaccines. Though we have made substantial progress in understanding the respiratory microbiome changes during pneumonia, there are still various knowledge gaps and challenges. Limitation in acquiring samples before CAP, difficulties is examining virome and mycobiome changes and lack of humanoid animal models to study micrbiome-host interactions are the most important limitations in this area that would require urgent research. Thus, widening the perspective from just pathogen based therapy to host based therapy is imperative. The use of advanced technologies to increase the current knowledge on respiratory microbiome, host- microbiome and immune system would potentially reduce indiscriminate use of antibiotics in infections such as pneumonia.

References

1. Hildebrandt G. Experimentelle Untersuchungen über das Eindringen pathogener Mikroorganismen von den Luftwegen und der Lunge aus. *Beitr Pathol Anat Physiol.* 1888;3:411–450.
2. Jones FS. The source of the microorganisms in the lungs of normal animals. *J Exp Med.* 1922;36:317–328.
3. Thomson SC, Hewlett RT. The fate of micro-organisms in inspired air. *Lancet.* 1896;147:86–87.
4. Cleary DW, Morris DE, Anderson RA, et al. The upper respiratory tract microbiome of indigenous orang Asli in north-eastern peninsular Malaysia. *NPJ Biofilms Microbiomes.* 2021;7(1). https://doi.org/10.1038/s41522-020-00173-5.
5. Bassis CM, Erb-Downward JR, Dickson RP, et al. Analysis of the upper respiratory tract microbiotas as the source of the lung and gastric microbiotas in healthy individuals. *MBio.* 2015;6(2):e00037. https://doi.org/10.1128/mBio.00037-15.
6. Welp AL, Bomberger JM. Bacterial community interactions during chronic respiratory disease. *Front Cell Infect Microbiol.* 2020;14(10):213. https://doi.org/10.3389/fcimb.2020.00213.
7. Patwa A, Shah A. Anatomy and physiology of respiratory system relevant to anaesthesia. *Indian J Anaesth.* 2015;59(9):533–541. https://doi.org/10.4103/0019-5049.165849.
8. Man W, de Steenhuijsen PW, Bogaert D. The microbiota of the respiratory tract: Gatekeeper to respiratory health. *Nat Rev Microbiol.* 2017;15:259–270. https://doi.org/10.1038/nrmicro.2017.14.
9. Sommariva M, Le Noci V, Bianchi F, et al. The lung microbiota: Role in maintaining pulmonary immune homeostasis and its implications in cancer development and therapy. *Cell Mol Life Sci.* 2020;77(14):2739–2749. https://doi.org/10.1007/s00018-020-03452-8.
10. Yun Y, Srinivas G, Kuenzel S, et al. Environmentally determined differences in the murine lung microbiota and their relation to alveolar architecture. *PLoS One.* 2014;9:e113466.
11. Aagaard K, Ma J, Antony KM, Ganu R, Petrosino J, Versalovic J. The placenta harbors a unique microbiome. *Sci Transl Med.* 2014;6(237):237ra65. https://doi.org/10.1126/scitranslmed.3008599.
12. Bosch AATM, Levin E, van Houten MA, et al. Development of upper respiratory tract microbiota in infancy is affected by mode of delivery. *EBioMedicine.* 2016;9:336–345. https://doi.org/10.1016/j.ebiom.2016.05.031.
13. Dominguez-Bello MG, Costello EK, Contreras M, et al. Delivery mode shapes the acquisition and structure of the initial microbiota across multiple body habitats in newborns. *Proc Natl Acad Sci U S A.* 2010;107:11971–11975.
14. Biesbroek G, Bosch AA, Wang X, et al. The impact of breastfeeding on nasopharyngeal microbial communities in infants. *Am J Respir Crit Care Med.* 2014;190(3):298–308. https://doi.org/10.1164/rccm.201401-0073OC [PMID: 24921688].
15. Pettigrew MM, Laufer AS, Gent JF, Kong Y, Fennie KP, Metlay JP. Upper respiratory tract microbial communities, acute otitis media pathogens, and antibiotic use in healthy and sick children. *Appl Environ Microbiol.* 2012;78(17):6262–6270. https://doi.org/10.1128/AEM.01051-12.
16. Duijts L, Jaddoe VWV, Hofman A, Moll HA. Prolonged and exclusive breastfeeding reduces the risk of infectious diseases in infancy. *Pediatrics.* 2010;126:e18–e25.
17. Jakobsson HE, Jernberg C, Andersson AF, Sjölund-Karlsson M, Jansson JK, Engstrand L. Short-term antibiotic treatment has differing long-term impacts on the human throat and gut microbiome. *PLoS One.* 2010;5(3):e9836. https://doi.org/10.1371/journal.pone.0009836.

18. Charlson ES, Chen J, Custers-Allen R, et al. Disordered microbial communities in the upper respiratory tract of cigarette smokers. *PLoS One*. 2010;5(12):e15216. https://doi.org/10.1371/journal.pone.0015216.
19. Lim MY, Yoon HS, Rho M, et al. Analysis of the association between host genetics, smoking, and sputum microbiota in healthy humans. *Sci Rep*. 2016;6:23745. https://doi.org/10.1038/srep23745.
20. Brook I, Gober AE. Recovery of potential pathogens and interfering bacteria in the nasopharynx of smokers and nonsmokers. *Chest*. 2005;127:2072–2075.
21. Stearns JC, Davidson CJ, McKeon S, et al. Culture and molecular-based profiles show shifts in bacterial communities of the upper respiratory tract that occur with age. *ISME J*. 2015;9(5):1246–1259. https://doi.org/10.1038/ismej.2014.250 [Erratum in: ISME J. 2015 9(5):1268].
22. Bogaert D, Keijser B, Huse S, et al. Variability and diversity of nasopharyngeal microbiota in children: A metagenomic analysis. *PLoS One*. 2011;6(2):e17035. https://doi.org/10.1371/journal.pone.0017035.
23. Eidi S, Kamali SA, Hajari Z, et al. Nasal and indoors fungal contamination in healthy subjects. *Health Scope*. 2016;5:e30033.
24. Bogaert D, De Groot R, Hermans PWM. *Streptococcus pneumoniae* colonisation: The key to pneumococcal disease. *Lancet Infect Dis*. 2004;4:144–154.
25. Bomar L, Brugger SD, Yost BH, Davies SS, Lemon KP. *Corynebacterium accolens* releases antipneumococcal free fatty acids from human nostril and skin surface triacylglycerols. *MBio*. 2016;7:[e01725-15].
26. Iwase T, Uehara Y, Shinji H, et al. *Staphylococcus epidermidis* Esp inhibits *Staphylococcus aureus* biofilm formation and nasal colonization. *Nature*. 2010;465(7296):346–349. https://doi.org/10.1038/nature09074.
27. Dickson RP, Erb-Downward JR, Freeman CM, et al. Spatial variation in the healthy human Lung microbiome and the adapted island model of Lung biogeography. *Ann Am Thorac Soc*. 2015;12(6):821–830. https://doi.org/10.1513/AnnalsATS.201501-029OC.
28. Marsh RL, Kaestli M, Chang AB, et al. The microbiota in bronchoalveolar lavage from young children with chronic lung disease includes taxa present in both the oropharynx and nasopharynx. *Microbiome*. 2016;4(1):37. https://doi.org/10.1186/s40168-016-0182-1.
29. van Woerden HC, Gregory C, Brown R, Marchesi JR, Hoogendoorn B, Matthews IP. Differences in fungi present in induced sputum samples from asthma patients and non-atopic controls: A community based case control study. *BMC Infect Dis*. 2013;13:69. https://doi.org/10.1186/1471-2334-13-69.
30. Young JC, Chehoud C, Bittinger K, et al. Viral metagenomics reveal blooms of anelloviruses in the respiratory tract of lung transplant recipients. *Am J Transplant*. 2015;15(1):200–209. https://doi.org/10.1111/ajt.13031.
31. Paun A, Danska JS. Immuno-ecology: How the microbiome regulates tolerance and autoimmunity. *Curr Opin Immunol*. 2015;37:34–39. https://doi.org/10.1016/j.coi.2015.09.004.
32. Gollwitzer ES, Marsland BJ. Impact of early-life exposures on immune maturation and susceptibility to disease. *Trends Immunol*. 2015;36:684–696.
33. Mosser DM, Edwards JP. Exploring the full spectrum of macrophage activation. *Nat Rev Immunol*. 2008;8:958–969.
34. Lai HY, Rogers DF. Mucus hypersecretion in asthma: Intracellular signalling pathways as targets for pharmacotherapy. *Curr Opin Allergy Clin Immunol*. 2010;10:67–76.
35. Howie SRC, Murdoch DR. Global childhood pneumonia: The good news, the bad news, and the way ahead. *Lancet Glob Health*. 2019;7(1):e4–e5.
36. WHO; 2019. https://www.who.int/news-room/fact-sheets/detail/pneumonia.

37. Chen JC, Jenkins-Marsh S, Flenady V, et al. Early-onset group B streptococcal disease in a risk factor-based prevention setting: A 15-year population-based study. *Aust N Z J Obstet Gynaecol*. 2019;59(3):422–429.
38. Lim WS, Macfarlane JT, Boswell TC, et al. Study of community acquired pneumonia aetiology (SCAPA) in adults admitted to hospital: Implications for management guidelines. *Thorax*. 2001;56(4):296–301. https://doi.org/10.1136/thorax.56.4.296.
39. Brandenburg JA, Marrie TJ, Coley CM, et al. Clinical presentation, processes and outcomes of care for patients with pneumococcal pneumonia. *J Gen Intern Med*. 2000;15 (9):638–646. https://doi.org/10.1046/j.1525-1497.2000.04429.x.
40. Cillóniz C, Torres A, Niederman M, et al. Community-acquired pneumonia related to intracellular pathogens. *Intensive Care Med*. 2016;42(9):1374–1386. https://doi.org/10.1007/s00134-016-4394-4.
41. Restrepo MI, Babu BL, Reyes LF, et al. Burden and risk factors for *Pseudomonas aeruginosa* community-acquired pneumonia: A multinational point prevalence study of hospitalised patients. *Eur Respir J*. 2018;52(2):1701190. https://doi.org/10.1183/13993003.01190-2017.
42. Vanoni NM, Carugati M, Borsa N, et al. Management of Acute Respiratory Failure due to community-acquired pneumonia: A systematic review. *Med Sci (Basel)*. 2019;7(1):10. https://doi.org/10.3390/medsci7010010.
43. Woodhead M, Blasi F, Ewig S, et al. Guidelines for the management of adult lower respiratory tract infections- -full version. *Clin Microbiol Infect*. 2011;17(Suppl 6): E1–59. https://doi.org/10.1111/j.1469-0691.2011.03672.x.
44. Falguera M, Ruiz-González A, Schoenenberger JA, et al. Prospective, randomised study to compare empirical treatment versus targeted treatment on the basis of the urine antigen results in hospitalised patients with community-acquired pneumonia. *Thorax*. 2010;65(2):101–106. https://doi.org/10.1136/thx.2009.118588.
45. Metlay JP, Waterer GW, Long AC, et al. (2019). Diagnosis and treatment of adults with community-acquired pneumonia. An official clinical practice guideline of the American Thoracic Society and Infectious Diseases Society of America. *Am J Respir Crit Care Med*. 2019;200(7):e45–e67. https://doi.org/10.1164/rccm.201908-1581ST.
46. Papazian L, Klompas M, Luyt CE. Ventilator-associated pneumonia in adults: A narrative review. *Intensive Care Med*. 2020;46(5):888–906. https://doi.org/10.1007/s00134-020-05980-0.
47. Blonz G, Kouatchet A, Chudeau N, et al. Epidemiology and microbiology of ventilator-associated pneumonia in COVID-19 patients: A multicenter retrospective study in 188 patients in an un-inundated French region. *Crit Care*. 2021;25 (1):72. https://doi.org/10.1186/s13054-021-03493-w.
48. Giacobbe DR, Battaglini D, Enrile EM, et al. Incidence and prognosis of ventilator-associated pneumonia in critically ill patients with COVID-19: A multicenter study. *J Clin Med*. 2021;10(4):555. https://doi.org/10.3390/jcm10040555.
49. Wu D, Wu C, Zhang S, Zhong Y. Risk factors of ventilator-associated pneumonia in critically III patients. *Front Pharmacol*. 2019;10:482. https://doi.org/10.3389/fphar.2019.00482.
50. Ewig S, Kolditz M, Pletz MW, Chalmers J. Healthcare-associated pneumonia: Is there any reason to continue to utilize this label in 2019? *Clin Microbiol Infect*. 2019;25 (10):1173–1179. https://doi.org/10.1016/j.cmi.2019.02.022.
51. Kalil AC, Metersky ML, Klompas M, et al. Executive summary: Management of Adults with Hospital-acquired and Ventilator-associated Pneumonia: 2016 clinical practice guidelines by the Infectious Diseases Society of America and the American Thoracic Society. *Clin Infect Dis*. 2016;63(5):575–582.
52. Endimiani A, Hujer KM, Hujer AM, et al. Are we ready for novel detection methods to treat respiratory pathogens in hospital-acquired pneumonia? *Clin Infect Dis*. 2011;52 (Suppl 4):S373–S383. https://doi.org/10.1093/cid/cir054.

53. Lung M, Codina G. Molecular diagnosis in HAP/VAP. *Curr Opin Crit Care*. 2012;18 (5):487–494. https://doi.org/10.1097/MCC.0b013e3283577d37.
54. Alvarez-Lerma F. Modification of empiric antibiotic treatment in patients with pneumonia acquired in the intensive care unit. ICU-acquired pneumonia study group. *Intensive Care Med*. 1996;22(5):387–394. https://doi.org/10.1007/BF01712153.
55. Zakharkina T, Martin-Loeches I, Matamoros S, et al. The dynamics of the pulmonary microbiome during mechanical ventilation in the intensive care unit and the association with occurrence of pneumonia. *Thorax*. 2017;72(9):803–810. https://doi.org/10.1136/thoraxjnl-2016-209158.
56. Woo S, Park SY, Kim Y, Jeon JP, Lee JJ, Hong JY. The dynamics of respiratory microbiota during mechanical ventilation in patients with pneumonia. *J Clin Med*. 2020;9 (3):638.
57. Emonet S, Lazarevic V, Leemann Refondini C, et al. Identification of respiratory microbiota markers in ventilator-associated pneumonia. *Intensive Care Med*. 2019;45 (8):1082–1092. https://doi.org/10.1007/s00134-019-05660-8.
58. Sommerstein R, Merz TM, Berger S, Kraemer JG, Marschall J, Hilty M. Patterns in the longitudinal oropharyngeal microbiome evolution related to ventilator-associated pneumonia. *Antimicrob Resist Infect Control*. 2019;8:81.
59. Fernandez-Barat L, Ben-Aicha S, Motos A, et al. Assessment of in vivo versus in vitro biofilm formation of clinical methicillin-resistant *Staphylococcus aureus* isolates from endotracheal tubes. *Sci Rep*. 2018;8(1):11906.
60. Hotterbeekx A, Xavier BB, Bielen K, et al. The endotracheal tube microbiome associated with *Pseudomonas aeruginosa* or *Staphylococcus epidermidis*. *Sci Rep*. 2016;6:36507. https://doi.org/10.1038/srep36507.
61. Kalanuria AA, Ziai W, Mirski M. Ventilator-associated pneumonia in the ICU. *Crit Care*. 2014;18(2):208.
62. Bao L, Zhang C, Dong J, Zhao L, Li Y, Sun J. Oral microbiome and SARS-CoV-2: Beware of Lung co-infection. *Front Microbiol*. 2020;11:1840. https://doi.org/10.3389/fmicb.2020.01840.
63. de Steenhuijsen Piters WA, Huijskens EG, Wyllie AL, et al. Dysbiosis of upper respiratory tract microbiota in elderly pneumonia patients. *ISME J*. 2016;10 (1):97–108. https://doi.org/10.1038/ismej.2015.99.
64. Segal LN, Alekseyenko AV, Clemente JC, et al. Correction: Enrichment of lung microbiome with supraglottic taxa is associated with increased pulmonary inflammation. *Microbiome*. 2013;2:21. https://doi.org/10.1186/2049-2618-2-21. Erratum for: Microbiome. 1:19.
65. Segal LN, Clemente JC, Tsay JC, et al. Enrichment of the lung microbiome with oral taxa is associated with lung inflammation of a Th17 phenotype. *Nat Microbiol*. 2016;1:16031. https://doi.org/10.1038/nmicrobiol.2016.31.
66. van't Hof W, Veerman EC, Nieuw Amerongen AV, Ligtenberg AJ. Antimicrobial defense systems in saliva. *Monogr Oral Sci*. 2014;24:40–51.
67. Marsland BJ, Trompette A, Gollwitzer ES. The gut-lung axis in respiratory disease. *Ann Am Thorac Soc*. 2015;12(Suppl. 2):S150–S156. https://doi.org/10.1513/AnnalsATS.201503-133AW.
68. Antosca KM, Chernikova DA, Price CE, et al. Altered stool microbiota of infants with cystic fibrosis shows a reduction in genera associated with immune programming from birth. *J Bacteriol*. 2019;201(16). https://doi.org/10.1128/JB.00274-19.
69. Felix KM, Jaimez IA, Nguyen TV, et al. Gut Microbiota Contributes to Resistance Against Pneumococcal Pneumonia in Immunodeficient $Rag^{-/-}$ Mice. *Front Cell Infect Microbiol*. 2018;8:118. https://doi.org/10.3389/fcimb.2018.00118.
70. Macia L, Thorburn AN, Binge LC, et al. Microbial influences on epithelial integrity and immune function as a basis for inflammatory diseases. *Immunol Rev*. 2012;245 (1):164–176. https://doi.org/10.1111/j.1600-065X.2011.01080.x.

71. Stokholm J, Blaser MJ, Thorsen J, et al. Maturation of the gut microbiome and risk of asthma in childhood. *Nat Commun.* 2018;9(1):141. https://doi.org/10.1038/s41467-017-02573-2 [Erratum in: Nat Commun. 9(1):704].
72. Abrahamsson TR, Jakobsson HE, Andersson AF, Bjorksten B, Engstrand L, Jenmalm MC. Low gut microbiota diversity in early infancy precedes asthma at school age. *Clin Exp Allergy.* 2014;44:842–850. https://doi.org/10.1111/cea.12253.
73. Qian LJ, Kang SM, Xie JL, et al. Early-life gut microbial colonization shapes Th1/Th2 balance in asthma model in BALB/c mice. *BMC Microbiol.* 2017;17(1):135. https://doi.org/10.1186/s12866-017-1044-0.
74. Demirci M, Tokman HB, Uysal HK, et al. Reduced *Akkermansia muciniphila* and *Faecalibacterium prausnitzii* levels in the gut microbiota of children with allergic asthma. *Allergol Immunopathol (Madr).* 2019;47(4):365–371. https://doi.org/10.1016/j.aller.2018.12.009.
75. Charlson ES, Bittinger K, Haas AR, et al. Topographical continuity of bacterial populations in the healthy human respiratory tract. *Am J Respir Crit Care Med.* 2011;184(8):957–963. https://doi.org/10.1164/rccm.201104-0655OC.
76. Luo M, Liu Y, Wu P, et al. Alternation of gut microbiota in patients with pulmonary tuberculosis. *Front Physiol.* 2017;8:822. https://doi.org/10.3389/fphys.2017.00822.
77. Negi S, Pahari S, Bashir H, Agrewala JN. Gut microbiota regulates mincle mediated activation of lung dendritic cells to protect against *mycobacterium tuberculosis. Front Immunol.* 2019;10:1142. https://doi.org/10.3389/fimmu.2019.01142.
78. Grayson MH, Camarda LE, Hussain SA, et al. Intestinal microbiota disruption reduces regulatory T cells and increases respiratory viral infection mortality through increased IFNγ production. *Front Immunol.* 2018;9:1587. https://doi.org/10.3389/fimmu.2018.01587.
79. Du X, Meng Q, Sharif A, et al. Surfactant proteins SP-A and SP-D ameliorate pneumonia severity and intestinal injury in a murine model of *Staphylococcus aureus* pneumonia. *Shock.* 2016;46(2):164–172. https://doi.org/10.1097/SHK.0000000000000587 [PMID: 26849628].
80. Enaud R, Hooks KB, Barre A, et al. Intestinal inflammation in children with cystic fibrosis is associated with Crohn's-like microbiota disturbances. *J Clin Med.* 2019;8(5):645. https://doi.org/10.3390/jcm8050645.
81. Debyser G, Mesuere B, Clement L, et al. Faecal proteomics: A tool to investigate dysbiosis and inflammation in patients with cystic fibrosis. *J Cyst Fibros.* 2016;15(2):242–250. https://doi.org/10.1016/j.jcf.2015.08.003.
82. van Werkhoven CH, Huijts SM. Vaccines to prevent pneumococcal community-acquired pneumonia. *Clin Chest Med.* 2018;39(4):733–752. https://doi.org/10.1016/j.ccm.2018.07.007.
83. Casadevall A, Scharff MD. Serum therapy revisited: Animal models of infection and development of passive antibody therapy. *Antimicrob Agents Chemother.* 1994;38(8):1695–1702. https://doi.org/10.1128/AAC.38.8.1695.
84. Wip C, Napolitano L. Bundles to prevent ventilator-associated pneumonia: How valuable are they? *Curr Opin Infect Dis.* 2009;22(2):159–166. https://doi.org/10.1097/QCO.0b013e3283295e7b.
85. Boltey E, Yakusheva O, Costa DK. 5 nursing strategies to prevent ventilator-associated pneumonia. *Am Nurse Today.* 2017;12(6):42–43.
86. Lacherade JC, Azais MA, Pouplet C, Colin G. Subglottic secretion drainage for ventilator-associated pneumonia prevention: An underused efficient measure. *Ann Transl Med.* 2018;6(21):422. https://doi.org/10.21037/atm.2018.10.40.
87. Ceccarelli G, Borrazzo C, Pinacchio C, et al. Oral bacteriotherapy in patients with COVID-19: A retrospective cohort study. *Front Nutr.* 2021;7:613928. https://doi.org/10.3389/fnut.2020.613928.

88. Batra P, Soni KD, Mathur P. Efficacy of probiotics in the prevention of VAP in critically ill ICU patients: An updated systematic review and meta-analysis of randomized control trials. *J Intensive Care*. 2020;8:81. https://doi.org/10.1186/s40560-020-00487-8.
89. Su M, Jia Y, Li Y, Zhou D, Jia J. Probiotics for the prevention of ventilator-associated pneumonia: A meta-analysis of randomized controlled trials. *Respir Care*. 2020;65 (5):673–685. https://doi.org/10.4187/respcare.07097.
90. Wang H, Lian P, Niu X, et al. TLR4 deficiency reduces pulmonary resistance to *Streptococcus pneumoniae* in gut microbiota-disrupted mice. *PLoS One*. 2018;13(12): e0209183. https://doi.org/10.1371/journal.pone.0209183.
91. Azad MAK, Sarker M, Wan D. Immunomodulatory effects of probiotics on cytokine profiles. *Biomed Res Int*. 2018;23:8063647.
92. Barbieri N, Herrera M, Salva S, Villena J, Alvarez S. *Lactobacillus rhamnosus* CRL1505 nasal administration improves recovery of T-cell mediated immunity against pneumococcal infection in malnourished mice. *Benef Microbes*. 2017;8(3):393–405. https://doi.org/10.3920/BM2016.0152.
93. De Boeck I, van den Broek MFL, Allonsius CN, et al. Lactobacilli have a niche in the human nose. *Cell Rep*. 2020;31(8), 107674. https://doi.org/10.1016/j.celrep.2020.107674 [PMID: 32460009].
94. Río ME, Zago Beatriz L, Garcia H, Winter L. The nutritional status change the effectiveness of a dietary supplement of lactic bacteria on the emerging of respiratory tract diseases in children. *Arch LatinoamNutr*. 2002;52:29–34.
95. de Vrese M, Winkler P, Rautenberg P, et al. Effect of *lactobacillus gasseri* PA 16/8, *Bifidobacterium longum* SP 07/3, B. bifidum MF 20/5 on common cold episodes: A double blind, randomized, controlled trial. *Clin Nutr*. 2005;24(4):481–491. https://doi.org/10.1016/j.clnu.2005.02.006.
96. CoboSanz JM, Mateos JA, Muñoz CA. Effect of lactobacillus casei on the incidence of infectious conditions in children. *Nutr Hosp*. 2006;21:547–551.
97. Puccio G, Cajozzo C, Meli F, Rochat F, Grathwohl D, Steenhout P. Clinical evaluation of a new starter formula for infants containing live *Bifidobacterium longum* BL999 and prebiotics. *Nutrition*. 2007;23(1):1–8. https://doi.org/10.1016/j.nut.2006.09.007.
98. Kloster Smerud H, Ramstad Kleiveland C, Roll Mosland A, et al. Effect of a probiotic milk product on gastrointestinal and respiratory infections in children attending daycare. *Microb Ecol Health Dis*. 2008;20:80–85.
99. Pregliasco F, Anselmi G, Fonte L, Giussani F, Schieppati S, Soletti L. A new chance of preventing winter diseases by the administration of symbiotic formulations. *J Clin Gastroenterol*. 2008;42(Suppl 3 Pt 2):S224–S233.
100. Moyad MA, Robinson LE, Zawada Jr ET, et al. Effects of a modified yeast supplement on cold/flu symptoms. *Urol Nurs*. 2008;28:50–55.
101. Leyer GJ, Li S, Mubasher ME, Reifer C, Ouwehand AC. Probiotic effects on cold and influenza-like symptom incidence and duration in children. *Pediatrics*. 2009;124(2): e172–e179. https://doi.org/10.1542/peds.2008-2666.
102. Rautava S, Salminen S, Isolauri E. Specific probiotics in reducing the risk of acute infections in infancy- -a randomised, double-blind, placebo-controlled study. *Br J Nutr*. 2009;101(11):1722–1726. https://doi.org/10.1017/S0007114508116282.
103. Cazzola M, Pham-Thi N, Kerihuel JC, Durand H, Bohbot S. Efficacy of a synbiotic supplementation in the prevention of common winter diseases in children: A randomized, double-blind, placebo-controlled pilot study. *Ther Adv Respir Dis*. 2010;4 (5):271–278. https://doi.org/10.1177/1753465810379010.
104. Hojsak I, Snovak N, Abdović S, Szajewska H, Misak Z, Kolacek S. *Lactobacillus* GG in the prevention of gastrointestinal and respiratory tract infections in children who attend day care centers: A randomized, double-blind, placebo-controlled trial. *Clin Nutr*. 2010;29(3):312–316. https://doi.org/10.1016/j.clnu.2009.09.008.

105. Makino S, Ikegami S, Kume A, Horiuchi H, Sasaki H, Orii N. Reducing the risk of infection in the elderly by dietary intake of yoghurt fermented with *lactobacillus delbrueckii* ssp. bulgaricus OLL1073R-1. *Br J Nutr.* 2010;104:998–1006.
106. Cáceres P, Montes S, Vega N, et al. Effects of *lactobacillus rhamnosus* HN001 on acute respiratory infections and intestinal secretory IgA in children. *Pediatr Infect Dis J.* 2010;5:353–362.
107. Guillemard E, Tanguy J, Flavigny A, de la Motte S, Schrezenmeir J. Effects of consumption of a fermented dairy product containing the probiotic *lactobacillus casei* DN-114 001 on common respiratory and gastrointestinal infections in shift workers in a randomized controlled trial. *J Am Coll Nutr.* 2010;29(5):455–468. https://doi.org/10.1080/07315724.2010.10719882.
108. Merenstein D, Murphy M, Fokar A, et al. Use of a fermented dairy probiotic drink containing *lactobacillus casei* (DN-114 001) to decrease the rate of illness in kids: The DRINK study. A patient-oriented, double-blind, cluster-randomized, placebo-controlled, clinical trial. *Eur J Clin Nutr.* 2010;64:669–677.
109. Berggren A, LazouAhren I, Larsson N, Onning G. Randomised, double-blind and placebo-controlled study using new probiotic lactobacilli for strengthening the body immune defence against viral infections. *Eur J Nutr.* 2011;50:203–210.
110. Kumpu M, Kekkonen RA, Kautiainen H, et al. Milk containing probiotic *lactobacillus rhamnosus* GG and respiratory illness in children: A randomized, double-blind, placebo-controlled trial. *Eur J Clin Nutr.* 2012;66(9):1020–1023. https://doi.org/10.1038/ejcn.2012.62.
111. Maldonado J, Cañabate F, Sempere L, et al. Human milk probiotic *lactobacillus fermentum* CECT5716 reduces the incidence of gastrointestinal and upper respiratory tract infections in infants. *J Pediatr Gastroenterol Nutr.* 2012;54(1):55–61. https://doi.org/10.1097/MPG.0b013e3182333f18 [Erratum in: J Pediatr Gastroenterol Nutr. 2012 Apr; 54(4):571].
112. Rerksuppaphol S, Rerksuppaphol L. Randomized controlled trial of probiotics to reduce common cold in schoolchildren. *Pediatr Int.* 2012;54(5):682–687. https://doi.org/10.1111/j.1442-200X.2012.03647.x.
113. Fujita R, Iimuro S, Shinozaki T, et al. Decreased duration of acute upper respiratory tract infections with daily intake of fermented milk: A multicenter, double-blinded, randomized comparative study in users of day care facilities for the elderly population. *Am J Infect Control.* 2013;41:1231–1235.
114. Smith TJ, Rigassio-Radler D, Denmark R, Haley T, Touger-Decker R. Effect of *Lactobacillus rhamnosus* LGG(R) and *Bifidobacterium animalis* ssp. lactis BB-12(R) on health-related quality of life in college students affected by upper respiratory infections. *Br J Nutr.* 2013;109:1999–2007.
115. Gerasimov SV, Ivantsiv VA, Bobryk LM, et al. Role of short-term use of *L. acidophilus* DDS-1 and *B. lactis* UABLA-12 in acute respiratory infections in children: A randomized controlled trial. *Eur J Clin Nutr.* 2016;70:463–469.
116. Chong HX, Yusoff NAA, Hor YY, et al. Lactobacillus plantarum DR7 improved upper respiratory tract infections via enhancing immune and inflammatory parameters: A randomized, double-blind, placebo-controlled study. *J Dairy Sci.* 2019;102(6):4783–4797. https://doi.org/10.3168/jds.2018-16103.

Further reading

117. Wang J, Liu KX, Ariani F, Tao LL, Zhang J, Qu JM. Probiotics for preventing ventilator-associated pneumonia: A systematic review and meta-analysis of high-quality randomized controlled trials. *PLoS One.* 2013;8(12):e83934. https://doi.org/10.1371/journal.pone.0083934. PMID: 24367620; PMCID: PMC3867481.

CHAPTER FIVE

Gut microbiome dysbiosis in neonatal sepsis

Jyoti Verma[a], M. Jeeva Sankar[b], Krishnamohan Atmakuri[c], Ramesh Agarwal[b], and Bhabatosh Das[a,*]

[a]Molecular Genetics Laboratory, Infection and Immunology Division, Translational Health Science and Technology Institute, NCR Biotech Science Cluster, Faridabad, India
[b]Department of Pediatrics, All India Institute of Medical Sciences, New Delhi, India
[c]Bacterial Pathogenesis Lab, Infection and Immunology Division, Translational Health Science and Technology Institute, Faridabad, India
*Corresponding author: e-mail address: bhabatosh@thsti.res.in

Contents

Progress in Molecular Biology and Translational Science, Volume 192
ISSN 1877-1173
https://doi.org/10.1016/bs.pmbts.2022.07.010

Abstract

Sepsis is a highly heterogeneous, life-threatening organ dysfunction primarily caused by a dysregulated immune response to counter bacterial, viral, or fungal infections, resulting in haemodynamic changes and significant morbidity and mortality across all ages. In recent times, it has become one of the foremost causes of morbidity and mortality among newborns globally. The neonates, particularly the preterm neonates, due to their immature immune systems and non-canonical microbial community acquisition in the gastrointestinal tract and other body habitats, are adversely affected compared to the elderly with immunocompromised conditions. The neonates could acquire microbiota in utero or during delivery from the mother's genital tract or postnatally from contact with hospital personnel and the immediate hospital environment after the birth. Other factors that may enhance the risk include early colonization of microbiota by pathogens that trigger dysbiosis of the gut microbiome accompanied by a dysregulated immune response, organ dysfunction, and potential death. The sepsis-linked mortality could be prevented by timely diagnosis, selective antibiotic therapy, and supportive postnatal care. Infections due to antibiotic-resistant bacteria severely restrict possible therapeutic options, thus extending hospital stays. A comprehensive analysis of the infecting pathogens, cognate host responses, and the microbiota present would certainly help formulate appropriate interventions.

Abbreviations

coNS coagulase-negative *Staphylococcus*
EOS early onset of sepsis
GBS group B *Streptococcus*
LOS late-onset of sepsis
rRNA ribosomal RNA
SIRS systemic inflammatory immune response

1. Introduction

Neonatal sepsis is a clinical syndrome, which is characterized by blood-borne infection caused by invasive bacteria, fungi and viruses in the first month of life. It accounts for ~2.9 million deaths among children less than 5 years and is the major cause of mortality among neonates followed by lower respiratory infections, and diarrheal diseases.[1] Most neonates who survive sepsis develop neurocognitive sequelae that impair their later growth and development. Globally, out of the three million annual cases of neonatal sepsis (~2200 per 1,00,000 live births), India accounts for the highest occurrence of sepsis among neonates (~17,000 per 1,00,000 live births).[2] In India the case fatality rate for neonatal sepsis varies from 25% to 65%.[3] There is paucity of accurate data that leads to underestimation of the actual cases. There is currently no agreed-upon definition of neonatal sepsis for low- and middle-income countries (LMICs)[4] and broadly includes pneumonia, sepsis, meningitis, arthritis, osteomyelitis and urinary tract infections.[5] Due to limited blood culture-testing facilities in most secondary hospitals of India and the non-specific features of clinical sepsis, quick and accurate identification of neonates with early onset of sepsis (EOS) is often a huge challenge.[6] Moreover, the culture testing results usually take around 48 h and are liable for many false-positive results after antenatal antibiotic exposure.[7]

The microbial communities that reside in the infant gut play an important role in development, overall health, and protection against several infectious and metabolic diseases. The gut microbiome is associated with the maturation of the host immune system, shaping of the gut microbiota by usage of different nutrients or due to production of various metabolites as by-products, and is involved in colonization resistance by selectively permitting the growth of specific bacteria and preventing the growth of pathogens.[8] The infant gut microbiome is not as complex as that of an adult gut and comprises uncomplicated flora from bacteria, viruses, and fungi domains. Most of the research studies have targeted the bacterial communities. However, viruses and fungi have been gaining importance lately. There is a simultaneous evolution of the communities acquired during infancy till childhood. The mode of delivery is the significant factor determining the early colonizers of the infant gut.[9] Furthermore, diet and other environmental factors influence the emergence of the mature gut microbial community.[10]

Advancement in the field of genomics has helped in deciphering the complex ecosystem of microbes in the human gut.[11] The preterm neonate's gut microbiota is vulnerable to perturbation, which is linked to life-threatening consequences such as necrotizing enterocolitis (inflammation of the intestinal tissues), late-onset of sepsis (LOS) and disorders in childhood and adult life such as asthma, obesity or neurological disorders.[12] The gut microbiome of premature infants shows a lower microbial diversity in contrast to full-term infant.[13] However, there still is a significant gap in knowledge with reference to the gut microbiota of neonates at birth or early postnatal dysbiosis being implicated in morbidity and mortality of the preterm neonates. Research studies with regard to the emerging world of viruses and fungi are still in their infancy and require a comprehensive understanding to have a holistic view of the complexities of the human gut microbiome. This would help in the formulation of interventions or the development of biomarker-based tests for the detection of any disorder at the earliest.

2. Human neonatal gut microbiome

Human microbiota constitute all of the microorganisms that live in the different parts of human body. Their number is estimated to be in trillions whose coordinated functions are important for the homeostasis of the human body. The complex microbial community residing in the gastrointestinal tract constitutes the gut microbiota. The members of the gut microbiota belong to different domains of life: Archaea, Bacteria, Eukarya and Viruses.[14] The gut microbiome serves an important interface to the external environment, maintenance of homeostasis, regulation of immune function, supports digestion and protection against pathogens.[15] The acquired gut microbiota in neonates evolves with time, which is driven by nutritional, hormonal, immunological, prebiotic effects of mother's milk and other environmental factors. The gut microbiome is believed to be a key determinant for the regulation of immunological, neural and endocrine pathways in an infant. The complete colonization of the gut occurs within 3 years of life. The dysbiosis in the early life of an infant is critical for the health and development of any disease complication. From birth to adulthood, the gut microbiome changes dramatically. Average number of microbes inhabiting the human gut is around 10 to 100 trillion.[16] However, the number of bacterial taxa in infants at birth is relatively low. Colonization in the infant gut occurs through the formation of complex microbial communities by de novo assembly, which is affected by various environmental and host factors. Earlier, it was assumed that the fetus develops in a sterile environment and the colonization with bacteria was presumed to occur after delivery. However, this incorrect belief was caused by the inability to culture microbes from amniotic fluid or infant skin surfaces.

The recent progress in 16S ribosomal RNA (rRNA) gene sequencing questions this notion and presented a clear picture of the commensal bacteria from the phylum Firmicutes, Proteobacteria, Bacteroidetes and Actinobacteria inhabiting the placental basal plate.[17] Surprisingly, this study also found a peculiar similarity between the microbial population residing in the placenta and the oral microbiome of the non-pregnant females. An intriguing study by Jimnez et al.,[18] depicted the presence of *Enterococcus*, *Staphylococcus* and *Streptococcus* species in low numbers within the umbilical cord blood of scheduled and elective human cesarean deliveries. The meconium of the term infants tested within 2 h of birth and before the initiation of breastfeeding, showed the presence of *Escherichia, Enterococcus and Staphylococcus.*[19] Collectively all these studies depict a bacterial community in placenta, umbilical blood stream and in the first infant stool. These studies suggest the initial colonization of the intestinal tract during utero-fetal development. The infant gut microbiota core is divided into six groups mainly: group 1, *Bifidobacteriales, Anaerostipes, Lactobacillales* and *Faecalibacterium*; group 2, *Bacteroidales* and *Verrucomicrobiales*; group 3, *Clostridiales*; group 4, *Enterobacteriales*; group 5, *Pasteurellales*; and group 6, *Selenomonadales.*[20] The development of the gut virome in newborns is dynamic, with initial colonization just after the bacterial colonization during the first weeks of life. However, comprehensive studies are limited in reference to virome, archaeome or fungi in the gut microbiome of neonates.[21]

2.1 Gut microbiome of normal infants

The gut microbiota is a critical component of the human microbiome. Just after birth, there is a rapid expansion of the infant's gut microbiome and other flora. During fetal life, the gut microbiome begins to evolve and eventually matures into a complex microbial ecosystem. The gut microbiota primarily includes six bacterial phyla: Firmicutes, Bacteroidetes, Actinobacteria, Proteobacteria, Tenericutes and Fusobacteria in an adult human.[22] The members from the phylum Firmicutes and Bacteroidetes dominate the adult gut microbiota. However, the infant intestine has members from the phylum Actinobacteria, specifically the genus *Bifidobacterium* in abundance.[23] The initial colonizers of normal infants include *E. coli, Enterobacter, Streptococcus* and *Staphylococcus.*[24] The gut microbiota in its early infancy is dominated by facultative and aerotolerant species. With time, these species reduce oxygen levels in the intestine leading to proliferation of diverse and complex communities with an abundance of anaerobic bacteria[25] and finally there is growth of strict anaerobes. *Proteobacteria* (mainly *Enterobacteriaceae*) has been observed to dominate the infant gut microbiota till the first week, followed by *Bifidobacteria, Clostridia* species and members of Firmicutes phylum.

2.2 Gut microbiome of premature infants

Globally 5–13% of all infants are born prematurely (i.e., before 37 weeks of gestation) every year.[26] The preterm birth along with many physiologic and environmental factors affects the normal maturation of the gut microbiome. Most of the preterm infants are enterally fed with formula feeds that are fortified for increased nutrient density. They possess an immature immune system; have longer hospital stays with indwelling orogastric or nasogastric tubes with a common exposure to antibiotics. These factors are responsible for the perturbation of microbial diversity leading to an initial overall decreased diversity and variation of species.[27] The preterm infant's microbiome is dominated by facultative anaerobes (*Enterobacteriaceae, Enterococcaceae, Lactobacillus* and *Weissella*), pathogenic bacteria (*E. coli, Klebsiella* and *Staphylococcus* species) and have a decreased proportion of obligate anaerobes (*Bacteroides* and *Bifidobacterium*).[28] Preterm infants were found to have a delay in the acquisition of normal commensal species in the intestine.[29] In due course of time their intestinal microbiome gets enriched with the addition of species towards a Clostridia-dominated biome.[30] Bacilli, Gammaproteobacteria and Clostridia are the predominant classes. Altogether, the gradual decrease in beneficial bacteria and the growth of harmful bacteria as well as the premature gut barrier & the immune system pose a high risk for infant gut pathology.[31]

3. Dysbiosis of the neonatal gut microbiome

A balanced gut microbiome of a neonate is required for the maintenance of physiological functions and homeostasis. Any change in the microbiome profiles is termed as dysbiosis. The neonatal gut microbiome is vulnerable to perturbation due to several factors including the delivery mode,[32,33] diet,[34,35] antibiotics[36,37] and the existence of older siblings at home.[38] Dysbiotic conditions may lead to invasion of pathogenic species and their growth further disrupts the immune system's important regulatory circuits that regulate pro- and anti-inflammatory checkpoints and balances. The naive neonatal microbiome is highly fragile and vulnerable to various external factors that can have long or short-term effects on the host's health. Any modification of the neonatal gut microbiome affects the development of systemic inflammatory response, which might result in sepsis, necrotizing enterocolitis or multi-organ failure leading to subsequent death. The disruption of the gut microbiota referred to as intestinal dysbacteriosis puts the neonates at the risk for the development of enterocolitis[39] or nosocomial infections further increasing the risk for onset of sepsis.[40] The decrease in abundance of Bifidobacteriaceae along with an increase in Enterobacteriaceae and Clostridiaceae family members mark the beginning of dysbiosis in the early months of life[12] (Fig. 1).

The consequences of neonatal gut microbiome dysbiosis have been linked to several diseases such as necrotizing colitis, neonatal sepsis and severe acute malnourishment (Table 1). The dysbiosis changes the

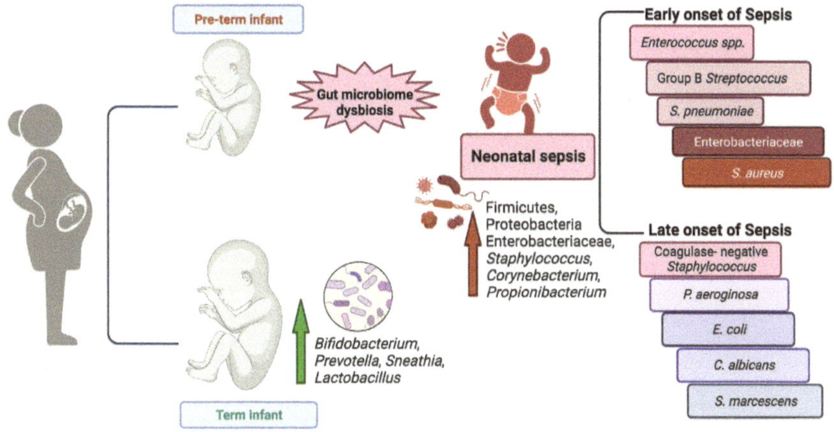

Fig. 1 Dysbiosis of neonatal gut microbiome leading to the onset of sepsis. A decrease in the abundance of Bifidobacteriaceae along with an increase in Enterobacteriaceae and Clostridiaceae family members marks the onset of neonatal sepsis. The microbes linked with the Early (EOS) and late-onset of sepsis (LOS) have been shown.

permeability of the intestines,[41] disrupts the local and systemic immune response[42] and alters the maturation of the gut microbiome as well as the metabolite profile. Infants with a relatively low abundance of bacterial genera are at the risk of developing asthma in later life.[43]

4. Factors modulating the neonatal microbiome

4.1 Mode of delivery

The mode of delivery is critical in the early colonization of the neonatal gut. The neonates born vaginally are colonized by maternal vaginal flora such as *Prevotella* and *Lactobacillus* species whereas, neonates delivered via C-section are colonized by skin flora such as *Staphylococcus* and *Corynebacterium*.[44,45] Those born by C-section have a delayed onset of Bacteroides and Bifidobacterium colonization and greater colonization by the Enterobacteriaceae family.[46] *Lactobacillus,* the dominant species toward the end of the gestation is vertically transmitted to vaginally born neonates.[47] The C-section delivered babies were reported to have *Proteobacteria* and *Firmicutes* as the major phyla in the initial days of life and *Actinobacteria* was reported in their feces at day 7 to 15 following birth.[25] These infants have more heterogeneous microbiota as compared to vaginally delivered infants till 12 months of life.[38] According to Dominguez-Bello et al., exposing C-section delivered neonates to maternal vaginal fluids at the time of birth can redirect the microbiome similar to the vaginally delivered neonates.[44]

4.2 Gestational age

The gestational age at birth is another critical factor that shapes the infant's gut microbiota. Preterm infants have an initial immature gut, are vulnerable to various health issues, and require longer hospital stay that result in aberrant gut microbiota composition. Their gut is predominated by Gram-positive bacteria including *Clostridia*, *Enterococcus* and *Staphylococcus* and other pathogenic microorganisms. The late colonization of the gut by commensal species such as *Bacteroides* or *Bifidobacterium* underlines the increased time taken for the gut microbiota composition to attain the signature composition of a healthy infant.[48] The variation is observed not only in composition but also in the functionality. Premature neonates showed an altered lipid metabolism[49] and the level of short-chain fatty acids was lower as compared to full-term infants (Table 1).[64]

Table 1 Microbiological causes of newborn sepsis.

Pathogen	Source	References
Gram-positive bacteria		
Coagulase-negative staphylococci	Vertical transmission and Hospital environment	50
Streptococci	Mother's vaginal microbiota	51
Enterococcus	Mother's gut and vaginal microbiota	52
Streptococcus pneumoniae	Mother's vaginal microbiota	53
Bacillus spp.	Hospital environment	54
Gram-negative bacteria		
Acinetobacter spp.	Hospital environment	55
Klebsiella pneumoniae	Mother's milk, hospital environment	56,57
Escherichia coli	Mother's gut, vaginal microbiota or hospital environment	58
Enterobacter aerogenes	Hospital environment	59
Pseudomonas aeruginosa	Hospital environment	60
Fungi		
Candida albicans	Hospital environment	61
Viruses		
Influenza virus	Hospital/home environment	62

Table 1 Microbiological causes of newborn sepsis.—cont'd

Pathogen	Source	References
Varicella zoster virus	Hospital/home environment	63
Cytomegalovirus	Hospital/home environment	63
Respiratory syncytial virus	Hospital/home environment	63
Epstein-Barr virus	Hospital/home environment	63
Herpes simplex virus	Hospital/home environment	63

4.3 Diet of infant

Diet is an important factor that has a significant impact on the development of the neonatal microbiome. The differences in the composition of gut microbiota of formula fed and breast-fed infants have been well-documented.[38] The breast milk components including nutrients, proteins, sugars, immunoglobulins, hormones and growth factors nurture the neonate. Human milk oligosaccharides (HMOs), an abundant component of breast milk, promote the growth of commensal microbes by acting as a specific substrate for growth. These act as prebiotics and promote growth of commensal bacterial species- *Bifidobacteria* and *Bacteroides*. The commensal species ward off any pathogenic organisms and have a protective effect on neonatal health. In addition to the modulation of the neonatal microbiome, breast milk comprises its own microbiome that changes over time. The 'sterile' colostrum contained *Staphylococcus*, *Streptococcus*, *Lactobacillus* and *Weissella* species; however, the breast milk collected after 1–6 months of delivery had a consistent population of microflora.[65] *Bacteroides*, *Clostridium difficile*, and *Enterobacteriaceae* were more prevalent in stool samples of formula-fed infants as compared to infants who received breast-milk.[66] The changeover from mother's milk to formula (i.e., bovine) milk and further transition to solid foods helps in the formation of mature microbiota.

4.4 Antibiotics

The usage of antibiotics in the intrapartum stages attributes to the decrease in bacterial diversity in the neonate's first stool and a supposedly reduced abundance of *Bifidobacteria* and *Lactobacilli* species. A comparable association was seen upon antibiotic administration to the neonates directly after birth.[67] There is paucity of data depicting the role of prenatal antibiotic usage on infants' health.

5. Neonatal sepsis

Neonatal sepsis is a systemic inflammatory response syndrome caused by an infection, and it is one of the leading causes of pediatric morbidity and mortality in infancy. It is one of the leading reasons of newborn mortality, attributing for 3,36,300 total deaths per year, and the 16th leading cause of death across all age groups.[68] The case fatality rate in term infants is 2%, 20% in premature infants, and 30% in those with concurrent meningitis.[69] It is clinically defined as a syndrome marked by infection and a systemic inflammatory immune response (SIRS),[70] which includes tachycardia or bradycardia, increase in leucocyte count, core temperature instability etc. The cascade of oxidative damage and inflammatory response leads to further progression of sepsis, which is independent of the presence of pathogens.[71] Excess inflammatory cytokine production, such as IL-1, IL-17, and TNF-alpha, causes a cytokine storm, which is responsible for the pathophysiology of sepsis.[72]

Neonatal sepsis is classified depending on the mode of acquisition and the time of onset: early and late-onset sepsis. EOS emerges due to the vertical transmission of bacteria from the mother during the perinatal period or at the time of birth bacteria can reach out to the fetus from the vagina, placenta, or in rare circumstances through the fallopian tubes.[73] Group B *Streptococcus* (GBS) and *E. coli* causes 70% of EOS infections in high-income countries. However, other pathogens like *Enterobacter* spp., *Enterococcus* spp., *Listeria monocytogens*, *Streptococcus pneumoniae* and *Staphylococcus aureus* may also be responsible.[51,74] *E. coli* is the principal causative pathogen in preterm infants and the second most common pathogen in term infants, causing severe infections and meningitis.[75] However, LOS occurs during the postnatal environmental exposure to various pathogenic bacteria. The usual cut off points considered between EOS and LOS are 72 h and 7 days.[76] LOS is more common in premature infants[77] and can happen at any time during the first three months. The main pathogens are mainly Coagulase-negative staphylococcus (CoNS) (48%) [62,60] and the highest mortality rates occur due to *Pseudomonas aeruginosa*, *E. coli*, *Candida albicans* and *Serratia marcescens*.[75] However, a cohort study from India did not find any difference in the pathogens between the EOS and LOS.[78] When compared to neonates infected with Gram-positive pathogens, the case fatality rate in neonates with sepsis caused by Gram-negative pathogens was higher. The Gram-negative pathogens included *Acinetobacter* spp., *Escherichia coli*, *Pseudomonas* spp., *Klebsiella* spp. and *Enterobacter* spp. The most common Gram-positive pathogens included Coagulase-negative staphylococcus, *Staphylococcus aureus*, and *Enterococcus* spp., however, Group B streptococci were isolated from only few newborns.

Clinically, EOS and LOS share very similar features of an early stage of mild and easily missed warning signs that, if untreated, may escalate into severe illness with instability of vital signs, irritability, lethargy, or seizures, and, eventually, multi-organ system dysfunction and failure. The time between the onset of disease and the final manifestations can be brief (a few hours) or the infant may experience mild symptoms that last a day or more before becoming severe. The parents and health practitioners often miss the initial signs due to their variability. The most common clinical features include abnormal vital signs, poor feeding, fever and respiratory distress.

5.1 Diagnosis

As a medical emergency, neonatal sepsis necessitates immediate medical and supportive care. A blood culture obtained after skin preparation and prior to antibiotic administration is the mainstay of the sepsis investigation. Automated blood culture detection systems are useful for rapid diagnosis, which requires a minimum of 1–2 viable bacterial colonies in the sample.[79] Other diagnostic tests that do not examine blood culture have poor diagnostic accuracy and are ineffective in determining antibiotic regimen. A complete blood count, particularly, the leucocyte and platelet count is recommended.[80] A lumbar puncture is required when (1) the blood culture tests positive; (2) the clinical results points to bacterial sepsis; (3) there is no improvement seen in the infant despite appropriate antimicrobial therapy; or (4) when there are symptoms related to the central nervous system including lethargy, inconsolable crying, fever, or seizures. Urine cultures might be considered for the evaluation of infants with LOS. C-reactive protein, Interleukin-6, and Procalcitonin are also used as biomarkers but lack sensitivity and specificity.[81] Recent research has revealed tumor necrosis factor-alpha (TNF-alpha),[82] neutrophil CD64,[83] toll-like receptors[84] and microRNAs as early diagnostic markers.[85] Presepsin (P-SEP) is a 13 KDa N-terminal fragment of the soluble form of CD14 (sCD14) that is cleaved in plasma by cathepsin D and simultaneously activates the innate immune system. It has emerged as a novel biomarker of sepsis and has been shown to increase during bacterial sepsis and decrease upon appropriate therapy. Its plasma levels can be measured within 17 min, which renders it suitable in emergency and critical care.[86] Another proinflammatory adipocyte-derived factor known as apelin, may be useful for diagnosis as its level was observed to elevate in the early onset sepsis neonates.[87]

5.2 Treatment

The principal treatment options for neonatal sepsis include the broad-spectrum antibiotics administered intravenously, which are narrowed down after the characterization of antibiotic sensitivity. Antibiotic combos commonly used to treat EOS pathogens include ampicillin plus gentamicin and vancomycin combined with an aminoglycoside or third generation cephalosporin. Treatment durations vary depending on the causative microbe or other underlying infection.[88] In the case of strains resistant against the first line of antibiotics, high spectrum antibiotics including carbapenems and piperacillin-tazobactam are used. Supportive care including intravenous fluids, vasopressor medications, phototherapy, invasive and non-invasive respiratory support, and occasionally steroids are required for the support.[69]

5.3 Risk factors for neonatal sepsis

The risk factors for neonatal sepsis are broadly classified into three categories: maternal factors, neonatal factors, and hospital-related factors, however, other factors as shown in Fig. 2 may act as potential risk factors. Various medical interventions, such as amniocentesis and cervical cerclage, disrupt the amniotic cavity, increasing the risk of intra-amniotic infection and neonatal sepsis.

5.3.1 Maternal risk factors

Premature rupture of membranes (PROM) and maternal infections [UTIs (62.1%), vulvovaginitis (24.2%), and chorioamnionitis (4.2%)] are the most likely causes of neonatal sepsis.[89] Other conditions such as fever at home,[90] bleeding disorder,[91] history of last gestation with neonatal infection,[89] instrumental labor,[92] prolonged labor,[91] poor cord care and poor feeding[93] were all found to be risk factors for neonatal sepsis.

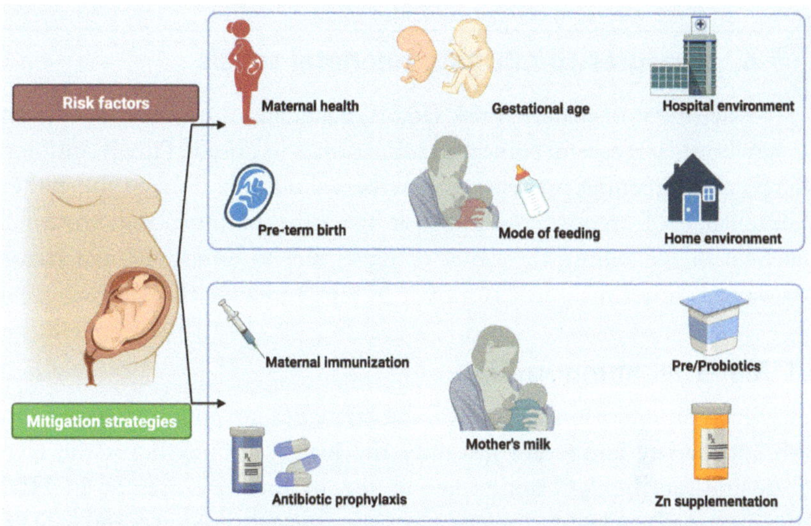

Fig. 2 Potential determinants of newborn sepsis risk and their mitigation strategies. The important risk factors for neonatal sepsis are majorly divided into maternal risk factors, neonatal risk factors and hospital-associated factors. Various measures can be utilized for the mitigation that includes maternal immunization, breast-feeding of the newborn, antibiotic prophylaxis, Zn supplementation and use of pre/probiotics.

5.3.2 Neonatal risk factors

Prematurity and low birth weight are the most significant neonatal risk factors. Low APGAR (Appearance, Pulse, Grimace, Activity and Respiration) score measured in the first and fifth minute just after the birth, perinatal asphyxia,[94,95] meconium-stained amniotic fluid (MSAF)[96] and lack of crying just after the birth[97] are also considered as the predisposing factors for neonatal sepsis. Abnormalities in the amniotic fluid and the underlying comorbidities in the infants are also found to have association with the onset of sepsis.

5.3.3 Hospital associated risk factors

The risk factors associated with neonatal care include association of the central venous catheter (CVC) insertion, mechanical ventilation, parenteral nutrition, surgery,[95,98] longer duration of NICU stay[91,97] and use of tocolytic drugs[99] are identified as risk factors. Other less studied risk factors include the mother's socioeconomic status, poor or delayed prenatal care, poor maternal nutrition, male sex, and infants born to African American mothers.[100]

6. Measures to mitigate neonatal sepsis

Assessment of maternal risk factors, adequate care of the mother and the newborn, and careful clinical examination is required. Timely antibiotic therapy and screening for sepsis biomarkers should be done at the earliest. Maintenance of aseptic conditions at the time of birth and postnatally will help in preventing the onset of sepsis due to environmental factors. The various other measures as shown in Fig. 2 include the following:

6.1 Maternal immunization

Maternal immunization is one of the strategies to endow the neonates with appropriate antibodies just after the birth.[101] They boost the levels of vaccine-specific IgG antibodies in the mother. At present in HICS (High income countries), three vaccines are routinely used in pregnancy that includes influenza, diarrhea and tetanus.[102] The vaccination may provide protection in the term infants till the infant completes the routine vaccinations, however in preterm infants, the maternal vaccines may not provide adequate neonatal immunity due to decreased transplacental transport of antibodies resulting in infants with lower antibody concentrations.[103]

6.2 Topical applications and antibiotic prophylaxis

According to research studies from Nepal and Malawi, topical vaginal chlorhexidine application during labor may have an antisepsis effect[104]; however, later studies did not show any impact and the intervention is therefore not recommended. Plant derivatives such as sunflower seed oil have been shown to provide protection from nosocomial infections in preterm neonates.[105] Studies from south Asia have shown the application of chlorhexidine to umbilical cord to reduce neonatal sepsis.[106]In developed countries, intrapartum antibiotic prophylaxis was shown to be efficient in decreasing newborn sepsis cases.[107]

6.3 Human milk feeding

Mother's milk is the best source of a diverse range of bioactive molecules, including antimicrobial proteins and peptides that safeguard against infection and inflammation. Lactoferrin present in the human milk inhibits bacterial adhesion and biofilm formation,[108] blocks the receptors for invasion of microbes, prevents binding of endotoxins from the intestinal pathogens[109] and inhibits the translocation of bacteria.[110] Moreover, lactoferrin promotes anti-inflammatory effects by interference in the TLR expression pathway.[110,111]

6.4 Pre- and probiotics

Prebiotics are nondigestible food constituents that enhance the growth of anaerobic commensals in mammalian colons. Lactoferrin is a naturally occurring prebiotic component of human milk as it stimulates the colonization by favorable commensal species of *Bifidobacterium* and *Lactobacillus*. Higher levels of lactoferrin have been shown to affect the immune system development and overall wellness of the neonate.[112] Lactulose, inulin, galactooligosaccharides and short-chain fructo-oligosaccharides include the prebiotics that have been studied extensively in humans, but their safety and effectiveness in the neonate population have received little attention.

Probiotics as defined by WHO are live microorganisms that, when administered in sufficient quantities, provide a health benefit to the host. The supplementation of probiotics in the preterm neonates confers the protective effect against dysbiotic conditions by promoting the bloom of normal commensal gut flora. *Lactobacillus* and *Bifidobacterium* spp. are commonly used probiotics for neonates. The usage of probiotics increases the local and systemic immunity, decreases the permeability of the gut to pathogens, and decreases the production of pro-inflammatory cytokines.[113]

The supplementation of probiotics *B. longum* subsp. *infantis* BB-02, *B. animalis* subsp. *lactis* BB-12 and *S. thermophilus* TH-4 were observed to enhance the probiotic species in preterm infant gut. Another study of the potential benefits of synbiotics, which is a combination of prebiotics and probiotics was carried out in 4566 enrolled infants with a body weight >2000 g or gestation period of 35 weeks in rural India. The results were a significant decrease in sepsis cases and number of deaths (risk ratio 0.60, 95% confidence interval 0.48–0.74).[114]

6.5 Zinc supplementation

The supplementation of zinc decreases the oxidative stress markers and targets Nuclear Factor Kappa B, thereby limiting the production of proinflammatory cytokine.[115,116] Preterm infants have low zinc stores by birth as well as reduced ability for its absorption and retention. A randomized controlled trial of 352 infants aged 17–120 days who had severe bacterial infection were supplemented with zinc and were observed to have less treatment failure.[117]

7. Future directions

Understanding the structure and functions of the basal microbiome in different body habitats in healthy neonates, as well as its differences in neonates suffering from sepsis, can aid in the early prediction of risk factors and the implementation of appropriate strategies to reduce disease burden. Interdisciplinary research will undoubtedly aid in bridging knowledge gaps in this field. Studies based on genomics, metagenomics, and artificial intelligence (AI) will surely pave the way for the early detection of sepsis, reducing morbidity and mortality among neonates. Preclinical animal models are required in sepsis research to produce a novel diagnostic biomarker; thus, new study models are needed as early as possible for the development of interventions against neonatal sepsis. The emergence of AMR has been another challenge that has jeopardized available treatment options and increased antibiotic misuse. To safely reduce neonatal antibiotic exposure, randomized trials should be set up to compare different antibiotic combinations, affordability, and mortality benefit. Point-of-care devices for detecting sepsis, as well as multi-omics studies to unravel the disease's complexities, should be prioritized.

8. Conclusion

One of the leading causes of neonatal mortality in the entire world is sepsis. Apart from mortality, it also results in long-term effects, including impaired neurodevelopment and growth. Sepsis is a systemic condition that occurs with dysregulated immune reactions in response to a pathogen. *Escherichia coli, Klebsiella pneumoniae, Staphylococcus aureus, Streptococcus* and Coagulase-negative staphylococci. Fungal, mainly *Candida*, infections are also emerging as potential causes of morbidity and mortality in neonatal sepsis. Early detection, systematic reporting, proper treatment, and effective measures to reduce infections and predict sepsis can reduce the burden of sepsis. Moreover, hospital care and associated facilities should be in compliance with good health care of neonates and mothers. Raising awareness among the general public could perhaps be useful for early diagnosis of sepsis in neonates. Effective management to lessen the burden of newborn sepsis requires robust research, enhanced testing, and reporting on risk factors from all the districts of India. Multicenter studies will aid in the evaluation of potential risk factors and the development of preventive measures. Furthermore, a thorough examination of the microbiome and metabolites in neonates with sepsis will undoubtedly pave the way for the identification of biomarkers and the development of any potential interventions.

Acknowledgments

The authors thank their respective institutions for their encouragement and assistance. The Biorender software was used for the preparation of figures.

Conflict of interests

The authors declare that there is no conflict of interests.

References

1. Rudd KE, Johnson SC, Agesa KM, et al. Global, regional, and national sepsis incidence and mortality, 1990–2017: Analysis for the global burden of disease study. *Lancet.* 2020;395(10219):200–211.
2. Roy S, Patil R, Apte A, et al. Feasibility of implementation of simplified management of young infants with possible serious bacterial infection when referral is not feasible in tribal areas of Pune district, Maharashtra, India. *Plos One.* 2020;15(8):e0236355. https://doi.org/10.1371/journal.pone.0236355.
3. Fleischmann-Struzek C, Goldfarb DM, Schlattmann P, Schlapbach LJ, Reinhart K, Kissoon N. The global burden of paediatric and neonatal sepsis: A systematic review. *Lancet Respir Med.* 2018;6(3):223–230.
4. Popescu CR, Cavanagh MMM, Tembo B, et al. Neonatal sepsis in low-income countries: Epidemiology, diagnosis and prevention. *Expert Rev Anti Infect Ther.* 2020;18(5):443–452.
5. Aggarwal R, Sarkar N, Deorari AK, Paul VK. Sepsis in the newborn. *Indian J Pediatr.* 2001;68(12):1143–1147.
6. Benitz WE, Wynn JL, Polin RA. Reappraisal of guidelines for management of neonates with suspected early-onset sepsis. *J Pediatr.* 2015;166(4):1070–1074.
7. Tewari VV, Jain N. Monotherapy with amikacin or piperacillin-tazobactum empirically in neonates at risk for early-onset sepsis: A randomized controlled trial. *J Trop Pediatr.* 2014;60(4):297–302.
8. Dalby MJ, Hall LJ. Recent advances in understanding the neonatal microbiome. *F1000Res.* 2020;9. https://doi.org/10.12688/f1000research.22355.1.
9. Ouyang R, Korpela K, Liu X, Xu G, de Vos WM, Kovatcheva-Datchary P. Early life microbiota—Impact of delivery mode and infant feeding. *Comprehensive Gut Microbiota.* 2022;25–38. https://doi.org/10.1016/b978-0-12-819265-8.00064-4.
10. Roswall J, Olsson LM, Kovatcheva-Datchary P, et al. Developmental trajectory of the healthy human gut microbiota during the first 5 years of life. *Cell Host Microbe.* 2021;29(5):765–776.e3.
11. Wu G, Zhao N, Zhang C, Lam YY, Zhao L. Guild-based analysis for understanding gut microbiome in human health and diseases. *Genome Med.* 2021;13(1):22.
12. Underwood MA, Mukhopadhyay S, Lakshminrusimha S, Bevins CL. Neonatal intestinal dysbiosis. *J Perinatol.* 2020;40(11):1597–1608.

13. Lee JKF, Tan LTH, Ramadas A, Mutalib NSA, Lee LH. Exploring the role of gut bacteria in health and disease in preterm neonates. *Int J Environ Res Public Health*. 2020;17(19):6963. https://doi.org/10.3390/ijerph17196963.
14. Baquero F, Nombela C. The microbiome as a human organ. *Clin Microbiol Infect*. 2012;18:2–4. https://doi.org/10.1111/j.1469-0691.2012.03916.x.
15. Patel RM, Denning PW. Intestinal microbiota and its relationship with necrotizing enterocolitis. *Pediatr Res*. 2015;78(3):232–238.
16. Qin J, Li R, Raes J, et al. A human gut microbial gene catalogue established by metagenomic sequencing. *Nature*. 2010;464(7285):59–65.
17. Aagaard K, Ma J, Antony KM, Ganu R, Petrosino J, Versalovic J. The placenta harbors a unique microbiome. *Sci Transl Med*. 2014;6(237). 237ra65.
18. Jiménez E, Fernández L, Marín ML, et al. Isolation of commensal bacteria from umbilical cord blood of healthy neonates born by cesarean section. *Curr Microbiol*. 2005;51 (4):270–274.
19. Jiménez E, Marín ML, Martín R, et al. Is meconium from healthy newborns actually sterile? *Res Microbiol*. 2008;159(3):187–193.
20. Vallès Y, Artacho A, Pascual-García A, et al. Microbial succession in the gut: Directional trends of taxonomic and functional change in a birth cohort of Spanish infants. *PLoS Genet*. 2014;10(6):e1004406. https://doi.org/10.1371/journal.pgen.1004406.
21. George S, Aguilera X, Gallardo P, et al. Bacterial gut microbiota and infections during early childhood. *Front Microbiol*. 2021;12:793050.
22. Rinninella E, Raoul P, Cintoni M, et al. What is the healthy gut microbiota composition? A changing ecosystem across age, environment, diet, and diseases. *Microorganisms*. 2019;7(1):14. https://doi.org/10.3390/microorganisms7010014.
23. Turroni F, Peano C, Pass DA, et al. Diversity of bifidobacteria within the infant gut microbiota. *PLoS One*. 2012;7(5):e36957.
24. Palmer C, Bik EM, DiGiulio DB, Relman DA, Brown PO. Development of the human infant intestinal microbiota. *PLoS Biol*. 2007;5(7):e177.
25. Del Chierico F, Vernocchi P, Petrucca A, et al. Phylogenetic and metabolic tracking of gut microbiota during perinatal development. *PLoS One*. 2015;10(9):e0137347.
26. Chawanpaiboon S, Vogel JP, Moller AB, et al. Global, regional, and national estimates of levels of preterm birth in 2014: A systematic review and modelling analysis. *Lancet Glob Health*. 2019;7(1):e37–e46.
27. Schwiertz A, Gruhl B, Löbnitz M, Michel P, Radke M, Blaut M. Development of the intestinal bacterial composition in hospitalized preterm infants in comparison with breast-fed, full-term infants. *Pediatr Res*. 2003;54(3):393–399. https://doi.org/10.1203/01.pdr.0000078274.74607.7a.
28. Arboleya S, Binetti A, Salazar N, et al. Establishment and development of intestinal microbiota in preterm neonates. *FEMS Microbiol Ecol*. 2012;79(3):763–772. https://doi.org/10.1111/j.1574-6941.2011.01261.x.
29. Arboleya S, Sánchez B, Milani C, et al. Intestinal microbiota development in preterm neonates and effect of perinatal antibiotics. *J Pediatr*. 2015;166(3):538–544. https://doi.org/10.1016/j.jpeds.2014.09.041.
30. La Rosa PS, Warner BB, Zhou Y, et al. Patterned progression of bacterial populations in the premature infant gut. *Proc Natl Acad Sci U S A*. 2014;111 (34):12522–12527.
31. Lacroix C, Collado MC, Chassard C, Corsetti A. *Infant Gut Microbiota Colonization and Food Impact*. Frontiers Media SA; 2019.
32. Stokholm J, Thorsen J, Blaser MJ, et al. Delivery mode and gut microbial changes correlate with an increased risk of childhood asthma. *Sci Transl Med*. 2020;12 (569). https://doi.org/10.1126/scitranslmed.aax9929.

33. Frese SA, Mills DA. Birth of the infant gut microbiome: Moms deliver twice! *Cell Host Microbe*. 2015;17(5):543–544.
34. Pannaraj PS, Li F, Cerini C, et al. Association between breast Milk bacterial communities and establishment and development of the infant gut microbiome. *JAMA Pediatr*. 2017;171(7):647–654.
35. Azad MB, Konya T, Maughan H, et al. Gut microbiota of healthy Canadian infants: Profiles by mode of delivery and infant diet at 4 months. *CMAJ*. 2013;185(5):385–394.
36. Bokulich NA, Chung J, Battaglia T, et al. Antibiotics, birth mode, and diet shape microbiome maturation during early life. *Sci Transl Med*. 2016;8(343). https://doi.org/10.1126/scitranslmed.aad7121.
37. Zeissig S, Blumberg RS. Life at the beginning: Perturbation of the microbiota by antibiotics in early life and its role in health and disease. *Nat Immunol*. 2014;15(4):307–310.
38. Martin R, Makino H, Cetinyurek Yavuz A, et al. Early-life events, including mode of delivery and type of feeding, siblings and gender, shape the developing gut microbiota. *PLoS One*. 2016;11(6):e0158498.
39. Claud EC, Keegan KP, Brulc JM, et al. Bacterial community structure and functional contributions to emergence of health or necrotizing enterocolitis in preterm infants. *Microbiome*. 2013;1(1):20.
40. Madan JC, Salari RC, Saxena D, et al. Gut microbial colonisation in premature neonates predicts neonatal sepsis. *Arch Dis Child Fetal Neonatal Ed*. 2012;97(6):F456–F462.
41. Iacob S, Iacob DG. Infectious threats, the intestinal barrier, and its trojan horse: Dysbiosis. *Front Microbiol*. 2019;10:1676.
42. Deshmukh HS, Liu Y, Menkiti OR, et al. The microbiota regulates neutrophil homeostasis and host resistance to Escherichia coli K1 sepsis in neonatal mice. *Nat Med*. 2014;20(5):524–530.
43. Arrieta MC, Stiemsma LT, Dimitriu PA, et al. Early infancy microbial and metabolic alterations affect risk of childhood asthma. *Sci Transl Med*. 2015;7(307). https://doi.org/10.1126/scitranslmed.aab2271.
44. Dominguez-Bello MG, Costello EK, Contreras M, et al. Delivery mode shapes the acquisition and structure of the initial microbiota across multiple body habitats in newborns. *Proc Natl Acad Sci*. 2010;107(26):11971–11975. https://doi.org/10.1073/pnas.1002601107.
45. Biasucci G, Benenati B, Morelli L, Bessi E, Boehm G. Cesarean delivery may affect the early biodiversity of intestinal bacteria. *J Nutr*. 2008;138(9):1796S–1800S.
46. Dogra S, Sakwinska O, Soh SE, et al. Rate of establishing the gut microbiota in infancy has consequences for future health. *Gut Microbes*. 2015;6(5):321–325.
47. Aagaard K, Riehle K, Ma J, et al. A metagenomic approach to characterization of the vaginal microbiome signature in pregnancy. *PLoS One*. 2012;7(6):e36466. https://doi.org/10.1371/journal.pone.0036466.
48. Milani C, Duranti S, Bottacini F, et al. The first microbial colonizers of the human gut: Composition, activities, and health implications of the infant gut microbiota. *Microbiol Mol Biol Rev*. 2017;81(4). https://doi.org/10.1128/MMBR.00036-17.
49. Hill CJ, Lynch DB, Murphy K, et al. Erratum to: Evolution of gut microbiota composition from birth to 24 weeks in the INFANTMET cohort. *Microbiome*. 2017;5(1):21.
50. Marchant EA, Boyce GK, Sadarangani M, Lavoie PM. Neonatal sepsis due to coagulase-negative staphylococci. *Clinical and Developmental Immunology*. 2013;2013:1–10. https://doi.org/10.1155/2013/586076.
51. Simonsen KA, Anderson-Berry AL, Delair SF, Davies HD. Early-onset neonatal sepsis. *Clin Microbiol Rev*. 2014;27(1):21–47.

52. Hufnagel M, Liese C, Loescher C, et al. Enterococcal colonization of infants in a neonatal intensive care unit: Associated predictors, risk factors and seasonal patterns. *BMC Infect Dis*. 2007;7(1). https://doi.org/10.1186/1471-2334-7-107.
53. Soto-Noguerón A, Carnalla-Barajas MN, Solórzano-Santos F, et al. Streptococcus pneumoniae as cause of infection in infants less than 60 days of age: Serotypes and antimicrobial susceptibility. *Int J Infect Dis*. 2016;42:69–73.
54. Cormontagne D, Rigourd V, Vidic J, Rizzotto F, Bille E, Ramarao N. Bacillus cereus induces severe infections in preterm neonates: Implication at the hospital and human milk bank level. *Toxins*. 2021;13(2):123. https://doi.org/10.3390/toxins13020123.
55. Zarrilli R, Bagattini M, Esposito EP, Triassi M. Acinetobacter infections in neonates. *Curr Infect Dis Rep*. 2018;20(12). https://doi.org/10.1007/s11908-018-0654-5.
56. Dorota P, Chmielarczyk A, Katarzyna L, et al. Klebsiella pneumoniae in breast Milk-a cause of sepsis in neonate. *Arch Med*. 2017;09(01). https://doi.org/10.21767/1989-5216.1000189.
57. Gupta A. Hospital-acquired infections in the neonatal intensive care unit- -Klebsiella pneumoniae. *Semin Perinatol*. 2002;26(5):340–345.
58. Bettelheim KA, Goldwater PN. Escherichia coli and sudden infant death syndrome. *Front Immunol*. 2015;6:343.
59. Ramachandran VLAKPG, Loiwal V, Kumar A, Ramachandran PG. Enterobacter aerogenes outbreak in a neonatal intensive care unit. *Pediatr Int*. 1999;41(2):157–161. https://doi.org/10.1046/j.1442-200x.1999.4121033.x.
60. Jefferies JMC, Cooper T, Yam T, Clarke SC. Pseudomonas aeruginosa outbreaks in the neonatal intensive care unit- -a systematic review of risk factors and environmental sources. *J Med Microbiol*. 2012;61(Pt 8):1052–1061.
61. Arsenault AB, Bliss JM. Neonatal candidiasis: New insights into an old problem at a unique host-pathogen Interface. *Curr Fungal Infect Rep*. 2015;9(4):246–252. https://doi.org/10.1007/s12281-015-0238-x.
62. Alexander-Miller MA. Challenges for the newborn following influenza virus infection and prospects for an effective vaccine. *Front Immunol*. 2020;11, 568651.
63. Gupta N, Richter R, Robert S, Kong M. Viral sepsis in children. *Front Pediatr*. 2018;6. https://doi.org/10.3389/fped.2018.00252.
64. Arboleya S, Martinez-Camblor P, Solís G, et al. Intestinal microbiota and weight-gain in preterm neonates. *Front Microbiol*. 2017;8:183.
65. Cabrera-Rubio R, Collado MC, Laitinen K, Salminen S, Isolauri E, Mira A. The human milk microbiome changes over lactation and is shaped by maternal weight and mode of delivery. *Am J Clin Nutr*. 2012;96(3):544–551.
66. Penders J, Vink C, Driessen C, London N, Thijs C, Stobberingh EE. Quantification ofBifidobacteriumspp.,Escherichia coliandClostridium difficilein faecal samples of breast-fed and formula-fed infants by real-time PCR. *FEMS Microbiol Lett*. 2005;243 (1):141–147. https://doi.org/10.1016/j.femsle.2004.11.052.
67. Reyman M, van Houten MA, Watson RL, et al. Effects of early-life antibiotics on the developing infant gut microbiome and resistome: A randomized trial. *Nat Commun*. 2022;13(1):893.
68. GBD 2015 Mortality and Causes of Death Collaborators. Global, regional, and national life expectancy, all-cause mortality, and cause-specific mortality for 249 causes of death, 1980–2015: A systematic analysis for the global burden of disease study 2015. *Lancet*. 2016;388(10053):1459–1544.
69. Kim F, Polin RA, Hooven TA. Neonatal sepsis. *BMJ*. 2020. https://doi.org/10.1136/bmj.m3672. m3672.
70. Wynn JL, Wong HR, Shanley TP, Bizzarro MJ, Saiman L, Polin RA. Time for a neonatal-specific consensus definition for sepsis. *Pediatr Crit Care Med*. 2014;15(6): 523–528.

71. Poggi C, Dani C. Sepsis and oxidative stress in the newborn: From pathogenesis to novel therapeutic targets. *Oxid Med Cell Longev*. 2018;2018:9390140.
72. Delano MJ, Ward PA. The immune system's role in sepsis progression, resolution, and long-term outcome. *Immunol Rev*. 2016;274(1):330–353.
73. Kim CJ, Romero R, Chaemsaithong P, Chaiyasit N, Yoon BH, Kim YM. Acute chorioamnionitis and funisitis: Definition, pathologic features, and clinical significance. *Am J Obstet Gynecol*. 2015;213(4 Suppl):S29–S52.
74. Cailes B, Kortsalioudaki C, Buttery J, et al. Epidemiology of UK neonatal infections: The neonIN infection surveillance network. *Arch Dis Child Fetal Neonatal Ed*. 2018;103(6):F547–F553.
75. Shah BA, Padbury JF. Neonatal sepsis: An old problem with new insights. *Virulence*. 2014;5(1):170–178.
76. Dong Y, Speer CP. Late-onset neonatal sepsis: recent developments. *Arch Dis Child Fetal Neonatal Ed*. 2015;100(3):F257–F263.
77. Giannoni E, Agyeman PKA, Stocker M, et al. Neonatal sepsis of early onset, and hospital-acquired and community-acquired late onset: A prospective population-based cohort study. *J Pediatr*. 2018;201:106–114.e4.
78. Investigators of the Delhi Neonatal Infection Study (DeNIS) collaboration. Characterisation and antimicrobial resistance of sepsis pathogens in neonates born in tertiary care centres in Delhi, India: A cohort study. *Lancet Glob Health*. 2016;4(10): e752–e760.
79. Garcia-Prats JA, Cooper TR, Schneider VF, Stager CE, Hansen TN. Rapid detection of microorganisms in blood cultures of newborn infants utilizing an automated blood culture system. *Pediatrics*. 2000;105(3 Pt 1):523–527.
80. Puopolo KM, Benitz WE, Zaoutis TE. COMMITTEE ON FETUS AND NEWBORN, COMMITTEE ON INFECTIOUS DISEASES. Management of Neonates Born at $\geq$35 0/7 Weeks' gestation with suspected or proven early-onset bacterial sepsis. *Pediatrics*. 2018;142(6). https://doi.org/10.1542/peds.2018-2894.
81. Rashwan NI, Hassan MH, Mohey El-Deen ZM, Ahmed AEA. Validity of biomarkers in screening for neonatal sepsis - a single center -hospital based study. *Pediatr Neonatol*. 2019;60(2):149–155.
82. Lv B, Huang J, Yuan H, Yan W, Hu G, Wang J. Tumor necrosis factor-α as a diagnostic marker for neonatal sepsis: A meta-analysis. *ScientificWorldJournal*. 2014;2014, 471463.
83. Dai J, Jiang W, Min Z, et al. Neutrophil CD64 as a diagnostic marker for neonatal sepsis: Meta-analysis. *Adv Clin Exp Med*. 2017;26(2):327–332.
84. Rohsiswatmo R, Azharry M, Sari TT, Bahasoan Y, Wulandari D. TLR2 and TLR4 expressions in late-onset neonatal sepsis: Is it a potential novel biomarker? *J Neonatal Perinatal Med*. 2021;14(3):361–367.
85. Fatmi A, Rebiahi SA, Chabni N, et al. miRNA-23b as a biomarker of culture-positive neonatal sepsis. *Mol Med*. 2020;26(1):94.
86. Piccioni A, Santoro MC, de Cunzo T, et al. Presepsin as early marker of sepsis in emergency department: A narrative review. *Medicina*. 2021;57(8):770. https://doi.org/10.3390/medicina57080770.
87. ELMeneza SAEN, ELMeneza SAE. Department of Pediatrics, et al. role of serum apelin in the diagnosis of early-onset neonatal sepsis. *Turk Arch Pediatr*. 2021;56 (6):563–568. https://doi.org/10.5152/turkarchpediatr.2021.21108.
88. Polin RA, Papile LA, Baley JE, et al. Management of Neonates with Suspected or proven early-onset bacterial sepsis. *Pediatrics*. 2012;129(5):1006–1015. https://doi.org/10.1542/peds.2012-0541.
89. Goulart AP, Valle CF, Dal-Pizzol F, Cancelier ACL. Fatores de risco para o desenvolvimento de sepse neonatal precoce em hospital da rede pública do Brasil. *Revista Brasileira de Terapia Intensiva*. 2006;18(2):148–153. https://doi.org/10.1590/s0103-507x2006000200008.

90. Pinheiro RdS, de Souza Pinheiro R, de Lima Ferreira LC, Brum IR, Guilherme JP, Monte RL. Estudo dos fatores de risco maternos associados à sepse neonatal precoce em hospital terciário da Amazônia brasileira. *Rev Bras Ginecol Obstet.* 2007;29 (8):387–395. https://doi.org/10.1590/s0100-72032007000800002.
91. Adatara P, Afaya A, Salia SM, et al. Risk factors associated with neonatal sepsis: A case study at a specialist Hospital in Ghana. *Scientific World Journal.* 2019;2019:1–8. https://doi.org/10.1155/2019/9369051.
92. Woldu MA. Assessment of the incidence of neonatal sepsis, its risk factors, antimicrobials use and clinical outcomes in Bishoftu general hospital, neonatal intensive care unit, Debrezeit-Ethiopia. *Pediatrics & Therapeutics.* 2014;04(04). https://doi.org/10.4172/2161-0665.1000214.
93. Onyedibe K, Toma B. Proportions and odds ratio analyses of maternal and perinatal factors associated with neonatal sepsis in a developing country. *Open Forum Infect Dis.* 2016;3(suppl_1). https://doi.org/10.1093/ofid/ofw172.569.
94. Ogunlesi TA, Ogunfowora OB, Osinupebi O, Olanrewaju DM. Changing trends in newborn sepsis in Sagamu, Nigeria: Bacterial aetiology, risk factors and antibiotic susceptibility. *J Paediatr Child Health.* 2011;47(1–2):5–11.
95. Leal YA, Álvarez-Nemegyei J, Velázquez JR, et al. Risk factors and prognosis for neonatal sepsis in southeastern Mexico: Analysis of a four-year historic cohort follow-up. *BMC Pregnancy Childbirth.* 2012;12:48.
96. Utomo MT. Risk factors of neonatal sepsis: A preliminary study in Dr. Soetomo hospital. *Indonesian J Trop Infect Dis.* 2010;1(1):23. https://doi.org/10.20473/ijtid.v1i1.3718.
97. Adatara P, Afaya A, Salia SM, et al. Risk factors for neonatal sepsis: A retrospective case-control study among neonates who were delivered by caesarean section at the trauma and specialist hospital, Winneba, Ghana. *Biomed Res Int.* 2018;2018:6153501.
98. Salem SY, Sheiner E, Zmora E, Vardi H, Shoham-Vardi I, Mazor M. Risk factors for early neonatal sepsis. *Arch Gynecol Obstet.* 2006;274(4):198–202.
99. Pereira H, Grilo E, Cardoso P, Noronha N, Resende C. Risk factors for healthcare associated sepsis in very low birth weight infants. *Acta Med Port.* 2016;29(4):261–267.
100. Bohanon FJ, Lopez ON, Adhikari D, et al. Race, income and insurance status affect neonatal sepsis mortality and healthcare resource utilization. *Pediatr Infect Dis J.* 2018;37(7):e178–e184. https://doi.org/10.1097/inf.0000000000001846.
101. Healy CM, Mary Healy C, Baker CJ. Maternal immunization. *Pediatr Infect Dis J.* 2007;26(10):945–948. https://doi.org/10.1097/inf.0b013e318156c18c.
102. Jones CE, Calvert A, Le Doare K. Vaccination in pregnancy-recent developments. *Pediatr Infect Dis J.* 2018;37(2):191–193.
103. van den Berg JP, Westerbeek EAM, Smits GP, van der Klis FRM, Berbers GAM, van Elburg RM. Lower transplacental antibody transport for measles, mumps, rubella and varicella zoster in very preterm infants. *PLoS One.* 2014;9(4):e94714.
104. McClure EM, Goldenberg RL, Brandes N, Darmstadt GL, Wright LL. For the CHX working group. The use of chlorhexidine to reduce maternal and neonatal mortality and morbidity in low-resource settings. *Int J Gynecol Obstet.* 2007;97(2):89–94. https://doi.org/10.1016/j.ijgo.2007.01.014.
105. Kumar A, Mishra S, Singh S, et al. Effect of sunflower seed oil emollient therapy on newborn infant survival in Uttar Pradesh, India: A community-based, cluster randomized, open-label controlled trial. *PLoS Med.* 2021;18(9):e1003680.
106. Sunil Sazawal, Usha Dhingra, Said M Ali, Arup Dutta, Saikat Deb, Shaali M Ame, et al. Show all authors Open Access Published: September 29, 2016. https://doi.org/10.1016/S2214-109X(16)30223-6 PlumX Metrics.
107. World Health Organization. *WHO Recommendations for Prevention and Treatment of Maternal Peripartum Infections.* World Health Organization; 2016.

108. Kallapur SG, Nitsos I, Moss TJM, et al. IL-1 mediates pulmonary and systemic inflammatory responses to chorioamnionitis induced by lipopolysaccharide. *Am J Respir Crit Care Med*. 2009;179(10):955–961.
109. Liao J, Kapadia VS, Brown LS, et al. The NLRP3 inflammasome is critically involved in the development of bronchopulmonary dysplasia. *Nat Commun*. 2015;6:8977.
110. Elass-Rochard E, Legrand D, Salmon V, et al. Lactoferrin inhibits the endotoxin interaction with CD14 by competition with the lipopolysaccharide-binding protein. *Infect Immun*. 1998;66(2):486–491.
111. Ochoa TJ, Noguera-Obenza M, Ebel F, Guzman CA, Gomez HF, Cleary TG. Lactoferrin impairs type III secretory system function in enteropathogenic Escherichia coli. *Infect Immun*. 2003;71(9):5149–5155.
112. Mimouni FB, Koletzko B. *Human Milk for Preterm Infants, an Issue of Clinics in Perinatology*. E-Book. Elsevier Health Sciences; 2017.
113. Watson RR, Preedy VR. *Probiotics, Prebiotics, and Synbiotics: Bioactive Foods in Health Promotion*. Academic Press; 2015.
114. Panigrahi P, Parida S, Nanda NC, et al. A randomized synbiotic trial to prevent sepsis among infants in rural India. *Nature*. 2017;548(7668):407–412.
115. Gammoh N, Rink L. Zinc in infection and inflammation. *Nutrients*. 2017;9 (6):624. https://doi.org/10.3390/nu9060624.
116. Prasad AS. Zinc: Mechanisms of host defense. *J Nutr*. 2007;137(5):1345–1349.
117. Bhatnagar S, Wadhwa N, Aneja S, et al. Zinc as adjunct treatment in infants aged between 7 and 120 days with probable serious bacterial infection: A randomised, double-blind, placebo-controlled trial. *Lancet*. 2012;379(9831):2072–2078.

Further reading

118. Plummer EL, Danielewski JA, Garland SM, Su J, Jacobs SE, Murray GL. The effect of probiotic supplementation on the gut microbiota of preterm infants. *J Med Microbiol*. 2021;70(8). https://doi.org/10.1099/jmm.0.001403.

CHAPTER SIX

Diarrheal disease and gut microbiome

Thandavarayan Ramamurthy[a,*], Shashi Kumari[b], and Amit Ghosh[a]
[a]ICMR-National Institute of Cholera and Enteric Diseases, Kolkata, India
[b]DBT-Translational Health Science and Technology Institute, Faridabad, India
*Corresponding author: e-mail address: ramamurthy.t@icmr.gov.in

Contents

Abstract

Diarrheal disease remains a great public health problem in many countries. Enteric infections caused by several viral, bacterial and parasitic species not only affect the host, but also alter the gut microbiome. The host physiology dictates the intestinal milieu and decides the composition and richness of gut microbiota, which forms a homeostatic ecosystem with numerous functions and provide protection against invading pathogens. During diarrheal infection, patients are affected by gut microbial dysbiosis, which benefits the pathogenic and pro-inflammatory bacteria by enhancing

Progress in Molecular Biology and Translational Science, Volume 192
ISSN 1877-1173
https://doi.org/10.1016/bs.pmbts.2022.08.002

their colonization and proliferation. Gut microbes are associated with several pathophysiological mechanisms, including distorted motility, intestinal barrier dysfunction, malabsorption, immunity disorder, systemic inflammation and changes in the gut-organ axis. Several abiotic factors and childhood malnutrition have negative influences on the gut microbiota, including antibiotics that lead to antibiotic-associated diarrhea and persistent infection. DNA sequencing and bioinformatic analyses enhanced our perception of the gut microbiota, network of metabolic interdependence and their role in health and disease. However, the precise functions of microbiota in gut homeostasis are not well defined. In this chapter, we recapitulate the impact of gut microbiota on diarrheal pathogens, their importance in the immune system and how reshaping the gut microbiota can help during the recovery phase. Additionally, we discuss about impediments and influences beyond diarrhea, particularly on the nutritional status of children.

1. Introduction

Diarrhea is an important cause of morbidity and mortality in all age groups. Children below 5 years of age are the most vulnerable group for this disease.[1] Worldwide, about 1.3 million deaths occur due to diarrhea alone every year, affecting about 500,000 children. The source of diarrhea is associated with consumption of contaminated food/water, sanitation conditions, hygiene practices and malnutrition. Clinically, diarrhea is described as the excretion of watery or mucoid stools, at least three times in a day. Clinical categorization of diarrhea helps to guide and management of the disease. Based on its nature, diarrhea has been categorized into three types (i) acute watery diarrhea lasting several hours to days, (ii) bloody diarrhea, which also referred as dysentery, and (iii) persistent diarrhea continuing for 7–14 days or longer. In most of the cases, self-limiting diarrhea persists for a short period that could be resolved/ameliorated with or without medication. Prolonged acute diarrhea with excessive bowel movements may be fatal if the patients are not hospitalized and treated properly.

As shown in Table 1, the infectious diarrhea can be associated with several pathogenic bacteria, viruses or parasites. These pathogens can cause two syndromes, non-inflammatory and inflammatory/invasive diarrhea. The non-inflammatory or watery diarrhea is caused by enterotoxins expressed by the pathogens that attach to the gut mucosa and interrupt the absorptive and/or secretory functions of the enterocytes. Inflammatory or bloody diarrhea is caused by the cytotoxin-producing, invasive/non-invasive groups of bacteria. These pathogens adhere and invade into the intestinal cells, stimulate the mucoid layers and cytokines to induce cytotoxins to provoke an acute inflammatory infection.

Table 1 Clinical features associated with acute diarrhea and causative agent.

Types of diarrhea	Bacteria	Virus	Parasite
Watery diarrhea	ETEC, EPEC, *Clostridium perfringens, Bacillus cereus, Staphylococcus aureus, Vibrio cholerae*	Rotavirus Norovirus	*Giardia, Cryptosporidium*
Bloody diarrhea	*Salmonella enterica* (non-typhoidal), *Shigella, Campylobacter*, EHEC, EIEC, *Clostridium difficile, Vibrio parahaemolyticus, Yersinia enterocolitica*		*Entamoeba histolytica*

Rotavirus and different pathogroups of *Escherichia coli* are the common etiological agents related to mild or severe diarrhea. *E. coli* is a normal microflora in the large intestine and mostly remain non-pathogenic. Horizontal acquisition of virulence encoding genes makes them pathogenic to the host. Enterotoxigenic (ETEC), enterohemorrhagic (EHEC), enteropathogenic (EPEC), enteroinvasive (EIEC), diffused adherent (DAEC) and enteroaggregative (EAEC) represents the major pathogroups of *E. coli*. Human intestinal microbiota, which participates in vital physiological and immunological processes are typified by diverse functions with continued stability. Many investigations have shown that the diarrheal infection can replace the normal gut microbiota till the recovery phase.

The human gut microbiota is enriched with different species consisting of viruses/phages, bacteria, archaea, fungi and parasites. Among these, bacteria dominate the gut microbiome with five phyla (*Actinobacteria, Bacteroidetes, Firmicutes, Proteobacteria* and *Verrucomicrobia*), each containing different species.[2] *Cyanobacteria, Fusobacteria, Lentisphaerae*, and *Spirochetes* are comparatively less abundant.[3,4] Although the taxonomic configuration of human gut microbiota diverges considerably across individuals, their functional ability is sustained. These microbes form a homeostatic network with numerous functions crucial for the health and to defend the invading pathogens. The composition and functions of gut microbiota vary with age, nutrient conditions, nature of infection, and health status. Persons with restricted diversity are usually pre-disposed to recurrent diseases.

Gut microbiota efficiently limits infection by enteric pathogens by secretion of antimicrobials, a phenomenon termed colonization resistance. Evidence of commensals in the gut that provide colonization resistance against diarrheal pathogens came from studies with germ-free animals and antibiotic treated animals whose gut microbiota has been disrupted. There are numerous mechanisms by which microbiota provides colonization resistance and these can be grouped into two broad categories viz., direct inhibition of pathogens and indirect inhibition of pathogens via their effect on the hosts. An example of direct inhibition is competition for nutrients. A number of physiological, and immunological features are governed by synergistic networks of the multispecies bacterial population. Microbial digestive enzymes help the host by converting complex macromolecules and also by producing essential vitamins. Fermentation of complex carbohydrates by the microbes leads to the production of short chain fatty acids (SCFAs), which are essential for various gastrointestinal processes. Gut microbes help the host in the development of stable mucosal immunity, protect gut-barrier function by maintaining the epithelial cell integrity and thick mucus layer and gut homeostasis. They also regulate intestinal epithelial restoration, restructure the tight junctions after the infection, lower the expression of inflammatory genes of intestinal cells and influence the immune system. Further, they also help in the protection process by secreting proteases of their own.[5–9]

The detection methods of clinically important enteric pathogen include bacterial culture-based diagnostics, microscopy, serology and PCR-based tests. These approaches are not appropriate to detect the commensal aerobic and anaerobic bacterial communities, as there is no universal culture or identification based methods. High-throughput sequencing based methods enabled by the advancement of next-generation sequencing platforms supported by the different software are in use to analyze and detect the diversity and the relative abundance of microbes. The 16S rRNA gene-based sequencing can identify the diversity of gut microbial communities. Metagenome sequencing can precisely distinguish many microbial communities and their functional genes based on gene annotation and bacterial metabolic pathways. These molecular methods have advanced our understanding of bacterial community functions in highly diverse microenvironments. Here, we review the current understanding of the association between the intestinal microbiota and diarrhea.

2. Composition of gut microbiome during diarrhea

In the gut, no bacterial phyla are exclusively beneficial to the host. Several studies indicates that healthy human gut generally has some beneficial phyla such as *Firmicutes*, *Bacteroidetes* and *Actinobacteria* and less number of *Proteobacteria* and *Verrucomicrobia*.[2] These phyla may certain species as disease causing pathobionts. *Firmicutes* and *Actinobacteria* are involved in energy resorption and maintenance of the mucosal barrier, respectively. *Bacteroidetes* has a highly conserved polysaccharide utilization locus that act as a commensal colonization factor.[10] This genetic locus helps the bacteria to invade the mucus and dwell deep in the colon crypt and thereby provide resistance against enteric infections. *Bacteroides* provide several beneficial effects such as carbohydrate fermentation, polysaccharide production, growth of gut-associated lymphoid tissue and activation $CD4^+$ T-cells that direct lymphocytes to sites of infection.[11,12] The conformation of gut microbes is dependent on several factors, including enteric infections caused by other bacteria, virus and parasites, travel, consumption of antibiotics and immune status. All these factors influence the bacterial α-diversity (species diversity *within* a bacterial community at a small scale, which represent a single ecosystem). Change in the host environment (e.g., infection, dietary source, pH, etc.) results in shift in relative abundance of bacterial species (Figs. 1 and 2). It is now well documented that in disease conditions, relative abundance of Bifidobacterial species changes significantly.

Several findings have explored the effect of diarrhea on gut bacterial communities in children. A number of factors such as geographical location, etiology of the infection, age and nutritional status are significantly associated with the structure of gut microbiome. In diarrheal children, increased the levels of streptococci were found regardless to the bacterial etiology. This trend was found closely associated with the nutrient-driven consequence of glucose provided with oral rehydration solution.[13]

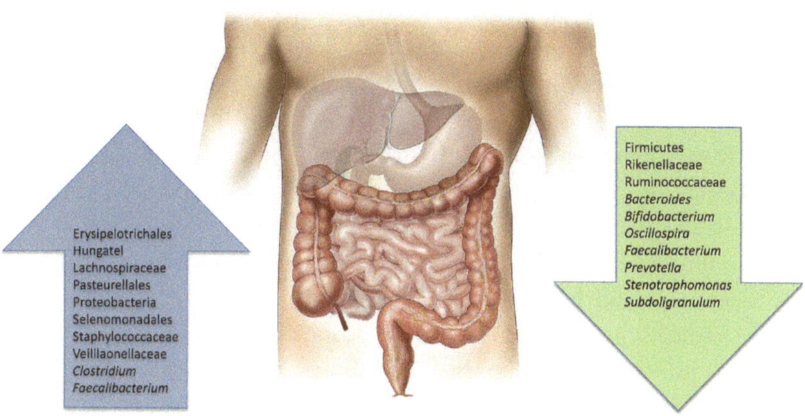

Fig. 1 Increase (light blue arrow) and decrease (pale green arrow) of gut microbiota during diarrheal infection.

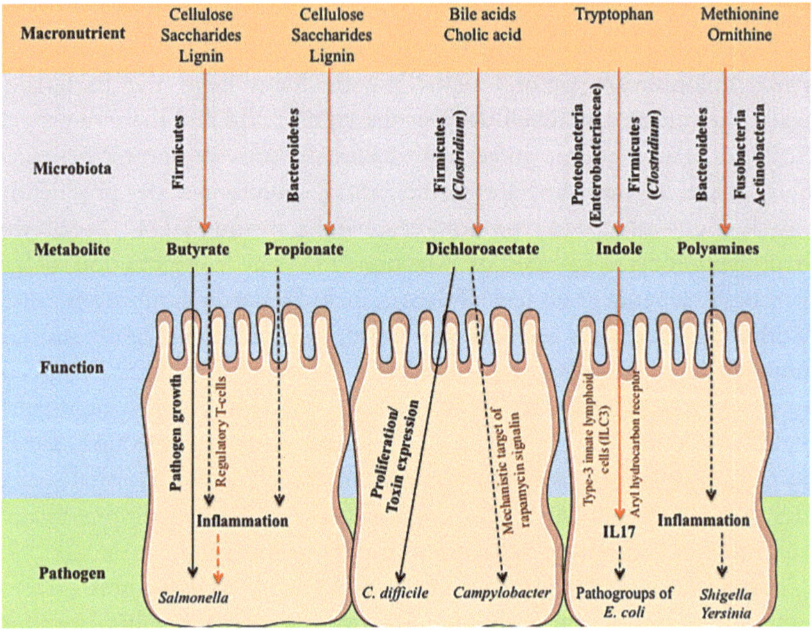

Fig. 2 Gut microbial activity and response of the pathogens with respect to dietary source. Red solid and dotted arrows indicate direct and indirect reaction initiation, respectively. Black solid arrows denote the direct inhibition and black dotted arrows indicate indirect inhibition.

Among the diarrheal children from Vietnam, steady rise in *Fusobacterium mortiferum*, and *E. coli* have been identified in the fecal microbiome during the early phase of infection. In dysenteric diarrhea, *Bifidobacterium pseudocatenulatum* was drastically declined irrespective of the pathogenic bacterial species.[14] Danish children, who had one episode of diarrhea, had about 10-fold increase of the genus *Prevotella* due its early colonization during the recovery period from diarrhea.[15] Similar re-colonization of *Prevotella* has also been demonstrated in *Vibrio cholerae*-induced childhood cholera in Bangladesh.[16]

Low abundance of *Bacteroides* and a higher abundance of *Clostridium* have been reported in Chinese children with diarrhea.[17] However, the healthy control group had a higher proportion of anaerobic bacteria than in the infected group, signifying that anaerobic bacteria play an essential part in stabilizing the gut function.[18] Among the taxonomic groups identified, Staphylococcaceae, Hungatella, Erysipelotrichales, Selenomonadales, and Pasteurellales were predominant in the infected group, whereas,

Lactobacillales, Clostridiaceae and Mycoplasmatales were higher in healthy children.[18] Association between gut bacterial composition and diarrhea in undernourished Peruvian children indicated an abundance of *Faecalibacterium prausnitzii*. On the other hand, *Bacteroides*, *Oscillospira*, Clostridiales, Ruminococcaceae and *Prevotella* were negatively influenced.[19]

In adult diarrheal patients, the richness of *Faecalibacterium*, *Stenotrophomonas*, and *Subdoligranulum* were significantly low. In addition, the composition of *Actinobacteria*, *Bacteroidetes* and *Firmicutes* were different in comparison with healthy controls.[20] Adult travelers suffering from diarrhea showed an enhanced level of Proteobacteria and reduced fractions of *Firmicutes* compared to non-diarrheal travelers.[21] In travelers' diarrhea (TD) cases, Bacteroidaceae, Rikenellaceae, and Ruminococcaceae were found to be low during infections, but the abundance of Lachnospiraceae and Veillonellaceae remained high.[22] In several studies, the importance of Ruminococcaceae in TD was investigated, as this group of bacteria produces SCFAs that help epithelial integrity that provides increased protection.[22,23]

3. Orchestrated mechanisms of commensals in preventing the pathogen colonization

3.1 Bile

Bile, the digestive secretion contains bile acids (BAs), cholesterol, phospholipid and IgA. Its primary function is to solubilize/emulsify dietary lipids and help in their absorption. Gut microbes play a central role in BAs homeostasis, participates in the BAs metabolism and convert primary BAs to secondary BAs in the gastrointestinal lumen. However, the detergent nature of the bile salts is likely to exert bacteriostatic activity on commensals, which overcome through the action of bile salt hydrolase enzyme.[24] Malabsorption or overproduction of BAs cause gut dysbiosis and diarrhea due to electrolyte imbalance. The BA-metabolizing *Lactobacilli* and *Bifidobacteria* improve the health by stimulating the host immune homeostasis.[25]

A typical condition which results from the reduction of BAs is the upsurge in the abundance of Prevotellaceae and Verrcomicrobiaceae with an increase in SCFAs (propionate and isocaproic acids) and decrease in caproic and heptanoic acids.[26] This in turn may cause low abundance of Lachnospiraceae, Ruminococcaceae, and Bacteroidaceae. Post-cholecystectomy diarrhea (PCD) develops in patients after the cholecystectomy. In such patients, bile acid directly enters into the duodenum and changes the composition of intestinal microbes that leads to diarrhea

(Fig. 2).[27] In the PCD group, a significant reduction in the microbial diversity was seen; especially, *Firmicutes/Bacteroidetes* ratio, and increase in *Bifidobacterium* and *Lactococcus*, and the other harmful microbiota (*Prevotella* and *Sutterella*). Low gut microbial abundance is associated with lipid metabolism pathway in PCD cases.

3.2 Gut function and immunity

Several gut microbial cell surface components and their signals are identified by specific receptors of epithelial and immune cells that can modify immune responses not only by inducing a range of immune cells exists in the lamina propria and also influence the host cell functions.[28] These immune cells are generally classified into innate and adaptive immune cells. Innate as well as adaptive responses are important to prevent colonization of invading pathogens. They also regulate systemic and local inflammatory responses that could be induced by dietary factors and antigens. Gut microbes are important in improving innate immunity and to modulate adaptive immune responses.[29] Gut related lymphoid tissue (GALT) caters large part of immune conditioning and protection. Experimental evidence obtained with germ-free animals has shown that commensal microbes are important for the GALT. Immune mediators stimulated by gut microbes are important in determining the disease outcome. Several immune receptors such as Toll-like receptors, IgA/IgG/IgM antibodies, $CD8^+$ T-cells, macrophages, and neutrophils are involved in this process.

Maintenance of immune homeostasis is a substantial activity, which requires discrimination between the harmless commensals and invading pathogens. This task is accomplished by the dynamic crosstalk between host immunity and microbiome. Microbial signals can also stimulate part of the regulatory T-cells (Tregs) in colonic tissue through the generation of SCFAs.[30,31] SCFAs are derived from fermentation of indigestible foods by the microbes (Fig. 2). Tregs are essential to sustain the immune homeostasis. Induction of Tregs reduces the Type 17 T-cells (Th17).[11] *Bacteroidetes* controls the inflammation by stimulating the T-helper cells (Th1 and Th17) and induce Tregs, which are important to resist immunity-associated diseases.[32]

The host gut cells actively support beneficial microbiota via generating of IgA, mucin, and antimicrobial peptides.[33,34] Intestinal epithelial cells use SCFAs for the regulation of different cellular activities such as gene expression, propagation, chemotaxis, and apoptosis.[35] In several animal models and human diarrheal cases, SCFAs are shown to prevent the proliferation and/or invasion of other bacterial by protecting the intestinal barrier functions, secretion of mucin, and countering the inflammation.[36,37] Colonization of the butyrate-producing *Firmicutes* many increase production of interleukins against invasive *E. coli* and *Salmonella* spp. Propionate produced in the human gut by *Bacteriodes* spp. and *Prevotella copri* protects the gut against several pathogroups of *E. coli.*[38] Reduction in the abundance commensal bacteria can lead to an unstable intestinal immune response, a condition that favors enteric pathogens by the activation of Th17 that regulates granulopoiesis and recruitment of neutrophils during infection.[39]

The innate immune system is normally activated by interruption of the intestinal barrier. Gut microbes improve the mucus barrier and *muc2 gene* expression, which collectively reduces inflammation by establishing a physical barrier between the lumen and the epithelium, regulation of mucosal dendritic cells, supports the production of plasma cells and IgA responses and sustains a balance between proinflammatory and regulatory outcomes.[40–42]

3.3 Intestinal iron transport

Transport of iron is important for the survival of most of the pathogens. It also influences the host–microbial interactions, including the expression of virulence. Iron is bound with high-affinity host proteins, which reduces the accessibility of iron by the microbes.[43] In humans the level of iron is $\sim 10^{-24}$ M, which is 8-fold less than the required amount for the bacterial growth.

Iron fortificants is required to prevent its deficiency and anemia in children. Iron-containing micronutrients and iron complements enhance the diarrheal risk. The excess iron provided in the form of food supplement in children decreases the level of *Lactobacillus* and *Bifidobacterium* and increases pathogen abundance.[44,45] This gut microbial imbalance often results in diarrhea and gut inflammation.[45–47] Various studies have indicated the existence of a strong iron/gut microbiota/host interaction. Iron supplementation was reported to increase *Clostridium, Dialister, Helicobacter, Lachnospira, Oscillibacter, Shigella, Slackia,* and *Xylanibacter,* while the abundance of others, such as *Acetanaerobacterium, Alloscardovia, Asaccharobacter, Barnesiella, Oscillospira, Phascolartobacterium,* and *Sporacetigenium,* was seen to be reduced in the intestine.[48] In the case of pathogenic bacteria, including *Salmonella, Clostridioides difficile* (formerly *Clostridium difficile*), *Clostridium perfringens,* and pathogenic *E. coli*, iron helps in their colonization ability and expression of virulence. A study conducted with diarrheal children in Africa, who were treated with and without iron and antibiotics displayed reduction in the abundance of *Bifidobacterium* along with a considerable rise in *C. difficile*. Children who received only the antibiotics had a greater abundance of *Bifidobacterium* and a decrease in pathogenic *E. coli* without affecting the *C. difficile* population.[49]

Most of the pathogenic bacteria has siderophores, which act as an outer membrane receptor for transferring iron-chelators. Enterobactin/enterochelin is a siderophore present in *E. coli* and *Salmonella enterica*. Presence of an iron active transport system and FeoB-mediated iron acquisition increase virulence of these pathogens.[50] During infection, siderophores regulate host cellular pathways, and establish a niche for bacterial replication. Beneficial gut microbes like members of the segmented filamentous bacteria produce an antimicrobial host protein (lipocalin-2, neutrophil gelatinase) that binds with enterobactin, to prevent the acquisition of iron.[51,52] Using the products of *iroN iroBCDE* gene cluster, *Salmonella* counters lipocalin-2 resistance by mediating the biosynthesis export and uptake of salmochelin, a linear glycosylated derivative of enterobactin that promotes iron acquisition during infection.[53]

3.4 Chemotherapy

In cancer patients, chemotherapy induced diarrhea (CID) is an unwarranted consequence after the administration of the cytotoxic drugs. CID occurs due to the damage of intestinal cells (gastroparesis) that initiates several inflammatory pathways, which causes lesions and displace commensal bacteria residing in the mucus layers. Orally administered small molecule tyrosine kinase inhibitor binds to the intracellular adenosine triphosphate domain of the tyrosine kinase and prevents downstream signaling followed by cell division and growth.

Symptoms of CID include loose stool with or without pain for a prolonged period of time, and poor response to treatment with gentamicin, berberine, and furoxone. Intestinal bacterial β-glucuronidase (BGUS) formed by several bacteria, which has been used as a carbon source.[54] BGUS contributes to the irinotecan (anti-cancer drug)-induced gut toxicity and induce diarrhea.[54] *C. perfringens, E. coli, Staphylococcus, Bacillus, Enterococcus Acinetobacter, Streptococcus* etc., play an important role in irinotecan-induced dysbiosis.[55,56] CID is associated with distinct changes in gut microflora with decreased levels of *Prevotella, Lactobacillus, Bifidobacterium, Bacteroides* and *Enterococcus* and an increase in *E. coli, Staphylococcus* and methanogenic archaea. This induces intestinal damage, altered gut function, and initiation of acute diarrhea.[57–59]

CID has also been correlated with the development of immune-related diarrhea. Treatment with anti-programed cell death protein-1 (PD-1) antibodies are one of the options for pulmonary cancer cases. However, patients administrated with anti-PD-1 antibody often suffers with immune-related diarrhea due to increase in the gut associated *Firmicutes* and *Proteobacteria.*[32] Bacteria belonging to these phyla increase gut epithelial permeability and cause inflammation leading to several gut-related disorders. Cytotoxic chemotherapy fluorouracil (5-FU) that is used to treat different cancers induces diarrhea and reduces tight junction protein occludin. Gastrointestinal mucositis (stomatitis), one of the manifestations of which is diarrhea caused by 5-FU was shown to reduce the abundance of *Bacteroidetes,* and *Firmicutes* enhance the populations of *Verrucomicrobia* and *Actinobacteria.*[60]

3.5 Malnutrition

In children, an undernourished condition causes less energy absorption, reduced abundance of gut microbiota and their metabolites, stunted growth and persistent diarrhea.[61] Intestinal villus blunting is a general trend in children with severe undernutrition that causes several metabolic perturbations.[62] Under such conditions, the abundance of anaerobic bacteria becomes particularly low, which causes reduction of energy levels, biosynthesis of vitamins, malabsorption, low immunity, and advancement of diarrheal and other infection.[63] In several studies it was shown that the malnutrition is closely related with transformation of gut microbiome.[64,65] Severely malnourished children with diarrhea have lower α-diversity with a poor abundance of Bacteroidaceae, Lachnospiraceae and enhanced levels of Enterobacteriaceae and Moraxellaceae.[66]

Investigation on the impact of susceptibility to rotavirus infection and microbiota using germ-free piglets transposed with infant's fecal microbiota (mimicking the symptoms of a malnourishment) showed that appropriate nourishment improved the quality of the microbiota. Severe diarrhea was evident with increased viral shedding in malnourished animals.[67] A study on the association between Norovirus and diarrhea in healthy and malnourished children showed a weaker association between the virus and diarrhea in malnourished children.[68] This was explained due the low expression of epithelial surface proteins in the malnourished children's gut that is required for viral entry. However, the processes by which the inadequate dietary nutrients disturb the intestinal homeostasis are not fully known.

4. Pathogen-mediated gut microbial modifications

4.1 *Clostridioides difficile*

C. difficile, which commonly present in the intestine causes drastic reduction in commensals in the gut. This pathogen causes healthcare-associated (iatrogenic) diarrhea due to excess use of antimicrobials, particularly fluroquinolones, penicillin, cephalosporin and clendamycin. Prolonged use of antibiotics increases predisposition of the patients to *C. difficile* infection (CDI) and progression of antibiotic associated diarrhea (AAD). The frequency of AAD depends on the antimicrobial agents, and other host related factors. Aging is closely associated with a decrease in gut microbial variation, metabolism, susceptibility to infections due to low immunity. Multidrug resistant *C. difficile* is one of the major enteropathogens of AAD, which results from a disequilibrium of the normal intestinal flora.

CDI is characterized by the displacement of the commensal microbiota in the gut, change in SCFA levels, increase in luminal carbohydrates and BAs, distorted water absorption, along with immune dysregulation, polyendocrinopathy and enteropathy. Decrease in B-cells, plasma cells, and IgA are the other outcomes reported in CDI cases.[69] One or many of these factors lead to life-threatening diarrhea. Manipulation of microbial composition by means of fecal microbiota transplantation (FMT) has been recommended as an effective measure to restore normal gut microbiota.[70] This implies displacement of microbiota from a healthy donor to the CDI patient. For example, *C. defficile* infection, which requires a breach in colonization resistance, can be successfully treated by FMT. The cure rate of single fecal microbiota transplantation of *C. difficile* infection is about 90%.[71] FMT also has been considered for restoring metabolic imbalances of SCFAs such as sialic acid, succinate and BAs.[72] In children with recurrent CDI, the FMT has shown significant improvement in the abundance and diversity of gut microbiota with the restoration of *Alistipes*, *Bifidobacterium*, *Dialister*, *Enterococcus*, *Flavonifractor*, *Oscillibacter*, and *Streptococcus*.[73]

Metagenomic analysis of patients with CDI in the intensive care units revealed abundances of Acetothermia, Cryptomycota, Deferribacteres, Enterobacteriaceae, Gammaproteobacteria, Porphyromonadaceae taxa. However, Saccharomycetes and species belonging to Clostridia remained significantly less.[74] Based on rDNA analysis, it was observed that a healthy adult male with AAD had decreased levels of clostridial groups.[75] Toxigenic *C. difficile*-positive diarrheal children showed higher levels of *Klebsiella* and *Proteobacteria. C. difficile* colonization accompanied by *Klebsiella* and *Ruminococcus* and *Bifidobacterium* appears to play a protective role in infants.[76] *Firmicutes* and *Bacteroidetes* remained high in *C. difficile*-negative and also among the carriers of non-toxigenic of *C. difficile* individuals.[77] Remarkably, several genera such as *Anaerostipes, Butyricimonas, Catenibacterium, Lachnospira, Odoribacter, Paraprevotella,* and *Phascolarctobacterium* were identified in most of the *C. difficile* infected cases.

It has been observed that CDI significantly reduces fecal BA levels, but increases the primary BA (cholic acid and taurocholate) concentrations.[78] Primary BAs support the growth of *C. difficile* spores, whereas the secondary BA (lithocholic acid) on the other hand prevent the growth and survival of *C. difficile.*[79] After FMT, metabolism of secondary BAs by members of the *Clostridium* XIVa clad, *Parasutterella, Ruminococcus* and *Bacteroides* showed resistance against CDI.[78,80] Several FMT studies have shown enhanced levels of anti-inflammatory bacteria, such as Desulfovibrionaceae, Lactobacillaceae, Porhyromonadeacae, Ruminococcaceae, and Sutturellaceae, and low abundance of Enterobacteriaceae and Veillonellaceae. In addition, a significant reduction in the inflammatory markers (proinflammatory cytokines) and normalization of C-reactive protein and fecal calprotectin were seen in recipients of FMT.[81]

Amino acid richness in a dysbiotic condition is an indication of less nutrient utilization by the gut microbes. Dybiosis is an ideal condition for the pathogenic *C. difficile.* In germ-free mice model, FMT with dysbiotic microbiota from diarrheal patients demonstrated an increase in proline has caused susceptibility to CDI. A mutation in the *prdB* gene of *C. difficile,* which codes for proline reductase showed reduced growth of this pathogen in the germ-free mice with displaced human gut microbiota.[70]

Toxin-producing *C. difficile* plays an important role in replacing several gut bacteria during infection. Binary AB enterotoxin is responsible for the *C. difficile*-associated diarrhea. In infants, bowel mucosa is immature and lack receptors for this toxin and hence *C. difficile* has been regarded as non-pathogenic in this age group.[82] In mouse model studies, it was established that *Saccharomyces* promotes intestinal IgA response to the toxin-A subunit of *C. difficile*.[83] In toxin B-positive *C. difficile* infected patients, the proportion of Enterobacteriaceae, Porphyromonadaceae, and Enterococcaceae, *Proteobacteria* persists significantly high, whereas *Firmicutes*, Lachnospiraceae, Ruminococcaceae, and Prevotellaceae remained very low.[84] For a more comprehensive information, the readers can refer to the review article by Ref. [85]

4.2 *V. cholerae*

During cholera infection, the efflux of secretory diarrhea, through the action of cholera toxin, removes the mucus layer along with the residing gut microbial community. Besides this, there are other mechanisms by which *V. cholerae* can regulate the level of commensals in the gut. The type-VI secretion system (T6SS) of *V. cholerae,* which delivers toxin to target cells to kill them, allows it to compete with the commensals by removing them directly through contact-dependent killing.[86] Antagonistic activity of *V. cholerae* and use of antibiotics during treatment also play their part in reducing the commensal bacteria from the gut. In *V. cholerae*-infected patients, *C. difficile* persisted in the gut for long periods, perhaps due to extensive gastrointestinal damage of mucus layer and intestinal epithelial cells.[86,87]

Many findings demonstrate the consequence of cholera on gut microbiome, especially on their influence during infection and recovery phases. *Streptococcus, Paracoccus*, *Prevotella*, and *Blautia* were found to be higher in infected individuals. Among these, *Blautia* reduced colonization of *V. cholerae*, whereas, *Paracoccus* supported the growth of the pathogen in vitro conditions.[88] Several molecular signals from commensal bacteria, including their metabolites are important determinants in shaping the disease outcomes and also resist against *V. cholerae* infection. Bile salt hydrolase enzyme encoded in the *bhs* gene of *B. obeum* degrades taurocholate, which is a host produced bile salt. As a consequence of this action, the gut colonization and activation virulence genes of *V. cholerae* are suppressed.[89] The metagenomic analysis of adult patients showed drastic reduction in the gut microbiome during the infection phase, but during the recovery phase it gets restored to a level similar to that of children.[90] It was also observed that autoinducers secreted by the commensal microbes during the recovery phase reduce the virulence expression of *V. cholerae*. *Prevotella* and *Bifidobacterium* communities are likely to give protection from cholera.[91]

Many metagenomic studies identified that a reduction in the abundance of *Bacteroidetes* and *Firmicutes* in the gut that makes a favorable condition for cholera infection.[89–91] During the early stages of convalescence, restoration of gut mucosa and the abundance of *Escherichia*, *Streptococcus* and *Enterococcus* has been observed.[16] This change was found to be linked to the aerobic condition in the gut. Colonization of carbohydrate metabolizing *Bacteroides*, obligate anaerobes and return of other normal bacterial flora are also seen during the recovery phase.[16] Maintenance of 'healthy' gut microbiome has been considered as one of the approaches in the intervention and prevention of cholera.

4.3 *Shigella* spp.

Production of virulence factors from the virulence plasmid (T3SS with many effector proteins), enterotoxins and exotoxin (Shiga toxin) are the important virulence factors of *Shigella* spp. Among the *Shigella* spp., *Shigella sonnei* express T6SS, which eliminates *E. coli*.[92] Diet-based fibers favors richness of *Prevotella* spp. in the gut. When the diet-based fibers are less, the gut microbes, especially the *Prevotella* uses the host's mucus glycoprotein that leads to weakening of the gut epithelial mucosal barrier.[93] In diarrheal children, abundance of *Prevotella* was negatively correlated with the invasion plasmid antigen encoding gene (*ipaH*) present in the *Shigella*/EIEC.[94] Overall, there is a paucity of information about bacterial communities that are protect against the *Shigella* infection.

4.4 Diarrheagenic *E. coli* (DEC)

Tin DEC, the virulence factors depend on the specific pathotype. One of the factors responsible for DEC virulence regulations is the presence of gut microbial metabolites.[95] Greater levels of histamine and low levels of ornithine in the DEC infected group is due to their adherence in the intestinal cells and also due to the higher abundance of *Streptococcus* and *Citrobacter*. Histamine is produced by immunogenic mast cells due to the presence of DEC. In contrast, higher production of L-ornithine is associated with healthy gut mucosa.[96]

Children infected with DEC display a distinct gut microbial composition. In DEC-positive diarrheal stool samples, a high fraction of *Bacteroidetes* and *Proteobacteria* and less abundance of *Firmicutes* have been reported.[97] In addition, *Escherichia albertii, Citrobacter werkmanii, Yersinia enterocolitica,* and *Haemophilus sputorum* were seemed to be significantly associated with DEC and recognized as indicator bacterial taxa to DEC-mediated diarrhea. In Ecuador, a case-control study showed the predominant occurrence of DAEC and ETEC among children with diarrhea. In a metagenomics analysis, *P. copri* was found in the non-diarrheal group, but *F. mortiferum* and *Campylobacter concisus* were found significantly higher in children with ETEC infection.[98] During DAEC infection, the levels of *Bifidobacterium longum* and *Alloprevotella tannerae* were increased substantially.[98]

Lactobacillales (*Enterococcus avium, Enterococcus faecalis,* and the *Streptococcus* spp.) dominated in Shiga toxin-producing *E. coli* (STEC)-positive cases,[99] but Bifidobacteriales and Clostridiales were remained low. *Bacteroides thetaiotaomicron* was shown to express the repressive effect on the expression of Shiga toxin (Stx) by the EHEC.[100] Vitamin B_{12} is a co-factor of ethanolamine utilization regulator protein (EutR), which is required for full activation of *eut* and locus of enterocyte effacement pathogenicity island operons in EHEC. Intake of vitamin B_{12} could control the expression of virulence in EHEC, as intake of vitamin B_{12} by the commensal *B. thetaiotaomicron* decrease in the production of Stx2.

4.5 Campylobacters

Campylobacter is a microaerophilic bacterium. Several virulence factors like adhesion proteins, flagellar secreted factors, serine proteas, invasion machinery and cytolethal distending toxin are associated with pathogenesis of this thermotolerant bacterium. In diarrheal children, the level of anaerobic *Bifidobacterium kashiwanohense* was found to be positively correlated with the presence of *Campylobacter.* This could be due to the modulation of *B. kashiwanohense* that reduce the intestinal homeostasis and facilitate the persistence of *Campylobacter* in the intestinal tract.[101] L-fucose is an essential microbial nutrient for the bacteria, which is acquired from the host or from microbes. Fucosylated structures are important bacterial adherence. *Campylobacter jejuni* do not have any homologs of fucosidase and hence depends on L-fucose of commensal microbes.[102] *Bacteroides, Gordonibacter,* and *Escherichia* are negatively correlated with *Campylobacter* as these bacteria produce lactic acids, SCFAs, enterocin, and urolithin molecules, which have the antimicrobial and anti-inflammatory properties. In addition, they modulate the microbial composition and host immune responses.[101]

4.6 Enteric viruses (rotavirus and norovirus)

For viral-mediated infections, the small intestine is the major location and hence the fecal microbiome might not correlate well with changes at this site.[103] Several structural proteins (VP) non-structural proteins (NS) and apoptosis of enterocytes in humans have shown to be involved in virulence and pathogenicity of these viruses. Infection of gut epithelial cells by the enteric viruses causes disarrangement of the gut microbial homeostasis and diversity. In children with rotaviral infection, microbial dysbiosis will be more significant with the increase in the abundance of Proteobacteria, Fusobacteria and *Streptococcus* populations with concomitant reduction in *Dialister* and Ruminococcaceae group.[20,104]

Convalescing phase of rotaviral infection has been influenced by mucin-degrading microbes. Incubation of cell-lines that produce mucin and intestinal enteroids of human origin with mucin-degrading *Bacteroides* or *Akkermansia* members along with rotavirus showed significant degradation of the glycan and altered binding ability of rotavirus.[103] It was also shown that one of the commensals, *Bacillus clausii* protect enterocytes against rotavirus with higher expression of mucin as well as the tight junction proteins. These proteins increase the mucosal barrier function, inhibition of reactive oxygen species production, secretions of pro-inflammatory cytokines and down-regulation of gene encoding the pro-inflammatory Toll-like receptor-3.[105]

In the mice model, *Bacteroides fragilis* stimulate IFN-β in the dendritic cells of colon lamina propria. This indicates that gut microbes could directly resist the viral infections.[106] The carbohydrate metabolism pathway was found to be significantly higher in rotavirus infected cases.[20] Rotavirus-damaged enterocytes of the small intestine villi cannot absorb carbohydrates and other nutrients and hence there will be an increase in the carbohydrate metabolism pathway.[107]

Norovirus is an RNA viruses, which causes acute gastroenteritis in humans. Studies on Norovirus infection in association with gut microbiota are limited. A recent work has shown the increased existence of *Corynebacterium, Holdemanella, Howardella, Massilia* and *Staphylococcus* in Norovirus infected individuals, however, the levels of Erysipelotrichaceae and Staphylococcaceae considerably differed, compared to the control group.[20] Norovirus bind with histo-blood group antigens (HBGA) and on the surface of bacterial membranes expressed with sialic acid residues.[108,109] Expression HBGA-like substance by some of the anaerobic bacteria like *Ruminococcus* and *Oxalobacter* reduce the viral infectivity.[110]

5. Severe acute respiratory syndrome coronavirus 2 (SARS-CoV-2)

The single-stranded RNA virus, SARS-CoV-2 is responsible for the greatest pandemic in recent times. Gastrointestinal manifestations, including diarrhea may contribute to the severity of this disease during the course of the infection. The main cause of diarrhea due to SARS-CoV-2 is its invasion into the angiotensin converting enzyme 2 (ACE2)-expressing epithelial cells of the small intestine. ACE2 plays an important role in the regulation of dietary amino acid and maintains gut homeostasis, innate immunity, and gut microbial ecosystem.[111] It also functions as a sodium-dependent neutral amino acid transporter (B^0AT1). The B^0AT1/ACE2 complex influences the gut microbial make-up and their functions, affecting the local and systemic immune-responses against many pathogens.[112]

In many findings, gut mycobial dysbiosis was demonstrated in SARS-CoV-2 patients along with a reduction in SCFA-producing bacteria.[113] In SARS-CoV-2 cases, opportunistic pathogens such as *Actinomyces, E. faecalis, Rothia, Saccharomyces cerevisiae, Streptococcus,* and *Veillonella* have been identified along with a reduction in the richness of *Bifidobacterium* spp.[114] SARS-CoV-2 also exhibited a positive correlation with members of Mucoromycota such as *Fusicatenibacter, Aspergillus niger.*[115] In non-severe patients, the microbial dysbiosis is associated with an uncharacteristic immune response.[113]

Based on the several reports, the gut microbial composition and their metabolic products are considered as indicators of SARS-CoV-2 infection. Pulmonary health is influenced by a crosstalk between the gut microbes and lungs. The "gut-lung axis", a bidirectional process involves microbial metabolites that regulate the lung through the blood can affect the gut microbiota during inflammation of the lungs. The gut microbial metabolic processes strongly control cytokine production and can enhance chronic phase protein and lung cell interferon signaling to protect from viral infection.[116] However, to confirm the role of gut microbes, more clinical evidences are required.

5.1 Intestinal parasites

Giardia-induces pathological changes such as malabsorption, shortening of the brush border microvilli, disturbances of biofilm composition and triggers dysbiosis.[117] *Giardia* infection leads to higher abundance of *Enterorhabdus*, which in turn increase deconjugation of bile salts and hydroxylated BAs such as glycochenodeoxycholic acid. This condition favors the growth of the parasites in the gut.[118] In the mice model, dysbiosis caused by *Giardia* increases *Proteobacteria* and reduces *Firmicutes* and *Melainabacteria* and cause diarrhea. This is facilitated directly through the anaerobic fermentative metabolism of the parasite and/or indirectly by the induction of gut inflammation.[119] Decreased abundance of the *Collinsella* and *Bacteroides* and *Lactobacillus* spp. exists in children infected with *Giardia lamblia* and *Entamoeba histolytica/dispar*, respectively.[120] A weak immune system under microbiome-depleted conditions influences the regulation of defensins, angiogenin-4 and low levels IgA intensify susceptibility to *G. duodenalis* infection.[121]

E. histolytica survives in the intestine by feeding on microbes and bacteria-induced mucin glycans, which exposes the intestinal epithelium.[122] Intestinal goblet cells increase the production of mucus to prevent parasitic infection.[123] As a consequence, the stability of commensal microbiota in the intestinal milieu gets distorted. Disruption of the host mucosa increases the abundance *Prevotella* species and an increase in host inflammatory responses.[124] Antimicrobial peptides induced by *E. histolytica* could also cause alterations in the microbial composition. In *E. histolytica* infected patients, dysbiotic state was seen to be associated with a reduction in *Bacteroides, Campylobacter, Clostridium, Eubacterium* and *Lactobacillus,* and an increase in *Bifidobacterium.*[125] It has been seen that increase in *E. coli* abundance safeguards the parasite from oxidative damage by generating malate dehydrogenase.[126] *E. histolytica* colonization has been significantly correlated with the composition of gut microbes, especially with the predominance of *Prevotellacopriis.*[127]

In several studies it was shown that stimulation of immunity by the gut microbiome protect against the infections caused by *E. histolytica.* In gnotobiotic mice, it was shown that colonization of *Clostridium scindens* is protected from *E. histolytica* infection through innate immunity. *C. scindens* and the microbially metabolized bile modulate hematopoietic precursors, which in turn provide protection against *E. histolytica.*[128] In children, increased level of *P. copri* was found to be related with

E. histolytica-mediated diarrhea.[129] This could be due to excessive immune activation triggered by *P. copri* and also to syndrome of environmental enteropathy.[129]

Cryptosporidium present in the intestinal epithelial cells impacts the gut microbiota and also damages the epithelium.[130] However, the short lived sporozoites, merozoites, microgametes and oocysts forms of the *Cryptosporidium* are extracellular and perturbs the gut microbiota of lumen and the mucosa. In the neonatal mice model, the gut microbiota synergize an immunostimulant polyinosinic:polycytidylic acid to increase intestinal immunity against *C. parvum*.[131] Fecal indole produced by *E. coli*, *Bacillus* spp., and *Clostridium* spp. helps *Cryptosporidium* to enhance the infection.[132]

6. Conclusion and future prospective

Gut microbiome is considered as an important component in the advancement of modern treatment alternatives for many diseases. However, compositional dynamics of microbiota are not fully understood, as it depends on the changing conditions of the gut milieu. Studies on gut microbiota, including the role of bacteriophages in disease manifestation is also needed focusing on the specific phase in which the microbes play a role in the disease progression. In addition, it would be advantageous to detect the mechanisms and signaling molecules implicated in the crosstalk between gut microbiota and enteric pathogens. Undefined bacterial species composition and the long-term health consequences of FMT are not fully explored. Hence, it is imperative to explore the microbial differences using high-throughput microbiome experimentations and in silico approaches. In the future, more attention has to be paid to these aspects in order to make this field of research as a part of personalized medicine.

Several commensal microbes in the gut influence anti-inflammatory and immunomodulatory properties through the caspases, gene expression of mucins, interleukins, nuclear factor-κB, and Toll-like receptors. Endorsing these assorted of functions, it is important to identify the potential gut microbial strains that can be used as probiotics. Identification of potential biomarkers that represent the onset and progression of persistent diarrhea caused by DEC and *C. difficile* are useful for further understanding the mechanisms associated with the gut microbiota, inflammation and morbidity. Considering the influence of micronutrients on gut microbiota, detailed investigations are needed in understanding the exact role in the physiological process and to find alternative approaches to combat their deficiencies.

Acknowledgments

This work was supported in part by the Indian National Science Academy (INSA) and the National Academy of Sciences (NASI), India.

Conflict of interest

The authors have no conflict of interest to declare.

References

1. GBD 2016 Diarrhoeal Disease Collaborators. Estimates of the global, regional, and national morbidity, mortality, and aetiologies of diarrhoea in 195 countries: a systematic analysis for the global burden of disease study 2016. *Lancet Infect Dis.* 2018;18 (11):1211–1228. https://doi.org/10.1016/S1473-3099(18)30362-1.
2. Kostic AD, Xavier RJ, Gevers D. The microbiome in inflammatory bowel disease: current status and the future ahead. *Gastroenterology.* 2014;146(6):1489–1499. https://doi.org/10.1053/j.gastro.2014.02.009.
3. Eckburg PB, Bik EM, Bernstein CN, et al. Diversity of the human intestinal microbial flora. *Science.* 2005;308(5728):1635–1638. https://doi.org/10.1126/science.1110591.
4. Tap J, Mondot S, Levenez F, et al. Towards the human intestinal microbiota phylogenetic core. *Environ Microbiol.* 2009;11(10):2574–2584. https://doi.org/10.1111/j.1462-2920.2009.01982.x.
5. Bron PA, Kleerebezem M, Brummer RJ, et al. Can probiotics modulate human disease by impacting intestinal barrier function? *Br J Nutr.* 2017;117(1):93–107. https://doi.org/10.1017/S0007114516004037.
6. Chichlowski M, De Lartigue G, German JB, Raybould HE, Mills DA. Bifidobacteria isolated from infants and cultured on human milk oligosaccharides affect intestinal epithelial function. *J Pediatr Gastroenterol Nutr.* 2012;55(3):321–327. https://doi.org/10.1097/MPG.0b013e31824fb899.
7. Mathewson ND, Jenq R, Mathew AV, et al. Gut microbiome-derived metabolites modulate intestinal epithelial cell damage and mitigate graft-versus-host disease. *Nat Immunol.* 2016;17(5):505–513. https://doi.org/10.1038/ni.3400.
8. Wickramasinghe S, Pacheco AR, Lemay DG, Mills DA. *Bifidobacteria* grown on human milk oligosaccharides downregulate the expression of inflammation-related genes in Caco-2 cells. *BMC Microbiol.* 2015;15:172. https://doi.org/10.1186/s12866-015-0508-3.
9. Yu LC, Wang JT, Wei SC, Ni YH. Host-microbial interactions and regulation of intestinal epithelial barrier function: from physiology to pathology. *World J Gastrointest Pathophysiol.* 2012;3(1):27–43. https://doi.org/10.4291/wjgp.v3.i1.27.
10. Lee SM, Donaldson GP, Mikulski Z, Boyajian S, Ley K, Mazmanian SK. Bacterial colonization factors control specificity and stability of the gut microbiota. *Nature.* 2013;501(7467):426–429. https://doi.org/10.1038/nature12447.
11. Round JL, Mazmanian SK. The gut microbiota shapes intestinal immune responses during health and disease. *Nat Rev Immunol.* 2009;9(5):313–323. https://doi.org/10.1038/nri2515.
12. Wexler HM. Bacteroides: the good, the bad, and the nitty-gritty. *Clin Microbiol Rev.* 2007;20(4):593–621. https://doi.org/10.1128/CMR.00008-07.
13. Kieser S, Sarker SA, Sakwinska O, et al. Bangladeshi children with acute diarrhoea show faecal microbiomes with increased *Streptococcus* abundance, irrespective of diarrhoea aetiology. *Environ Microbiol.* 2018;20(6):2256–2269. https://doi.org/10.1111/1462-2920.14274.

14. The HC, Florez de Sessions P, Jie S, et al. Assessing gut microbiota perturbations during the early phase of infectious diarrhea in Vietnamese children. *Gut Microbes*. 2018;9 (1):38–54. https://doi.org/10.1080/19490976.2017.1361093.
15. Mortensen MS, Hebbelstrup Jensen B, Williams J, et al. Stability and resilience of the intestinal microbiota in children in daycare - a 12 month cohort study. *BMC Microbiol.* 2018;18(1):223. https://doi.org/10.1186/s12866-018-1367-5.
16. David LA, Weil A, Ryan ET, et al. Gut microbial succession follows acute secretory diarrhea in humans. *MBio*. 2015;6(3):e00381–15. https://doi.org/10.1128/mBio.00381-15.
17. Niu J, Xu L, Qian Y, et al. Evolution of the gut microbiome in early childhood: a cross-sectional study of Chinese children. *Front Microbiol.* 2020;11:439. https://doi.org/10.3389/fmicb.2020.00439.
18. Wen H, Yin X, Yuan Z, Wang X, Su S. Comparative analysis of gut microbial communities in children under 5 years old with diarrhea. *J Microbiol Biotechnol.* 2018;28 (4):652–662. https://doi.org/10.4014/jmb.1711.11065.
19. Rouhani S, Griffin NW, Yori PP, et al. Diarrhea as a potential cause and consequence of reduced gut microbial diversity among undernourished children in Peru. *Clin Infect Dis*. 2020;71(4):989–999. https://doi.org/10.1093/cid/ciz905.
20. Mizutani T, Aboagye SY, Ishizaka A, et al. Gut microbiota signature of pathogen-dependent dysbiosis in viral gastroenteritis. *Sci Rep.* 2021;11(1):13945. https://doi.org/10.1038/s41598-021-93345-y.
21. Leo S, Lazarevic V, Gaïa N, et al. Andremont a The VOYAG-R study group, Schrenzel J, Ruppé E. The intestinal microbiota predisposes to traveler's diarrhea and to the carriage of multidrug-resistant Enterobacteriaceae after traveling to tropical regions. *Gut Microbes*. 2019;10(5):631–641. https://doi.org/10.1080/19490976.2018.1564431.
22. Walters WA, Reyes F, Soto GM, et al. Epidemiology and associated microbiota changes in deployed military personnel at high risk of traveler's diarrhea. *PLoS One*. 2020;15(8):e0236703. https://doi.org/10.1371/journal.pone.0236703.
23. Yutin N, Galperin MY. A genomic update on clostridial phylogeny: Gram-negative spore formers and other misplaced clostridia. *Environ Microbiol.* 2013;15 (10):2631–2641. https://doi.org/10.1111/1462-2920.12173.
24. De Smet I, Van Hoorde L, Vande Woestyne M, Christiaens H, Verstraete W. Significance of bile salt hydrolytic activities of lactobacilli. *J Appl Bacteriol.* 1995;79(3):292–301. https://doi.org/10.1111/j.1365-2672.1995.tb03140.x.
25. Vlasova AN, Chattha KS, Kandasamy S, et al. *Lactobacilli* and *Bifidobacteria* promote immune homeostasis by modulating innate immune responses to human rotavirus in neonatal gnotobiotic pigs. *PLoS One*. 2013;8(10):e76962. https://doi.org/10.1371/journal.pone.0076962.
26. Sagar NM, Duboc H, Kay GL, et al. The pathophysiology of bile acid diarrhoea: differences in the colonic microbiome, metabolome and bile acids. *Sci Rep.* 2020;10 (1):20436. https://doi.org/10.1038/s41598-020-77374-7.
27. Li YD, Liu BN, Zhao SH, Zhou YL, Bai L, Liu EQ. Changes in gut microbiota composition and diversity associated with post-cholecystectomy diarrhea. *World J Gastroenterol.* 2021;27(5):391–403. https://doi.org/10.3748/wjg.v27.i5.391.
28. Sansonetti PJ, Di Santo JP. Debugging how bacteria manipulate the immune response. *Immunity*. 2007;26(2):149–161. https://doi.org/10.1016/j.immuni.2007.02.004.
29. Omenetti S, Pizarro TT. The Treg/Th17 Axis: a dynamic balance regulated by the gut microbiome. *Front Immunol.* 2015;6:639. https://doi.org/10.3389/fimmu.2015.00639.
30. Arpaia N, Campbell C, Fan X, et al. Metabolites produced by commensal bacteria promote peripheral regulatory T-cell generation. *Nature*. 2013;504(7480):451–455. https://doi.org/10.1038/nature12726.

31. Lathrop SK, Bloom SM, Rao SM, et al. Peripheral education of the immune system by colonic commensal microbiota. *Nature.* 2011;478(7368):250–254. https://doi.org/10.1038/nature10434.
32. Liu T, Xiong Q, Li L, Hu Y. Intestinal microbiota predicts lung cancer patients at risk of immune-related diarrhea. *Immunotherapy.* 2019;11(5):385–396. https://doi.org/10.2217/imt-2018-0144.
33. Caballero S, Pamer EG. Microbiota-mediated inflammation and antimicrobial defense in the intestine. *Annu Rev Immunol.* 2015;33:227–256. https://doi.org/10.1146/annurev-immunol-032713-120238.
34. Shapiro H, Thaiss CA, Levy M, Elinav E. The cross talk between microbiota and the immune system: metabolites take center stage. *Curr Opin Immunol.* 2014;30:54–62. https://doi.org/10.1016/j.coi.2014.07.003.
35. Corrêa-Oliveira R, Fachi JL, Vieira A, Sato FT, Vinolo MA. Regulation of immune cell function by short-chain fatty acids. *Clin Transl Immunol.* 2016;5(4):e73. https://doi.org/10.1038/cti.2016.17.
36. Morrison DJ, Preston T. Formation of short chain fatty acids by the gut microbiota and their impact on human metabolism. *Gut Microbes.* 2016;7(3):189–200. https://doi.org/10.1080/19490976.2015.1134082.
37. Sun X, Jia Z. Microbiome modulates intestinal homeostasis against inflammatory diseases. *Vet Immunol Immunopathol.* 2018;205:97–105. https://doi.org/10.1016/j.vetimm.2018.10.014.
38. Louis P, Flint HJ. Formation of propionate and butyrate by the human colonic microbiota. *Environ Microbiol.* 2017;19(1):29–41. https://doi.org/10.1111/1462-2920.13589.
39. Pelletier M, Maggi L, Micheletti A, et al. Evidence for a cross-talk between human neutrophils and Th17 cells. *Blood.* 2010;115(2):335–343. https://doi.org/10.1182/blood-2009-04-216085.
40. Burgess SL, Buonomo E, Carey M, et al. Bone marrow dendritic cells from mice with an altered microbiota provide interleukin 17A-dependent protection against *Entamoeba histolytica* colitis. *MBio.* 2014;5(6), e01817. https://doi.org/10.1128/mBio.01817-14.
41. Macpherson AJ, McCoy KD. Stratification and compartmentalisation of immunoglobulin responses to commensal intestinal microbes. *Semin Immunol.* 2013;25(5):358–363. https://doi.org/10.1016/j.smim.2013.09.004.
42. Shan M, Gentile M, Yeiser JR, et al. Mucus enhances gut homeostasis and oral tolerance by delivering immunoregulatory signals. *Science.* 2013;342(6157):447–453. https://doi.org/10.1126/science.1237910.
43. Nairz M, Schroll A, Sonnweber T, Weiss G. The struggle for iron-a metal at the host-pathogen interface. *Cell Microbiol.* 2010;12(12):1691–1702. https://doi.org/10.1111/j.1462-5822.2010.01529.x.
44. Balmer SE, Wharton BA. Diet and faecal flora in the newborn: breast milk and infant formula. *Arch Dis Child.* 1989;64(12):1672–1677. https://doi.org/10.1136/adc.64.12.1672.
45. Jaeggi T, Kortman GA, Moretti D, et al. Iron fortification adversely affects the gut microbiome, increases pathogen abundance and induces intestinal inflammation in Kenyan infants. *Gut.* 2015;64(5):731–742. https://doi.org/10.1136/gutjnl-2014-307720.
46. Paganini D, Zimmermann MB. The effects of iron fortification and supplementation on the gut microbiome and diarrhea in infants and children: a review. *Am J Clin Nutr.* 2017;106(Suppl 6):1688S–1693S. https://doi.org/10.3945/ajcn.117.156067.
47. Paganini D, Uyoga MA, Kortman GAM, et al. Prebiotic galacto-oligosaccharides mitigate the adverse effects of iron fortification on the gut microbiome: a randomised controlled study in Kenyan infants. *Gut.* 2017;66(11):1956–1967. https://doi.org/10.1136/gutjnl-2017-314418.

48. Seyoum Y, Baye K, Humblot C. Iron homeostasis in host and gut bacteria-a complex interrelationship. *Gut Microbes*. 2021;13(1):1–19. https://doi.org/10.1080/19490976.2021.1874855.
49. Paganini D, Uyoga MA, Kortman GAM, et al. Iron-containing micronutrient powders modify the effect of oral antibiotics on the infant gut microbiome and increase post-antibiotic diarrhoea risk: a controlled study in Kenya. *Gut*. 2019;68(4):645–653. https://doi.org/10.1136/gutjnl-2018-317399.
50. Boyer E, Bergevin I, Malo D, Gros P, Cellier MF. Acquisition of Mn(II) in addition to Fe(II) is required for full virulence of *Salmonella enterica* serovar typhimurium. *Infect Immun*. 2002;70(11):6032–6042. https://doi.org/10.1128/IAI.70.11.6032-6042.2002.
51. Laan M, Cui ZH, Hoshino H, et al. Neutrophil recruitment by human IL-17 via C-X-C chemokine release in the airways. *J Immunol*. 1999;162(4):2347–2352.
52. Swanson BA, Taylor EV, Chew ME, et al. Specific commensal bacterium critically regulates gut microbiota osteoimmunomodulatory actions during normal postpubertal skeletal growth and maturation. *JBMR Plus*. 2020;4(3):e10338. https://doi.org/10.1002/jbm4.10338.
53. Raffatellu M, Bäumler AJ. *Salmonella*'s iron armor for battling the host and its microbiota. *Gut Microbes*. 2010;1(1):70–72. https://doi.org/10.4161/gmic.1.1.10951.
54. Wallace BD, Wang H, Lane KT, et al. Alleviating cancer drug toxicity by inhibiting a bacterial enzyme. *Science*. 2010;330(6005):831–835. https://doi.org/10.1126/science.1191175.
55. Chamseddine AN, Ducreux M, Armand JP, et al. Intestinal bacterial β-glucuronidase as a possible predictive biomarker of irinotecan-induced diarrhea severity. *Pharmacol Ther*. 2019;199:1–15. https://doi.org/10.1016/j.pharmthera.2019.03.002.
56. Leung JW, Liu YL, Leung PS, Chan RC, Inciardi JF, Cheng AF. Expression of bacterial beta-glucuronidase in human bile: an in vitro study. *Gastrointest Endosc*. 2001;54(3):346–350. https://doi.org/10.1067/mge.2001.117546 [PMID: 11522976].
57. Pal SK, Li SM, Wu X, et al. Stool bacteriomic profiling in patients with metastatic renal cell carcinoma receiving vascular endothelial growth factor-tyrosine kinase inhibitors. *Clin Cancer Res*. 2015;21(23):5286–5293. https://doi.org/10.1158/1078-0432.CCR-15-0724.
58. Secombe KR, Van Sebille YZA, Mayo BJ, Coller JK, Gibson RJ, Bowen JM. Diarrhea induced by small molecule tyrosine kinase inhibitors compared with chemotherapy: potential role of the microbiome. *Integr Cancer Ther*. 2020;19. https://doi.org/10.1177/1534735420928493.
59. Stringer AM, Al-Dasooqi N, Bowen JM, et al. Biomarkers of chemotherapy-induced diarrhoea: a clinical study of intestinal microbiome alterations, inflammation and circulating matrix metalloproteinases. *Support Care Cancer*. 2013;21(7):1843–1852. https://doi.org/10.1007/s00520-013-1741-7.
60. Li HL, Lu L, Wang XS, et al. Alteration of gut microbiota and inflammatory cytokine/chemokine profiles in 5-fluorouracil induced intestinal mucositis. *Front Cell Infect Microbiol*. 2017;7:455. https://doi.org/10.3389/fcimb.2017.00455.
61. Plovier H, Cani PD. Microbial impact on host metabolism: opportunities for novel treatments of nutritional disorders? *Microbiol Spectr*. 2017;5(3). https://doi.org/10.1128/microbiolspec.BAD-0002-2016.
62. Farràs M, Chandwe K, Mayneris-Perxachs J, et al. Characterizing the metabolic phenotype of intestinal villus blunting in Zambian children with severe acute malnutrition and persistent diarrhea. *PLoS One*. 2018;13(3), e0192092. https://doi.org/10.1371/journal.pone.0192092.
63. Million M, Diallo A, Raoult D. Gut microbiota and malnutrition. *Microb Pathog*. 2017;106:127–138. https://doi.org/10.1016/j.micpath.2016.02.003.

64. Julio-Pieper M, López-Aguilera A, Eyzaguirre-Velásquez J, et al. Gut susceptibility to viral invasion: contributing roles of diet, microbiota and enteric nervous system to mucosal barrier preservation. *Int J Mol Sci.* 2021;22(9):4734. https://doi.org/10.3390/ijms22094734.
65. Kane AV, Dinh DM, Ward HD. Childhood malnutrition and the intestinal microbiome. *Pediatr Res.* 2015;77(1–2):256–262. https://doi.org/10.1038/pr.2014.179.
66. Castro-Mejía JL, O'Ferrall S, Krych Ł, et al. Restitution of gut microbiota in Ugandan children administered with probiotics (*Lactobacillus rhamnosus* GG and *Bifidobacterium animalis* subsp. *lactis* BB-12) during treatment for severe acute malnutrition. *Gut Microbes.* 2020;11(4):855–867. https://doi.org/10.1080/19490976.2020.1712982.
67. Kumar A, Vlasova AN, Deblais L, et al. Impact of nutrition and rotavirus infection on the infant gut microbiota in a humanized pig model. *BMC Gastroenterol.* 2018;18(1):93. https://doi.org/10.1186/s12876-018-0810-2.
68. Tickell KD, Sharmin R, Deichsel EL, et al. The effect of acute malnutrition on enteric pathogens, moderate-to-severe diarrhoea, and associated mortality in the global enteric multicenter study cohort: a post-hoc analysis. *Lancet Glob Health.* 2020;8(2):e215–e224. https://doi.org/10.1016/S2214-109X(19)30498-X.
69. Johal SS, Lambert CP, Hammond J, James PD, Borriello SP, Mahida YR. Colonic IgA producing cells and macrophages are reduced in recurrent and non-recurrent *Clostridium difficile* associated diarrhoea. *J Clin Pathol.* 2004;57(9):973–979. https://doi.org/10.1136/jcp.2003.015875.
70. Battaglioli EJ, Hale VL, Chen J, et al. *Clostridioides difficile* uses amino acids associated with gut microbial dysbiosis in a subset of patients with diarrhea. *Sci Transl Med.* 2018;10(464):eaam7019. https://doi.org/10.1126/scitranslmed.aam7019.
71. Terveer EM, Vendrik KE, Ooijevaar RE, et al. Faecal microbiota transplantation for *Clostridioides difficile* infection: four years' experience of the Netherlands Donor Feces Bank. *United European Gastroenterol J.* 2020;8(10):1236–1247. https://doi.org/10.1177/2050640620957765.
72. Carlucci C, Jones CS, Oliphant K, et al. Effects of defined gut microbial ecosystem components on virulence determinants of *Clostridioides difficile. Sci Rep.* 2019;9(1):885. https://doi.org/10.1038/s41598-018-37547-x.
73. Li X, Gao X, Hu H, et al. Clinical efficacy and microbiome changes following fecal microbiota transplantation in children with recurrent *Clostridium difficile* infection. *Front Microbiol.* 2018;9:2622. https://doi.org/10.3389/fmicb.2018.02622.
74. Duan J, Meng X, Liu S, et al. Gut microbiota composition associated with *Clostridium difficile*-positive diarrhea and *C. difficile* type in ICU patients. *Front Cell Infect Microbiol.* 2020;10:190. https://doi.org/10.3389/fcimb.2020.00190.
75. Young VB, Schmidt TM. Antibiotic-associated diarrhea accompanied by large-scale alterations in the composition of the fecal microbiota. *J Clin Microbiol.* 2004;42(3):1203–1206. https://doi.org/10.1128/JCM.42.3.1203-1206.2004.
76. Rousseau C, Levenez F, Fouqueray C, Doré J, Collignon A, Lepage P. *Clostridium difficile* colonization in early infancy is accompanied by changes in intestinal microbiota composition. *J Clin Microbiol.* 2011;49(3):858–865. https://doi.org/10.1128/JCM.01507-10.
77. Lees EA, Carrol ED, Ellaby NAF, et al. Characterization of circulating *Clostridium difficile* strains, host response and intestinal microbiome in hospitalized children with diarrhea. *Pediatr Infect Dis J.* 2020;39(3):221–228. https://doi.org/10.1097/INF.0000000000002559 [PMID: 31876614].
78. Brown JR, Flemer B, Joyce SA, et al. Changes in microbiota composition, bile and fatty acid metabolism, in successful faecal microbiota transplantation for *Clostridioides difficile* infection. *BMC Gastroenterol.* 2018;18(1):131. https://doi.org/10.1186/s12876-018-0860-5.

79. Heeg D, Burns DA, Cartman ST, Minton NP. Spores of *Clostridium difficile* clinical isolates display a diverse germination response to bile salts. *PLoS One.* 2012;7(2): e32381. https://doi.org/10.1371/journal.pone.0032381.
80. Staley C, Kelly CR, Brandt LJ, Khoruts A, Sadowsky MJ. Complete microbiota engraftment is not essential for recovery from recurrent *Clostridium difficile* infection following fecal microbiota transplantation. *MBio.* 2016;7(6):e01965–16. https://doi.org/10.1128/mBio.01965-16.
81. Konturek PC, Koziel J, Dieterich W, et al. Successful therapy of *Clostridium difficile* infection with fecal microbiota transplantation. *J Physiol Pharmacol.* 2016;67(6):859–866.
82. Eglow R, Pothoulakis C, Itzkowitz S, et al. Diminished *Clostridium difficile* toxin a sensitivity in newborn rabbit ileum is associated with decreased toxin a receptor. *J Clin Invest.* 1992;90(3):822–829. https://doi.org/10.1172/JCI115957.
83. Qamar A, Aboudola S, Warny M, et al. *Saccharomyces boulardii* stimulates intestinal immunoglobulin a immune response to *Clostridium difficile* toxin a in mice. *Infect Immun.* 2001;69(4):2762–2765. https://doi.org/10.1128/IAI.69.4.2762-2765.2001.
84. Han SH, Yi J, Kim JH, Lee S, Moon HW. Composition of gut microbiota in patients with toxigenic *Clostridioides (Clostridium) difficile*: comparison between subgroups according to clinical criteria and toxin gene load. *PLoS One.* 2019;14(2):e0212626. https://doi.org/10.1371/journal.pone.0212626.
85. Seekatz AM, Young VB. *Clostridium difficile* and the microbiota. *J Clin Invest.* 2014;124(10):4182–4189. https://doi.org/10.1172/JCI72336.
86. Cho JY, Liu R, Macbeth JC, Hsiao A. The interface of *vibrio cholerae* and the gut microbiome. *Gut Microbes.* 2021;13(1):1937015. https://doi.org/10.1080/19490976.2021.1937015.
87. VanInsberghe D, Elsherbini JA, Varian B, Poutahidis T, Erdman S, Polz MF. Diarrhoeal events can trigger long-term *Clostridium difficile* colonization with recurrent blooms. *Nat Microbiol.* 2020;5(4):642–650. https://doi.org/10.1038/s41564-020-0668-2.
88. Midani FS, Weil AA, Chowdhury F, et al. Human gut microbiota predicts susceptibility to *vibrio cholerae* infection. *J Infect Dis.* 2018;218(4):645–653. https://doi.org/10.1093/infdis/jiy192.
89. Alavi S, Mitchell JD, Cho JY, Liu R, Macbeth JC, Hsiao A. Interpersonal gut microbiome variation drives susceptibility and resistance to cholera infection. *Cell.* 2020;181(7):1533–1546.e13. https://doi.org/10.1016/j.cell.2020.05.036.
90. Hsiao A, Ahmed AM, Subramanian S, et al. Members of the human gut microbiota involved in recovery from *vibrio cholerae* infection. *Nature.* 2014;515(7527):423–426. https://doi.org/10.1038/nature13738.
91. Levade I, Saber MM, Midani FS, et al. Predicting *Vibrio cholera*e infection and disease severity using metagenomics in a prospective cohort study. *J Infect Dis.* 2021;223(2):342–351. https://doi.org/10.1093/infdis/jiaa358.
92. Anderson MC, Vonaesch P, Saffarian A, Marteyn BS, Sansonetti PJ. *Shigella sonnei* encodes a functional T6SS used for interbacterial competition and niche occupancy. *Cell Host Microbe.* 2017;21(6):769–776.e3. https://doi.org/10.1016/j.chom.2017.05.004.
93. Desai MS, Seekatz AM, Koropatkin NM, et al. A dietary fiber-deprived gut microbiota degrades the colonic mucus barrier and enhances pathogen susceptibility. *Cell.* 2016;167(5):1339–1353.e21. https://doi.org/10.1016/j.cell.2016.10.043.
94. Lindsay B, Oundo J, Hossain MA, et al. Microbiota that affect risk for shigellosis in children in low-income countries. *Emerg Infect Dis.* 2015;21(2):242–250. https://doi.org/10.3201/eid2101.140795.

95. Gallardo P, Izquierdo M, Vidal RM, Soto F, Ossa JC, Farfan MJ. Gut microbiota-metabolome changes in children with diarrhea by diarrheagenic *E. coli*. *Front Cell Infect Microbiol*. 2020;10:485. https://doi.org/10.3389/fcimb.2020.00485.
96. O'Mahony L, Akdis M, Akdis CA. Regulation of the immune response and inflammation by histamine and histamine receptors. *J Allergy Clin Immunol*. 2011;128 (6):1153–1162. https://doi.org/10.1016/j.jaci.2011.06.051.
97. Gallardo P, Izquierdo M, Vidal RM, et al. Distinctive gut microbiota is associated with diarrheagenic *Escherichia coli* infections in Chilean children. *Front Cell Infect Microbiol*. 2017;7:424. https://doi.org/10.3389/fcimb.2017.00424.
98. Peña-Gonzalez A, Soto-Girón MJ, Smith S, et al. Metagenomic signatures of gut infections caused by different *Escherichia coli* pathotypes. *Appl Environ Microbiol*. 2019;85(24):e01820–19. https://doi.org/10.1128/AEM.01820-19.
99. Gigliucci F, von Meijenfeldt FAB, Knijn A, et al. Metagenomic characterization of the human intestinal microbiota in fecal samples from STEC-infected patients. *Front Cell Infect Microbiol*. 2018;8:25. https://doi.org/10.3389/fcimb.2018.00025.
100. Cordonnier C, Le Bihan G, Emond-Rheault JG, Garrivier A, Harel J, Jubelin G. Vitamin B12 uptake by the gut commensal bacteria *Bacteroides thetaiotaomicron* limits the production of Shiga toxin by enterohemorrhagic *Escherichia coli*. *Toxins (Basel)*. 2016;8(1):14. https://doi.org/10.3390/toxins8010014.
101. Terefe Y, Deblais L, Ghanem M, et al. Co-occurrence of *Campylobacter* species in children from eastern Ethiopia, and their association with environmental enteric dysfunction, diarrhea, and host microbiome. *Front Public Health*. 2020;8:99. https://doi.org/10.3389/fpubh.2020.00099.
102. Garber JM, Nothaft H, Pluvinage B, et al. The gastrointestinal pathogen *Campylobacter jejuni* metabolizes sugars with potential help from commensal *Bacteroides vulgatus*. *Commun Biol*. 2020;3(1):2. https://doi.org/10.1038/s42003-019-0727-5.
103. Engevik MA, Banks LD, Engevik KA, et al. Rotavirus infection induces glycan availability to promote ileum-specific changes in the microbiome aiding rotavirus virulence. *Gut Microbes*. 2020;11(5):1324–1347. https://doi.org/10.1080/19490976.2020.1754714.
104. Sohail MU, Al Khatib HA, Al Thani AA, Al Ansari K, Yassine HM, Al-Asmakh M. Microbiome profiling of rotavirus infected children suffering from acute gastroenteritis. *Gut Pathog*. 2021;13(1):21. https://doi.org/10.1186/s13099-021-00411-x.
105. Paparo L, Tripodi L, Bruno C, et al. Protective action of *Bacillus clausii* probiotic strains in an *in vitro* model of rotavirus infection. *Sci Rep*. 2020;10(1):12636. https://doi.org/10.1038/s41598-020-69533-7.
106. Stefan KL, Kim MV, Iwasaki A, Kasper DL. Commensal microbiota modulation of natural resistance to virus infection. *Cell*. 2020;183(5):1312–1324.e10. https://doi.org/10.1016/j.cell.2020.10.047.
107. Ramig RF. Pathogenesis of intestinal and systemic rotavirus infection. *J Virol*. 2004;78 (19):10213–10220. https://doi.org/10.1128/JVI.78.19.10213-10220.2004.
108. Bartnicki E, Cunha JB, Kolawole AO, Wobus CE. Recent advances in understanding noroviruses. *F1000Res*. 2017;6:79. https://doi.org/10.12688/f1000research.10081.1.
109. Walker FC, Baldridge MT. Interactions between noroviruses, the host, and the microbiota. *Curr Opin Virol*. 2019;37:1–9. https://doi.org/10.1016/j.coviro.2019.04.001.
110. Gozalbo-Rovira R, Rubio-Del-Campo A, Santiso-Bellón C, et al. Interaction of intestinal bacteria with human rotavirus during infection in children. *Int J Mol Sci*. 2021;22(3):1010. https://doi.org/10.3390/ijms22031010.
111. Hashimoto T, Perlot T, Rehman A, et al. ACE2 links amino acid malnutrition to microbial ecology and intestinal inflammation. *Nature*. 2012;487(7408):477–481. https://doi.org/10.1038/nature11228.

112. Viana SD, Nunes S, Reis F. ACE2 imbalance as a key player for the poor outcomes in COVID-19 patients with age-related comorbidities—role of gut microbiota dysbiosis. *Ageing Res Rev*. 2020;62:101123. https://doi.org/10.1016/j.arr.2020.101123.
113. Zhou Y, Shi X, Fu W, et al. Gut microbiota dysbiosis correlates with abnormal immune response in moderate COVID-19 patients with fever. *J Inflamm Res*. 2021;14:2619–2631. https://doi.org/10.2147/JIR.S311518.
114. Gu S, Chen Y, Wu Z, et al. Alterations of the gut microbiota in patients with coronavirus disease 2019 or H1N1 influenza. *Clin Infect Dis*. 2020;71(10):2669–2678. https://doi.org/10.1093/cid/ciaa709.
115. Lv L, Gu S, Jiang H, et al. Gut mycobiota alterations in patients with COVID-19 and H1N1 infections and their associations with clinical features. *Commun Biol*. 2021;4(1):480. https://doi.org/10.1038/s42003-021-02036-x.
116. Trottein F, Sokol H. Potential causes and consequences of gastrointestinal disorders during a SARS-CoV-2 infection. *Cell Rep*. 2020;32(3):107915. https://doi.org/10.1016/j.celrep.2020.107915.
117. Berry ASF, Johnson K, Martins R, et al. Natural infection with *Giardia* is associated with altered community structure of the human and canine gut microbiome. *mSphere*. 2020;5(4):e00670–20. https://doi.org/10.1128/mSphere.00670-20.
118. Riba A, Hassani K, Walker A, et al. Disturbed gut microbiota and bile homeostasis in *Giardia*-infected mice contributes to metabolic dysregulation and growth impairment. *Sci Transl Med*. 2020;12(565):eaay7019. https://doi.org/10.1126/scitranslmed.aay7019.
119. Barash NR, Maloney JG, Singer SM, Dawson SC. *Giardia* alters commensal microbial diversity throughout the murine gut. *Infect Immun*. 2017;85(6):e00948–16. https://doi.org/10.1128/IAI.00948-16.
120. von Huth S, Thingholm LB, Kofoed PE, et al. Intestinal protozoan infections shape fecal bacterial microbiota in children from Guinea-Bissau. *PLoS Negl Trop Dis*. 2021;15(3):e0009232. https://doi.org/10.1371/journal.pntd.0009232.
121. Maertens B, Gagnaire A, Paerewijck O, De Bosscher K, Geldhof P. Regulatory role of the intestinal microbiota in the immune response against *Giardia*. *Sci Rep*. 2021;11(1):10601. https://doi.org/10.1038/s41598-021-90261-z.
122. Ngobeni R, Samie A, Moonah S, Watanabe K, Petri Jr WA, Gilchrist C. *Entamoeba* species in South Africa: correlations with the host microbiome, parasite burdens, and first description of *Entamoeba bangladeshi* outside of Asia. *J Infect Dis*. 2017;216(12):1592–1600. https://doi.org/10.1093/infdis/jix535.
123. Leon-Coria A, Kumar M, Chadee K. The delicate balance between *Entamoeba histolytica*, mucus and microbiota. *Gut Microbes*. 2020;11(1):118–125. https://doi.org/10.1080/19490976.2019.1614363.
124. Larsen JM. The immune response to *Prevotella* bacteria in chronic inflammatory disease. *Immunology*. 2017;151(4):363–374. https://doi.org/10.1111/imm.12760.
125. Verma AK, Verma R, Ahuja V, Paul J. Real-time analysis of gut flora in *Entamoeba histolytica* infected patients of Northern India. *BMC Microbiol*. 2012;12:183. https://doi.org/10.1186/1471-2180-12-183.
126. Shaulov Y, Shimokawa C, Trebicz-Geffen M, et al. *Escherichia coli* mediated resistance of *Entamoeba histolytica* to oxidative stress is triggered by oxaloacetate. *PLoS Pathog*. 2018;14(10), e1007295. https://doi.org/10.1371/journal.ppat.1007295.
127. Morton ER, Lynch J, Froment A, et al. Variation in rural African gut microbiota is strongly correlated with colonization by *Entamoeba* and subsistence. *PLoS Genet*. 2015;11(11), e1005658. https://doi.org/10.1371/journal.pgen.1005658.
128. Burgess SL, Leslie JL, Uddin J, et al. Gut microbiome communication with bone marrow regulates susceptibility to amebiasis. *J Clin Invest*. 2020;130(8):4019–4024. https://doi.org/10.1172/JCI133605.

129. Gilchrist CA, Petri SE, Schneider BN, et al. Role of the gut microbiota of children in diarrhea due to the protozoan parasite *Entamoeba histolytica*. *J Infect Dis*. 2016;213 (10):1579–1585. https://doi.org/10.1093/infdis/jiv772.
130. Ras R, Huynh K, Desoky E, Badawy A, Widmer G. Perturbation of the intestinal microbiota of mice infected with *Cryptosporidium parvum*. *Int J Parasitol*. 2015;45 (8):567–573. https://doi.org/10.1016/j.ijpara.2015.03.005.
131. Lantier L, Drouet F, Guesdon W, et al. Poly(I:C)-induced protection of neonatal mice against intestinal *Cryptosporidium parvum* infection requires an additional TLR5 signal provided by the gut flora. *J Infect Dis*. 2014;209(3):457–467. https://doi.org/10.1093/infdis/jit432.
132. Chappell CL, Darkoh C, Shimmin L, et al. Fecal indole as a biomarker of susceptibility to *Cryptosporidium* infection. *Infect Immun*. 2016;84(8):2299–2306. https://doi.org/10.1128/IAI.00336-16.

CHAPTER SEVEN

Gut microbiome dysbiosis in inflammatory bowel disease

Shruti Lal[a], Bharti Kandiyal[a], Vineet Ahuja[b], Kiyoshi Takeda[c], and Bhabatosh Das[a,*]

[a]Molecular Genetics Laboratory, Infection and Immunology Division, Translational Health Science and Technology Institute, NCR Biotech Science Cluster, Faridabad, India
[b]Department of Gastroenterology and Human Nutrition, All India Institute of Medical Sciences, New Delhi, India
[c]Laboratory of Immune Regulation, Department of Microbiology and Immunology, Graduate School of Medicine, Osaka University, Suita, Japan
*Corresponding author: e-mail address: bhabatosh@thsti.res.in

Contents

Abstract

Inflammatory bowel disease (IBD) is a complex multi-factorial chronic relapsing disease of the digestive tract where dysbiosis of autochthonous intestinal microbiota, environmental factors and host genetics are implicated in the disease development, severity, course and treatment outcomes. The two clinically well-defined forms of IBD are Crohn's disease (CD) and ulcerative colitis (UC). The CD affects the local immune response of the entire gastrointestinal tract whereas the inflammation in UC is mainly restricted to the colonic mucosa. Prolong progressive inflammation due to CD and UC often lead to colonic cancer. In healthy individuals, the enormous taxonomic diversity and functional potency of gut microbiota including members from the bacterial and fungal microbiota

Progress in Molecular Biology and Translational Science, Volume 192
ISSN 1877-1173
https://doi.org/10.1016/bs.pmbts.2022.09.003

tune the host immunity and keep the gastric environment beneficial and protective. However, expansion of pathobionts, autochthonous microbes with the potency of pathogenicity in dysbiotic condition, in the gastrointestinal tract and subsequently enriched inflammatory microbial products in the gastrointestinal milieu attract different immune cells and activate aberrant host immune response which leads to excessive production and secretion of different cytokines that damage the colonic epithelial cells and manifest chronic inflammatory digestive disease. In the current chapter, we provided our updated understanding about the different bacterial and fungal pathobionts, their genomic and metabolic signatures, and geo-specific diversity of gut microbes linked with IBD across the globe at the molecular resolution. An improved understanding of IBD and the factors associated with the disease will be a boost for therapeutic development and disease management.

Abbreviations

ASC acute severe colitis
CD Crohn's disease
HDAC histone-deacetylase
IBD inflammatory bowel disease
ITS internal transcribed spacer
LPS lipopolysaccharides
MAM microbial anti-inflammatory molecule
PAMPs pathogen associated molecular patterns
SCFAs short chain fatty acids
UC ulcerative colitis

1. Introduction

Inflammatory bowel disease (IBD), a multi-factorial chronic relapsing disease of the digestive tract, comprising Crohn's disease (CD) and ulcerative colitis (UC), that leads to discrete inflammation in a specific part or all the part of digestive tract due to aberrant immune response.[1] IBD affects millions of people across the globe and imposes a huge burden on society by affecting the quality of life of the patients by impairing their ability to work and participate in social functioning.[2] IBD is defined by a multiple of gastrointestinal symptoms, including abdominal pain, diarrhea, constipation, or bloody stool. The IBD is a multifactorial disease caused by gut microbiome dysbiosis, environmental factors and mutations in the specific loci of the host genomes.[1] Perturbation of the gut microbiome, all the microbial taxa living in the gastrointestinal tract and the surrounding environment containing microbial and host derived molecules, appears to be one of the key risk factors for IBD.[3] Recent advances in the culture independent profiling of bacterial and fungal microbiota either by limited taxonomic resolution methods such as 16S ribosomal RNA (rRNA) gene sequencing or by high-resolution shotgun metagenome sequencing and in silico identification of functional potency of microbial genomes with pro-inflammatory or anti-inflammatory natures give a clue to the scientific community to proceed toward disease causality at the molecular level.[4] Species, subspecies and strain-level identification of microbial taxa living in the gastrointestinal tract shows signatures of bacterial and fungal taxa and their diversity in different parts of the globe.[5] Integrative analyses of metagenomes (collection of genomes of a microbial community), metatranscriptomes (the entire transcripts of microbial community living in a specific ecosystem), metaproteomes (pool of microbial proteins in a body habitat or environment), metabolomes (all the metabolites present in a specific samples) of human microbiome helps immensely in understanding at the molecular resolution the importance of gut microbiome dynamics over period and epidemiology of IBD in developed and developing countries. Reduce abundance of gut symbionts and expansion of disease specific bacterial and fungal pathobionts interrupt the immune homeostasis in the gut and active aberrant immune response.[6] Abundant metagenomes but little or no expression of biosynthetic functions with anti-inflammatory or pro-inflammatory potency was a major bottleneck in the identification of pseudo pathobionts in IBD. However, due to rapid expansion and use of proteomics and metabolomics in the clinical studies for decoding the genomic potency and functionality it has been well established that the keystone bacterial and fungal pathobionts associated with pathogenesis of IBD play a crucial role in disease severity and treatment outcome.

2. Global epidemiology of inflammatory bowel disease

The disease burden of IBD is very heterogeneous across the globe, and it varies significantly both within and between different regions of geography and ethnicity.[2] According to the Centres for Disease Control and Prevention, the prevalence of IBD in the adult population of the United States of America (USA) in 2015 was close to 1.5% (https://www.cdc.gov/ibd/data-and-statistics/prevalence.html). However, the disease burden of IBD is very heterogeneous in the USA and multiple factors including sociodemographic characteristics, ethnicity, and urbanity play an important role in IBD epidemiology.[7] IBD is very common in European descendants and the highest rates of IBD have been reported in Europe and North America. In comparison to Asian and African descendants, the prevalence of IBD in European descendants is very deep-rooted, particularly in the population living in industrialized areas. In the last three decades, the disease burden of CD in Denmark increased from 5.2 to 9.1 per 100,000, while the incidence of UC in Denmark rose from 10.7 to 18.6 per 100,000 from 1980 to 2013. In France, the incidence of CD increased from 4.2 to 9.5 per 100,000 while the incidence of UC rose from 1.6 per to 4.1 per 100,000. Recent studies have shown the rapid increase of the IBD burden in Asia, including India.[8] In Asia, the disease burden of IBD increased from about 166 thousand cases in 2017 to nearly 220 thousand cases in 2020.[9] A recent prediction indicated that the disease burden of IBD would be expanded to nearly a half-million cases in 2035. In China, the disease burden of IBD is highest in the Guangzhou region (3.4/100,000) followed by Hong Kong (3.06/100,000).[10] In Taiwan, IBD incidence doubled from 4.54/100,000 to 9.39/100,000 in 10 years (2000–2010). The disease burden of IBD in western Asia, including Kuwait and Israel, is 2.8/100,000 and 5.04/100,000, respectively. Among Asian countries, the disease burden of IBD is also very high in India (9.31/100,000). Different geo-epidemiological observations revealed that within India, the southern part has a higher prevalence of UC compared with the northern part of the country. In addition, socioeconomic status, sedentary life, and the increased use of westernized food in the urban population added extra fuel to the IBD burden.

3. Clinical features of inflammatory bowel disease

The integrity of the intestinal epithelial cells, recognition of microbial metabolites or dietary components, production of antimicrobial peptides, optimal permeability of the gastric epithelial barrier, and differentiation and renewal of the intestinal epithelial cells are key to maintaining the intestinal homeostasis and outlined healthy gut. Patients suffering from IBD couldn't maintain homeostasis in the gastrointestinal tract and manifested aberrant immune responses.[11] IBD is a chronic intermittent relapsing progressive inflammatory disorder of the gastrointestinal tract characterized by (i) gastric mucosal inflammation (ii) chronic diarrhea (iii) abdominal pain (iv) rectal bleeding and fatigue. Gastric inflammation in CD has a segmented pattern across the gastrointestinal tract, but the terminal ileum is the most frequently affected organ. In the case of UC, the most frequently affected organ is the rectum, but sometimes the inflammation spreads across the colon.

In addition to the above-mentioned common symptoms, spontaneous repeated flares and remissions of inflammation have also been reported in many adult patients suffering from severe IBD. Extra-intestinal manifestations of IBD are inflammations in the skin (erythema nodosum, pyoderma), eye (uveitis, iritis, conjunctivitis), gallbladder (cholecystolithiasis), liver and bile duct (primary sclerosing cholangitis, liver damage), kidney (nephrolithiasis), bone (osteoporosis) or different joints (arthralgias and arthritis).[12] Aberrant activation and excess production of cytokines and mediators from immune cells substantially contribute to the IBD-associated complications, including ulcer formation, cancer, fistulas, and stenosis.[13] Immune cells involved in innate immune responses like neutrophils, ILCs, T cells, neutrophils, macrophages, and DCs are often migrated to the inflamed region of the gastrointestinal tract and mediate tissue damage. In cases of extra-intestinal manifestations, mostly T cells migrate to the lymph nodes and bloodstream, causing tissue destruction and organ damage. However, the disease symptoms, degree of severity, and manifestations of IBD are heterogeneous, and multiple factors contribute to the heterogeneity.

4. Four major factors linked with inflammatory bowel disease

4.1 Gut microbiome

The human gastrointestinal tract harbors trillions of microbes that represent bacteria, fungi, viruses, archaea, and protozoa.[14,15] Microbes from each domain are very diverse, both in terms of their genomic repertoires and functional potency. Functions linked to xenobiotic metabolism, virulence, inflammation, and fitness are dynamics, often linked with mobile genetic elements, and can disseminate between different members through horizontal gene transfer.[16] With time, humans changed their lifestyles, food habits, and were exposed to different classes of xenobiotics depending on their living environment. Such factors have a direct impact on the composition, growth and functions of autochthonous and allochthonous microbiota including pathobionts that may be involved in disease development, progression, severity, and treatment outcomes.[6] Host genetics or environmental factors that provide selective advantages to the pathobionts over commensals and symbionts by selective recognition of pathogen associated molecular patterns (PAMPs) may be associated with the rising frequency of IBD in developing countries undergoing industrialization and westernization.

4.1.1 Gut bacteria

Hundreds of association studies in human and model organisms have shown varying degrees of gut microbial dysbiosis, more specifically the perturbation of the bacterial component of the gut microbiota, in patients suffering from IBD (Table 1; Fig. 1). Many clinical studies adopted targeted metagenomics and decoded partial DNA sequences of the 16S rRNA genes in the

Table 1 Bacterial and fungal dysbiosis in the gastrointestinal tract of IBD patients.

Name of the microbes	Relative abundance in IBD	Important molecules/functions	Reference
Faecalibacterium prausnitzii	Decreased	Short-chain fatty acids, Microbial anti-inflammatory molecule	17
Prevotella copri	Decreased	Butyrate and propionate	17
Ruminococcus, Roseburia, Coprococcus, Blautia, Eubacterium and *Dorea*	Decreased	Short-chain fatty acids	17
Bifidobacterium adolescentis	Decreased	Tryptophan production	18
Ruminococcus gnavus	Increased	Production of mucolytic enzymes	19
Alistipes massiliensis	Increased	Inflammation	17
Alistipes putredinis	Increased	Hydrolyze tryptophan to Indole	20
Escherichia coli	Increased	Lipopolysaccharides	18
Serratia marcescens	Increased	Lipopolysaccharides	19
Bacteroides fragilis	Increased	Polysaccharide A	21
Bacteroides vulgatus	Increased	Succinate	22
Streptococcus anginosus	Increased	Group G antigens	23
Aggregatibacter segnis	Increased	Short-chain fatty acids	24
Clostridium difficile	Increased	Exotoxins (toxin A and toxin B)	25
Helicobacter hepaticus	Increased	Through recruitment of neutrophils	26
Klebsiella pneumoniae	Increased	Haemolysin co-regulated protein	27
Saccharomyces cerevisiae	Decreased	Anti-inflammatory	19
Candida albicans	Increased	Candidalysin	28
Candida tropicalis	Increased	Associated with anti-*S. cerevisiae* antibodies	19

Table 1 Bacterial and fungal dysbiosis in the gastrointestinal tract of IBD patients.—cont'd

Name of the microbes	Relative abundance in IBD	Important molecules/functions	Reference
Aspergillus clavatus	Increased	Cytochalasin E	29
Cryptococcus neoformans	Increased	Laccase	30
Malassezia sympodialis	Increased	Ligand of CARD9	31

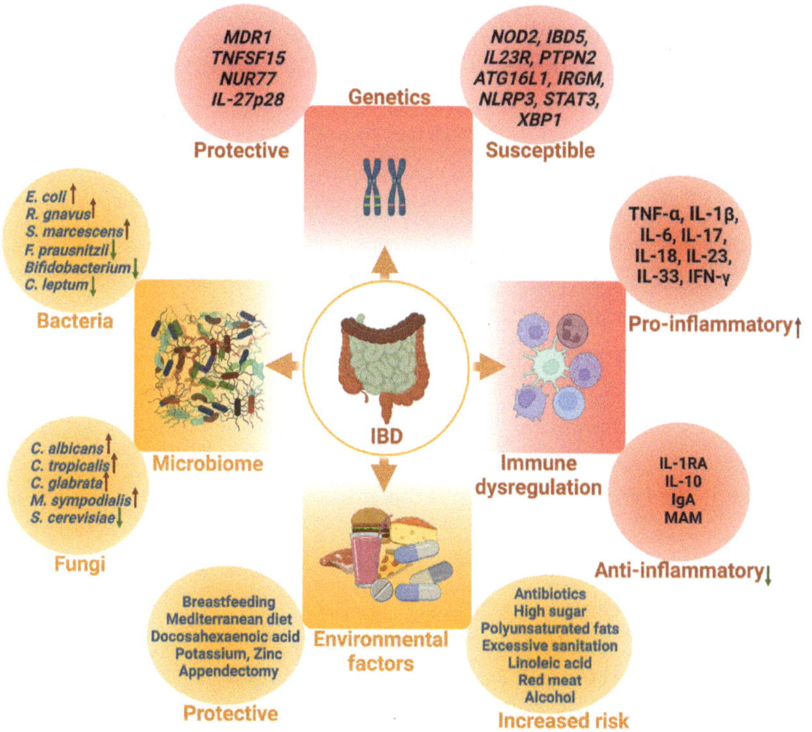

Fig. 1 Multiple factors that are linked with Inflammatory Bowel Disease. Microbiome dysbiosis at the taxonomic resolution are represented by up and downward arrows. Both bacterial and fungal dysbiosis are important in IBD progression, severity and treatment outcomes. Host genetics associated with innate immunity, adaptive immunity, T-cell regulation, mucosal immunity, immune tolerance, autophagy, Apoptosis, epithelial barrier functions are linked with disease susceptibility as well as resistance against IBD. Dietary components including refined rice, high- fat foods and dietary additives are also potential risk factors for IBD. B cell development and differentiation, and expansion of regulatory T (T_{reg}) cells, maintenance of mucosal integrity and barrier functions and production and activation of cytokines activation of dendritic cells and macrophages could also increase the susceptibility of IBD. IL, Interleukin; IFN, Interferon; TNF, Tumor necrosis factor; MAM, Microbial Anti-inflammatory Molecule. Microbial species: *R. gnavus, Ruminococcus gnavus; S. marcescens, Serratia marcescens; F. prausnitzii, Faecalibacterium prausnitzii; C. leptum, Clostridium leptum; C. albicans, Candida albicans; C. tropicalis, Candida tropicalis; C. glabrata, Candida glabrata; M. sympodialis, Malassezia sympodialis; S. cerevisiae, Saccharomyces cerevisiae.*

biological samples and predicted microbial richness, diversity, and dynamics during the course of treatments.[33] Phylogenetic Investigation of Communities by Reconstruction of Unobserved States (PICRUSt) algorithms are widely used to overcome the limited resolution of 16S rRNA based studies and understands the functional potency of the community of microbial taxa. Few association studies have used multi-omics platforms and different animal models and extended to identify the microbial and host functions linked with the IBD and their potential mechanisms in disease development and progression.

A gut microbiome study in acute severe colitis (ASC) and moderate ulcerative colitis (UC) from our group identified reduced richness and normalized alpha diversity, and high inter-individual heterogeneity in the patients' population compared to the healthy controls.[32] Reduced abundance of Firmicutes such as *Faecalibacterium prausnitzii* and other short-chain fatty acids (SCFA) producing bacteria, and increased abundance of Proteobacteria is a clear compositional signature in UC reported by several groups across the globe.[34] Reduced abundance of *F. prausnitzii* is related to a higher incidence in endoscopic recurrence of active CD in adult patients. In addition, patients suffering from IBD contain a lower number of *Prevotella, Alloprevotella, Dialister, Anaerovorax, Sporobacter, Slackia, Butyricococcus, Enterorhabdus, Howardella, Subdoligranulum, Bilophila, Ruminococcus, Lachnospiraceae, Roseburia, Oscillibacter,* and *Clostridium XIVb*. Bacterial taxa including *Escherichia-Shigella, Peptococcus, Enterobacter, Staphylococcus, Listonella, Asteroleplasma Pseudomonas,* and *Actinomyces* are often overcrowded in the intestine of UC patients.[35] Principal component analysis of microbiota in the fecal samples of UC, CD, and healthy controls clustered more than 150 bacterial species separately from each other's, with CD representing the most dysbiotic state compared to other groups.

The protective effects of gut microbiota are mostly due to anti-inflammation activity or modulation of host signaling through the binding of a specific receptor in the intestinal epithelial cells. *F. prausnitzii* synthesizes and secretes bioactive peptides derived from a single protein known as a microbial-anti-inflammatory molecule (MAM). The MAM inhibits nuclear factor (NF)-κB and reduces the production of IL-1β induced IL-8 by intestinal epithelial cells.[36] A similar MAM-like protein having a signature GGDEF domain has also been identified in the genome of another commensal Firmicute *Roseburia intestinalis*. *Lactobacillus rhamnosus* GG, commensal firmicutes, also secrete two bioactive proteins p75 and p40, which reduce epithelial cell apoptosis by inhibiting functions of pro-inflammatory cytokines. Other than bioactive peptides, anti-inflammatory small molecules, molecular weight <10 kDa, are also identified from *Saccharomyces boulardii* and *Bacteroides thetaiotaomicron,* which functions by interacting with host NF-κB pathway.

Pathobionts mostly induced inflammation by producing endotoxin and other metabolites that act as a ligand for different cell-surface receptors like TLR, G-PCR, enzyme-linked receptors and induce cytokine productions in lymphocytes, PMC, and other immune cells in the innate immune system as well as.

Pseudopathobionts: The gastrointestinal tract harbors several pathobionts such as *Helicobacter pylori, C. difficile, H. hepaticus*, SFB, EHEC, *P. mirabilis, Klebsiella pneumoniae*, and *Enterococcus* spp. that are directly involved in the pathogenesis of CD by inducing inflammation. However, metagenomic analysis of the microbial community in the fecal samples of IBD patients revealed that the abundance of some bacterial species like *Dialister invisus* is significantly high in the disease conditions, but functional analysis of the metabolic pathways by monitoring transcript level of relevant biosynthetic pathways shown with low or non-existent gene expression. Such bacterial taxon is either dormant or inactive and their numbers are increased because of the disease specific environment and could be recognized as pseudopathobionts.

4.1.2 Gut fungi

The fungal microbiota known as mycobiota is an important component of the human gut microbiota. Mycobiota coexist with other microbiota in the gastrointestinal tract with a mutual understanding and sharing complementary metabolic functions.[5] In addition, mycobiota plays an important role in host immune modulation by directly interacting with helper T cells (T_H17) and other lymphocytes and immune cells.[37] Deep sequencing of the internal transcribed spacer (ITS) region, genomics, culturomics, and functional analysis of isolated fungal strains tremendously helps in understanding the immunoreactivity of fungal-derived antigens and metabolites in inflammation. A recent study on the gut microbiome of healthy Indian and Japanese populations uncovered that *Candida albicans, C. tropicalis*, and *C. glabrata* are very common in the gut of healthy Indians while *Saccharomyces cerevisiae* is the most prevalent fungus in the gut of healthy Japanese population.[5]

Interestingly, multiple pieces of evidence indicate that increased abundance of several species of *Candida* (*C. albicans, C. tropicalis*, and *C. glabrata*) and *Malassezia* (*M. restricta* and *M. globosa*) have immune cell damaging potency and strongly induce local inflammation in the gastrointestinal tract through IL-1β dependent pathways (Table 1; Fig. 1). At the taxonomic resolution it was observed that an increased abundance of *Candida* and *Malassezia* is often associated with UC, although the association of *Malassezia* with CD has also recently been reported.[38] At the molecular resolution, it was observed that candidalysin, a peptide secreted by *C. albicans*, is majorly responsible for the pathogenicity of UC.[39] IL-1β is a crucial player in inflammation in the gastrointestinal tract and its increased level is consistently detected in the rectal biopsy of patients suffering from UC. Increased abundance of *C. albicans* in the colon and *a* higher number of IL-1β producing phagocytes are positively correlated with disease severity of UC. Increased abundance of *Malassezia* is observed in the surgical mucosal samples of patients suffering from CD. Polymorphism in the CARD9 is crucial for *Malassezia* mediated inflammation in CD patients. However, the *Malassezia* derived antigens or metabolites, which interact with CARD9 are yet to be identified. Anti-*S. cerevisiae* mannan antibodies (ASCA) have also been used as biomarkers to predict CD. Mannan, a natural polysaccharide derived from baker's yeast, is recognized by the Toll-like receptors (TLR) and macrophage mannose receptor (MR) and strongly induces inflammation through the innate immune system.[40] A recent study from our group (unpublished data) indicates a reduced abundance of specific strains of *S. cerevisiae/paradoxus* and an increased abundance of *Aspergillus* spp. and *Alternaria tenuissima* may link with the IBD pathogenesis of the Indian patients.

4.2 Host genetics

Genome-wide association studies (GWAS) have identified more than 200 genetic polymorphisms (SNPs and In-del) mostly linked with the functions of innate immunity (CCL11, CCR6, IL8RA, MST1), adaptive immunity (IL5, BACH2, IL7R, IRF5), T-cell regulation (NDFIP1, TNFSF8, TAGAP, IL2, IL2R, IL7R, PRDM1, IFNG, IL23), mucosal immunity (CARD9, NOD2, ITLN1, REL, SLC11A1), immune tolerance (IL10, IL27, IL1R1, NOD2, CREM), autophagy (ATG16L1, IRGM, NOD2, CUL2, DAP, PARK7), Apoptosis (FASLG, THADA, DAP, PUS10), epithelial barrier functions (GNA12, CDH1, MUC19, ITLN1, ERRFL1), etc., are linked with the increased risk of IBD.[1] The study of IBD genetics has made tremendous progress in the past decades with the availability of multiple deep sequencing platforms, which have substantially improved in variant or SNP identification in large cohorts of well-phenotyped patients and control samples. Genetic defects due to spontaneous mutation in the susceptible loci of IBD patients mostly modulate immune perturbation at the system-level in the mucosal immune system. Such mutations also affect the commensal ecosystem by enabling pathobiont expansion.

4.3 Host Immunobiology

One of the most important factors associated with chronic inflammation and gastric ailment in IBD is an abrupt immune response in the gastrointestinal tract. In an ailment-free host, there may be excellent stability between a shielding and a deleterious reaction of the immune system, which will become perturbed in patients with IBD.[41] The immune cells (T-helper 1 and T-helper 17 in CD and T-helper 2 in UC), cytokines (Tumor Necrosis Factor, transforming growth factor, and interleukins IL-12, IL-17, and IL23), chemokines, reactive oxygen species (ROS), neuropeptides, and non-immune cells (myeloid, epithelial, mesenchymal) are the ones responsible to mediate the inflammatory responses.[42]

4.3.1 Cytokines

In the most recent genetic and immunological research, cytokines have received special attention and been directly linked in IBD pathogenesis where they have an important role in controlling intestinal irritation and the associated clinical symptoms of IBD. Cytokines appear to have an essential role in the pathogenesis of progressive and unfavorable styles of IBD, which are related to complications such as intestinal stenosis, rectal bleeding, abscess and fistula formation, and the improvement of colitis-related neoplasias.[43,44] The GWASs have pointed to many IBD susceptibility loci including genes produce cytokine, chemokine receptor signaling, and T helper (TH) cell-mediated responses. Further research has identified IBD risk loci that include genes that encode cytokines (for example, IL-2, IL-21, interferon-γ (IFN-γ), IL-10, and IL-27), which highlighted the central role of cytokines in disease pathogenesis.[45] The outcome of therapeutic applications of anti-inflammatory cytokines, which include IFNβ, IL-10, and

IL-11, was not very encouraging. However, the first breakthrough came with the scientific use of a neutralizing antibody unique to TNF (infliximab) in sufferers with CD. Anti-TNF therapy led to significant scientific development and macroscopic restoration of the infected mucosa detected in endoscopy (mucosal restoration) in patients with CD. Since there have been comparable similarities in patients with UC, anti-TNF therapy with numerous chimeric, humanized, which includes adalimumab, certolizumab pegol, golimumab, and infliximab, is now taken into consideration as an essential backbone of biological therapeutics in IBD.[46,47] In patients suffering from IBD and in the experimental mouse models of colitis, pro-inflammatory and anti-inflammatory cytokines are produced with the help of numerous cells of the mucosal immune system in response to environmental triggers, which include commensal microorganisms. In particular, dendritic cells (DCs), neutrophils, macrophages, natural killer (NK) cells, intestinal epithelial cells (IECs), innate lymphoid cells (ILCs), mucosal effector T cells (T helper 1 (TH1), TH2 and TH17) and regulatory T (TReg) cells are the common producers of different cytokines. In intestinal homeostasis condition the appropriate balance between pro-inflammatory and anti-inflammatory cytokines are maintain by plethora of signaling systems.[48] Novel strategies to target cytokine pathways and cytokine signaling cascades are being evolved as keys for the treatment of IBD.

4.3.2 Barrier function

The mucosal barrier has three foremost components, the mucus layer, the epithelial glycocalyx, and the surface epithelium itself. The integrity of barrier function lies largely upon tight junction characteristics. In healthy subjects, there may be minimal direct interaction between the microbiome present in the lumen and the epithelium—the non-stop mucus layer in the colon is responsible to keep the surface epithelium out of contact with the microorganisms, and the microbe-free environment of the distal small intestine is guaranteed by the ileocaecal valve. While a few IBD patients can also have additional genetic elements responsible for the weakening of the mucosal barrier, it is probable that environmental elements can be even more critical. Some are diffuse and indirect, such as the consequence of pressure on the mucosa barrier, whereas others can be more apparent, such as the impact of pathogen-associated gastroenteritis, which is frequently recognized as a cause of IBD relapse.[49] The strategies adopted by intestinal pathogens to infect the intestinal epithelium are evidence of how the penetration of the epithelia by commensal organisms is inhibited by the mucus barrier. Intestinal pathogens have developed mechanisms to bypass or avoid the secreted mucus barrier. Being motile, they can pass through mucus, and often bind with mucin carbohydrates, encoding enzymes that damage the mucin polymer community facilitating the penetration of mucus.[50,51] The dome epithelium, however, is devoid of goblet cells and is consequently lacking a secreted mucus barrier along with altered expression of cell surface glycoproteins. Although the dome epithelium appears to lack host defenses, for mammals, especially given the potentially lethal nature of gastrointestinal infections, maintaining controlled immunity to common flora is a critical requirement. We can conclude that access has proven to be more important than completely blocking potential pathogens. In the assessment of the non-inflammatory sIgA response set up toward commensal flora, where the pathogens pass the barrier and destroy the epithelium, an inflammatory reaction is generated which is more comparable to the persistent inflammatory reaction in IBD.[52]

Disturbances in the barrier characteristics in human IBD include the reduction of the barrier and antimicrobial secretions, decreased abundance of secretory cells, elevated permeability, disabled tight junctions, vast reduction, or even absence of the epithelium in which ulceration occurs. While one can also infer that most of those are outcomes rather than reasons for irritation, the records from animal models show that barrier dysfunction should essentially be a primary disorder in a few cases of IBD. Even if this isn't the case, those functions will cause inflammation and complicate the resolution of lesions. The mucus layer in IBD seems to be more incredibly inhabited by microorganisms, specifically in regions in which there may be minimum or less excessive inflammation.[53] Therapeutic recovery of the barrier function should reduce the pathophysiology and clinical symptoms in both CD and UC. Immunomodulating capsules and biologics are likely to behave in elements without delay or in a roundabout way through the recovery of a few elements of barrier functions, and similarly, studies are necessary to comprehend the significance of those factors in their activity.[54]

4.3.3 Antimicrobial peptides

The human body secretes antimicrobial proteins (AMPs) as an innate immune response to killing or limiting the growth of microorganisms.

AMPs are a class of peptide antibiotics that function as key effectors of innate immunity. AMP is normally found in the intestinal mucosa where it is in constant contact with the gut microbiota.[55] Different varieties of AMPs produced have one-of-a-kind associations with IBD. A genetic variant within the TCF4 promoter region that's a form of α-defensin- is mentioned to be related to CD.[56] Some researchers say protease inhibitors are closely linked to the development of IBD. For example, an excessive elafin expression became notable in the intestinal tissues of sufferers of ulcerative colitis.[57] The colonic epithelial cells of patients with IBD have an accelerated Human hepatocarcinoma-gut-pancreas/pancreatitis-associated protein (HIP/PAP) expression.[58] Fecal lactoferrin concentration accelerated in patients with IBD and remained the same in patients with irritable bowel syndrome. Therefore, the lactoferrin stage is used for the diagnosis of IBD. Measurement of lactoferrin is simple, inexpensive, and very stable in fecal samples. Furthermore, decreased fecal lactoferrin concentrations were associated with healing responses to mucosal agents.[59,60] When lactoferrin was given orally to a model with DSS-induced colitis, the signs of colitis were increased in keeping with doses, and cytokine disbalance was resolved.[61] Increased levels of hepcidin in chronic inflammatory situations like IBD blocks the movement of iron from macrophages, resulting in the reduction of the iron level of the blood serum, leading to anemia. Since hepcidin reduces iron usage by microorganisms, the increase of pathogens can be prohibited.[62] IBD-associated anemia and colitis symptoms may be reduced by the inhibition of hepcidin expression.

AMPs expressed with the aid of pathogen contamination can damage the pathogen itself thus neutralizing immoderate immune reactions. However, the unusual expression of antimicrobial proteins ought to raise the chances of disturbances in inflammation regulation and autoimmune disease. Thus, we are able to say that interactions among AMPs and IBD are an essential key to the remedy of IBD and are presently understudied for the same.

4.3.4 Mycobacteria restriction functions

Mycobacterium avium paratuberculosis has been involved in Crohn's pathogenesis keeping Dalziel's original description of the condition in 1913 in consideration.[63] The link is primarily based upon the similarity of Crohn's and Johne's ailment (a bovine condition), signified by granulomatous inflammation of the intestines.[64] The literature covering MAP and Crohn's disease is considerable. There were clashing reviews regarding MAP DNA positivity from Crohn's biopsies; however, it's been assumed that this could be because of MAP DNA moving intracellularly in the human cell cytoplasm; as a result, the fake negativity might exist.[65]

Critics of the theory that MAP is the cause put forward many arguments. Most importantly, anti-TNF antibodies were highly effective in treating Crohn's disease, but have been implicated in latent tuberculosis reactivation.[66] Moreover, the considerable majority of research using numerous strategies has detected MAP DNA or cultured it at a better rate from tissues of patients suffering from Crohn's disease than from patients with ulcerative colitis and different disorders, even though the mentioned healing frequency in both Crohn's disease and ulcerative colitis ranges from 0% to 100%.[66] These consequences are constant with possibilities: either Crohn's disease is caused by MAP infection in a subset of patients who are both particularly vulnerable to this organism or who are genetically at risk of contamination, or this can additionally colonize the ulcerated mucosa of the patients of the Crohn's disease; however, it now no longer causes or perpetuates intestinal irritation. Even, if possible, the transmission of MAP occurs, a conceivable mechanism of tissue damage and induction of continuous intestinal irritation has now no longer been proposed. Even advocates of the notion that MAP explains Crohn's disease concede that infection, if present, lacks histochemical proof of acid-fast staining in Crohn's disease tissues and the bacterial load observed is very low. Although current records do not compulsorily blame MAP for Crohn's disease, they do not necessarily exclude this possibility.[66]

4.3.5 The role of short-chain fatty acids in IBD progression

The short-chain fatty acids (SCFAs) consist of 2–6 carbon atoms and are solely synthesized by the microorganisms that live in the human gastrointestinal tract.[67] The non-digestible starch, plant-derived inulin, white bran, oat bran, cellulose, pectin access carbohydrates serve as a major substrate for the synthesis of SCFA.[67] However, the most abundant SCFAs in the gut were acetate C2, propionate C3, and butyrate C4 in a 3:1:1 M ratio, accounting for approximately 95% of the SCFA population.[68,69] Within the gut microflora, acetate and propionate are mainly produced by Bacteroides, while firmicutes are responsible as a major contributor to butyrate production. A fewer gastrointestinal tract microflora, including *F. prausnitzii, Eubacterium rectale, Eubacterium hallii and Ruminococcus bromii*, are engaged in the majority of butyrate production.[70] Both of these are known as SCFAs producers at the phylum level.[71] SCFAs, mostly propionate and butyrate, are well known for their anti-inflammatory, anti-anxiety, and anticancer effects. Several studies have used SCFAs to establish its role in the prevention and mitigation of

IBD, hypertension, and coronary artery diseases. Furthermore, when comparing the level of SCFAs in the fecal samples between healthy subjects and IBD patients it was observed that their concentration has significantly decreased patient population.[70,72] A recent human-based study has shown a reduced abundance of *F. prauntizi* and *Roseburia hominis* species, the major producers of SCFAs, in the gastrointestinal tract could be used as characteristic markers of ulcerative colitis in IBD patients.[73]

By regulating the activity of the G protein-coupled receptors GPR-41, GPR-43, and GPR-109a, SCFAs including acetate, propionate, and butyrate contribute significantly to host physiology and gut homeostasis.[74] Due to the small size and simple structure, SCFAs can easily enter the nucleus and inhibit the activity of the histone-deacetylase (HDAC) enzyme that causes chromatin condensation, which results in the suppression of target genes implicated in the up-regulation of pro-inflammatory pathways.[75] GPCR-43 receptors are mainly found on various immune cells such as basophils, neutrophils, eosinophils, and dendritic cells and are reported to modulate the immune system via SCFA signaling.[76,77] In an experimental murine colitis model, inhibiting HDACs reduced the severity of colitis progression and improved pro-inflammatory cytokines such as IL-6, TNF-α, and IFN-γ.[78] In the DSS-colitis paradigm, GPR 43 and GPR 109a knockout mice had more severe colitis, demonstrating their combined function in promoting gut immunity.[79]

4.3.6 LPS: An intrinsic factor promoting the progression of IBD

Lipopolysaccharides (LPSs) are the carbohydrate-lipid conjugate molecules located on the outer membrane of Gram-negative bacteria. Two dominant bacterial phyla, Bacteroidetes and Proteobacteria, are the major sources of LPS in the gastrointestinal tract. However, the toxicity and inflammation levels of LPS are heterogeneous and mostly depend on the source bacterium. For example, the LPS of *Bacteroides* has minimal or no inflammatory property, while the LPS released by *E. coli* is highly toxic and can enhance the high level of secretion of calprotectin protein in the feces.[80] In addition to the source bacterium, the toxicity of LPS is also modulated by the surrounding gut microbiota by modulating host physiology and facilitating LPS entry into systemic circulation. Metabolites derived from a set of gut microbial taxa lead to the leakiness of the gut, which allows the translocation of the toxic LPS from the gut into the bloodstream and triggers the activation of a non-gut-associated population of immune cells.[81] It has been reported that in the IBD population, an elevated amount of LPS and

LPS-binding receptors were present in the serum in comparison to a healthy individual that can elicit strong immunological responses.[82] Innate immune cells were found to be the major interactors of LPS within the gut. Among the various TLR receptors, TLR-4 has been reported as the most potent receptor for LPS. However, the TLR-4 receptor has two major domains known as myeloid differentiation: primary response gene 88 and TIR domain-containing activator adaptor protein.[83] TLR-4 combined with the other moieties such as MD-2, LBP and CD-14 represent a new form of receptor called PRR (Pathogen recognition receptor). LPS interacts with LPS-binding protein (LBP), and CD14 then recognizes it. The TLR-4 signaling is activated by the LPS-LBP-CD14 complex via the adaptor proteins MyD88, IL-1R-associated kinase 4 (IRAF4), and TNF receptor-associated factor 6 (TRAF6). This route induces the generation of cytokines and activates NF-κb and mitogen-activated protein kinases.[84] TLR4's aberrant expression in the intestinal epithelium may be linked to the onset and development of IBD. The mucosa in the non-IBD control group lacks or expresses TLR minimally.[85] Increased TLR-4 expression in IBD patients may be caused by impaired host tolerance to luminal LPS antigens.[85,86] Tan et al.[76,77] found that in the inflamed and injured intestinal mucosa, TLR4-positive macrophages were restricted to regions of inflammatory intestinal mucosa of the UC patients, further enhancing hyperreactivity to LPS. There are strong associations in UC patients between disease activity, histopathological scores, fecal microbial score, and increased TLR-4 mRNA expression.[76,77]

4.3.7 Effects of diet: A major contributor in the alleviation of IBD

The gastrointestinal tract symptoms of nearly two-thirds of IBD patients are thought to be caused by irregular eating habits.[87] Gastrointestinal tract symptoms of IBD patients have been reported made worse by partially absorbed carbs (found in foods like lactose found in dairy products, beans, onion, cabbage, apples, and wheat), fatty meals, coffee, alcohol, and spicy foods.[87]

4.3.7.1 The specific carbohydrate diet (SCD)

The first time specific carbohydrate diets were devised for IBD was in the 1920s. In the treatment and management of IBD they were later utilized as a therapeutic agents.[88] Suskind[89] notes that monosaccharides (glucose, fructose, and galactose) found in fruits, nuts, honey, and completely fermented yogurt are the only carbs allowed in the diet.[89] In a study of 50 IBD patients who were self-managing with SCD, 66% of patients experienced a clinical remission after around 10 months of following the dietary plan. Additionally, several patients were able to stop receiving corticosteroid medication.[90] The authors particularly point out alterations in the gut microbiota they had previously noticed as a contributing factor to the beneficial outcomes they had noticed.

It is believed that SCD transforms the intestinal microbiota from a pro-inflammatory condition to a non-inflammatory state and reduces intestinal inflammation.[91] It is advised that patients follow the diet for up to 1 year while their condition is still active and for an additional year after their symptoms have subsided.

4.3.7.2 The low FODMAP diet

The term FODMAP refers to fermentable oligosaccharides, disaccharides, monosaccharides, and polyols.[92,93] concurs that nutritional therapy, particularly the low FODMAP diet, ought to be regarded as the first-line treatment for IBS patients. Moreover, the FODMAP diet has three major characteristic features that include: Poor absorption in the region of the small intestine; short-chain osmotically active, and fast fermentable.[92,93] Low FODMAP foods are those that lack lactose in dairy products, have low fructans and galactans in vegetables, and have a lower fructose content.[94]

4.3.8 Salt composition: Does it affect IBD progression?

The activity of Treg cells contributes to the maintenance of the peripheral immune system. Once the Treg cells are active, other immunosuppressive cytokines including IL-10 and TGF are produced, and their combined action inhibits the release of cytokines from other subsets of T cells.[95] Recent investigations have shown that the HSD (high salt diet) negatively impacts the gut immune system.[96,97] However, HSD increased the susceptibility of chemically generated models of colitis in mice while inducing a histologically evident inflammation by a process that was dependent on the generation of interleukin (IL)-17, most probably by both type 3 innate lymphoid cell (ILC3) and Th17 cells.[95] HSDs not only influence immune components, but also reshape the gut microbiota. The Wei et al.[98] work is highly focused on the impact of a high salt diet on a colitis model using trinitrobenzenesulfonic acid (TNBS). Their group has shown that mice fed a high salt diet exhibited an elevation in lamina propria IL-17A-producing cells. Furthermore, a high salt diet significantly decreased the release of IL-10 and the immunosuppressive effects of T-regs rather than having an impact on the overall population of Treg.

5. Microbiome based therapeutics for inflammatory bowel disease

Discovery, development, and clinical use of human microbiome-based therapeutics against infectious and metabolic diseases have made tremendous progress in the last 10 years.[99] Microbiomes, microbiota, microbial-derived proteins and metabolites were isolated from different body habitats of the healthy subjects or produced in laboratory conditions for immediate or future use. Fecal microbiome transplantation (FMT), which involves injecting fresh or previously collected fecal samples in the form of a solution from a healthy donor into a patient's gastrointestinal tract, has proven to be an effective treatment for drug-resistant *C. difficile* infections.[100] FMT is now widely used across the globe as an alternative to chemotherapy in mild to moderate UC. A combination therapy, FMT with an anti-inflammatory diet (FMT-AID), seems very promising in IBD treatment.[101] Substantial efforts have also been made to develop synthetic microbial communities that produce short-chain fatty acids (SCFAs), secondary bile acids, tryptophan metabolites, and vitamins with important roles in immune maturation and resistance against colonization by allochthonous microbiota for therapeutic implementation in IBD.[102] SCFAs suppress inflammation and reduce the risk of development and progression of IBD.

6. Perspectives

Although bacterial and fungal dysbiosis at the taxonomic level is well established,[103] the microbial functions and the precise host pathways that interact with the microbial-derived molecules remain poorly understood. In addition, such studies provided the microbial ecology of the disease state but did not demonstrate causality. Some studies have extended the initial findings to the next level where they used animal models to replicate the disease phenotypes, but the distinct microbial compositions in humans and animals raise several questions, although the phenotypes are partially replicated. Most of the studies conducted to understand the role of the gut microbiome in metabolic or infectious diseases have used fecal samples as representative biospecimens for decoding the composition, diversity, and functions of the gut microbiome. However, the large number of symbiotic microorganisms living close to the epithelial surface may not be accurately captured in those studies. As a result, the true dysbiosis in the intestinal milieu that leads to uncontrolled cytokine production and a harmful immune response in chronic inflammation is yet to be identified. Nevertheless, the studies that used targeted metagenomics (16S rRNA) have additional limitations in taxonomic resolution, and thus the strain-specific host-microbial interactions in the human gut are not established. Recent advances in multi-omics technologies, culturomics, and germ-free animals have revealed new insights into the human microbiome's functional potency and functional activity in health and disease. These discoveries enriched microbiome-based diagnostic and therapeutic research for chronic and acute diseases, including IBD.

7. Conclusion

IBD is a multi-factorial chronic inflammatory disease of the digestive tract. Analysis of the composition and diversity of the gastrointestinal tract microbiome by targeted (16S-rRNA, ITS) and shotgun metagenomics and decoding of community microbial functional potency and functional activities of the colonic microbiome by metatranscriptomics provided consistent evidence that taxonomic and/or functional dysbiosis in the gut microbiome is a hallmark of IBD. Microbes living in the human intestinal mucosa may not be accurately captured if fecal samples are being used to understand the composition of the gastrointestinal tract microbiota, but they do play a crucial role in modulating mucosal immunity and disease severity in individual patients. Multi-cohorts' studies have also provided evidence that in addition to gut microbiome, infectious microbes, environmental factors such as geography, smoking, excess sanitation, and hygiene, diets like high fat, excess simple carbohydrates, red meat, genetic susceptibility, ethnic origin, and a dysregulated immune system can lead to chronic mucosal inflammation and tissue damage. Both specific chemical moiety (drugs) and microbiome-based therapy are in various stages of development or clinical use and have shown promising effects in the prevention and treatment of IBD.

Acknowledgment

B.D. acknowledges all the members of MGL for critical reading of the chapter. S.L. and B.K. acknowledge Dept. of Biotechnology (DBT), Govt. of India for research fellowship. Images are created using BioRender.

Author contributions

B.D. designed the chapter. S.L., B.K., V.A., K.T., and B.D., wrote, reviewed and edited the chapter.

Funding

This study received financial support from the Dept. of Biotechnology (DBT), Govt. of India (Grant No. BT/PR38173/MED/97/474/2020 and No.102/IFD/SAN/5603/2018-2019).

Conflict of interest

Authors declare that there is no conflict of interest.

References

1. Khor B, Gardet A, Xavier RJ. Genetics and pathogenesis of inflammatory bowel disease. *Nature.* 2011 Jun 15;474(7351):307–317. https://doi.org/10.1038/nature10209. PMID: 21677747; PMCID: PMC3204665.
2. Kaplan GG. The global burden of IBD: from 2015 to 2025. *Nat Rev Gastroenterol Hepatol.* 2015 Nov;12(12):720–727. https://doi.org/10.1038/nrgastro.2015.150. Epub 2015 Sep 1. PMID: 26323879.
3. Lavelle A. and Sokol H., Gut microbiota-derived metabolites as key actors in inflammatory bowel disease. Nat Rev Gastroenterol Hepatol. 2020;17(4): 223-237. doi: 10.1038/s41575-019-0258-z. Epub 2020 Feb 19. PMID: 32076145.
4. Gilbert J, Blaser M, Caporaso J, et al. Current understanding of the human microbiome. *Nat Med.* 2018;24:392–400. https://doi.org/10.1038/nm.4517.
5. Pareek S, Kurakawa T, Das B, et al. Comparison of Japanese and Indian intestinal microbiota shows diet-dependent interaction between bacteria and fungi. *NPJ Biofilms Microbiomes.* 2019 Dec 20;5(1):37. https://doi.org/10.1038/s41522-019-0110-9. PMID: 31885873; PMCID: PMC6925221.
6. Lynch SV, Pedersen O. The human intestinal microbiome in health and disease. *N Engl J Med.* 2016 Dec 15;375(24):2369–2379. https://doi.org/10.1056/NEJMra1600266. PMID: 27974040.
7. Hammer T, Langholz E. The epidemiology of inflammatory bowel disease: balance between east and west? A narrative review. *Dig Med Res.* 2020;3:48.
8. Kedia S, Ahuja V. Epidemiology of inflammatory bowel disease in India: the Great Shift East. *Inflamm Intest Dis.* 2017 Nov;2(2):102–115. https://doi.org/10.1159/000465522. Epub 2017 Apr 8. PMID: 30018961; PMCID: PMC5988149.
9. Olfatifar M, Zali MR, Pourhoseingholi MA, et al. The emerging epidemic of inflammatory bowel disease in Asia and Iran by 2035: a modeling study. *BMC Gastroenterol.* 2021;21:204. https://doi.org/10.1186/s12876-021-01745-1.
10. Cui G, Yuan A. A systematic review of epidemiology and risk factors associated with Chinese inflammatory bowel disease. *Front Med (Lausanne).* 2018 Jun 19;5:183. https://doi.org/10.3389/fmed.2018.00183. PMID: 29971235; PMCID: PMC6018203.
11. Neurath MF. Targeting immune cell circuits and trafficking in inflammatory bowel disease. *Nat Immunol.* 2019;20:970–979. https://doi.org/10.1038/s41590-019-0415-0.
12. Levine JS, Burakoff R. Extraintestinal manifestations of inflammatory bowel disease. *Gastroenterol Hepatol (N Y).* 2011 Apr;7(4):235–241. PMID: 21857821; PMCID: PMC3127025.

13. Friedrich M, Pohin M, Powrie F. Cytokine networks in the pathophysiology of inflammatory bowel disease. *Immunity*. 2019 Apr 16;50(4):992–1006. https://doi.org/10.1016/j.immuni.2019.03.017. PMID: 30995511.
14. Das B, Ghosh TS, Kedia S, et al. Analysis of the gut microbiome of rural and urban healthy Indians living in sea level and high altitude areas. *Sci Rep*. 2018 Jul 4;8(1):10104. https://doi.org/10.1038/s41598-018-28550-3. PMID: 29973712; PMCID: PMC6031670.
15. The Human Microbiome Project Consortium. Structure, function and diversity of the healthy human microbiome. *Nature*. 2012;486:207–214. https://doi.org/10.1038/nature11234.
16. Bag S, Ghosh TS, Banerjee S, et al. Molecular insights into antimicrobial resistance traits of commensal human gut microbiota. *Microb Ecol*. 2019;77(2):546–557. https://doi.org/10.1007/s00248-018-1228-7. Epub 2018 Jul 16. PMID: 30009332.
17. Halfvarson J, Brislawn CJ, Lamendella R, et al. Dynamics of the human gut microbiome in inflammatory bowel disease. *Nat Microbiol*. 2017;2:17004. https://doi.org/10.1038/nmicrobiol.2017.4.
18. Nikolaus S, Schulte B, Al-Massad N, et al. Increased tryptophan metabolism is associated with activity of inflammatory bowel diseases. *Gastroenterology*. 2017;153:1504–1516.e2. https://doi.org/10.1053/j.gastro.2017.08.028.
19. Hoarau G, Mukherjee PK, Gower-Rousseau C, et al. Bacteriome and mycobiome interactions underscore microbial dysbiosis in familial crohn's disease. *MBio*. 2016;7. https://doi.org/10.1128/mBio.01250-16.
20. Parker BJ, Wearsch PA, Veloo ACM, Rodriguez-Palacios A. The genus alistipes: gut bacteria with emerging implications to inflammation, cancer, and mental health. *Front Immunol*. 2020;11:906. https://doi.org/10.3389/fimmu.2020.00906.
21. Round JL, Mazmanian SK. Inducible Foxp3+ regulatory T-cell development by a commensal bacterium of the intestinal microbiota. *Proc Natl Acad Sci USA*. 2010;107:12204–12209. https://doi.org/10.1073/pnas.0909122107.
22. Connors J, Dawe N, Van Limbergen J. The role of succinate in the regulation of intestinal inflammation. *Nutrients*. 2018;11. https://doi.org/10.3390/nu11010025.
23. Reissmann S, Friedrichs C, Rajkumari R, et al. Contribution of Streptococcus anginosus to infections caused by groups C and G streptococci, southern India. *Emerging Infect Dis*. 2010;16:656–663. https://doi.org/10.3201/eid1604.090448.
24. Choi J-S, Park N-H, Hwang S-Y, et al. The antibacterial activity of various saturated and unsaturated fatty acids against several oral pathogens. *J Environ Biol*. 2013;34:673–676.
25. Khanna S, Shin A, Kelly CP. Management of *Clostridium difficile* infection in inflammatory bowel disease: expert review from the Clinical Practice Updates Committee of the AGA Institute. *Clin Gastroenterol Hepatol*. 2017;15:166–174. https://doi.org/10.1016/j.cgh.2016.10.024.
26. Kim SC, Tonkonogy SL, Albright CA, et al. Variable phenotypes of enterocolitis in interleukin 10-deficient mice monoassociated with two different commensal bacteria. *Gastroenterology*. 2005;128:891–906. https://doi.org/10.1053/j.gastro.2005.02.009.
27. Schirmer M, Garner A, Vlamakis H, Xavier RJ. Microbial genes and pathways in inflammatory bowel disease. *Nat Rev Microbiol*. 2019;17:497–511. https://doi.org/10.1038/s41579-019-0213-6.
28. Li XV, Leonardi I, Putzel GG, et al. Immune regulation by fungal strain diversity in inflammatory bowel disease. *Nature*. 2022;603:672–678. https://doi.org/10.1038/s41586-022-04502-w.
29. Qiao K, Chooi Y-H, Tang Y. Identification and engineering of the cytochalasin gene cluster from Aspergillus clavatus NRRL 1. *Metab Eng*. 2011;13:723–732. https://doi.org/10.1016/j.ymben.2011.09.008.

30. Li Q, Wang C, Tang C, He Q, Li N, Li J. Dysbiosis of gut fungal microbiota is associated with mucosal inflammation in Crohn's disease. *J Clin Gastroenterol.* 2014;48: 513–523. https://doi.org/10.1097/MCG.0000000000000035.
31. Sokol H, Leducq V, Aschard H, et al. Fungal microbiota dysbiosis in IBD. *Gut.* 2017;66:1039–1048. https://doi.org/10.1136/gutjnl-2015-310746.
32. Kedia S., Ghosh T.S., Jain S., Desigamani A., Kumar A., Gupta V., Bopanna S., Yadav D.P., Goyal S., Makharia G., Travis S.P.L., Das B., Ahuja V., Gut microbiome diversity in acute severe colitis is distinct from mild to moderate ulcerative colitis. J Gastroenterol Hepatol. 2021; 36(3):731-739. doi: 10.1111/jgh.15232. Epub 2020 Sep 14. PMID: 32870508.
33. Senapati T, Kothidar A, Banerjee SK, DAS B. Insights into the gastrointestinal tract microbiomes of Indian population. *J Biosci.* 2019 Oct;44(5):113. PMID: 31719222 (web archive link).
34. Caruso R, Lo BC, Núñez G. Host–microbiota interactions in inflammatory bowel disease. *Nat Rev Immunol.* 2020;20:411–426. https://doi.org/10.1038/s41577-019-0268-7.
35. Ryan FJ, Ahern AM, Fitzgerald RS, et al. Colonic microbiota is associated with inflammation and host epigenomic alterations in inflammatory bowel disease. *Nat Commun.* 2020;11:1512. https://doi.org/10.1038/s41467-020-15342-5.
36. Quévrain E, Maubert MA, Michon C, et al. Identification of an anti-inflammatory protein from *Faecalibacterium prausnitzii*, a commensal bacterium deficient in Crohn's disease. *Gut.* 2016 Mar;65(3):415–425. https://doi.org/10.1136/gutjnl-2014-307649. Epub 2015 Jun 4. PMID: 26045134; PMCID: PMC5136800.
37. Li XV, Leonardi I, Iliev ID. Gut mycobiota in immunity and inflammatory disease. *Immunity.* 2019 Jun 18;50(6):1365–1379. https://doi.org/10.1016/j.immuni.2019.05.023. PMID: 31216461; PMCID: PMC6585451.
38. Limon JJ, Tang J, Li D, et al. Malassezia is associated with crohn's disease and exacerbates colitis in mouse models. *Cell Host Microbe.* 2019 Mar 13;25(3). https://doi.org/10.1016/j.chom.2019.01.007. 377–388.e6. Epub 2019 Mar 5. PMID: 30850233; PMCID: PMC6417942.
39. Naglik JR, Gaffen SL, Hube B. Candidalysin: discovery and function in Candida albicans infections. *Curr Opin Microbiol.* 2019;52:100–109. https://doi.org/10.1016/j.mib.2019.06.002.
40. Wang M, Chen Y, Zhang Y, Zhang L, Lu X, Chen Z. Mannan-binding lectin directly interacts with Toll-like receptor 4 and suppresses lipopolysaccharide-induced inflammatory cytokine secretion from THP-1 cells. *Cell Mol Immunol.* 2011 May;8(3):265–275. https://doi.org/10.1038/cmi.2011.1. Epub 2011 Mar 7. PMID: 21383675; PMCID: PMC4012877.
41. Dave M, Papadakis KA, Faubion Jr WA. Immunology of inflammatory bowel disease and molecular targets for biologics. *Gastroenterol Clin North Am.* 2014 Sep;43(3): 405–424. https://doi.org/10.1016/j.gtc.2014.05.003. PMID: 25110250; PMCID: PMC4480636.
42. Neurath MF. Animal models of inflammatory bowel diseases: illuminating the pathogenesis of colitis, ileitis and cancer. *Dig Dis.* 2012;30(Suppl 1):91–94.
43. Becker C, Fantini MC, Schramm C, et al. TGF-beta suppresses tumor progression in colon cancer by inhibition of IL-6 trans-signaling. *Immunity.* 2004 Oct;21(4):491–501. https://doi.org/10.1016/j.immuni.2004.07.020. PMID: 15485627.
44. Peyrin-Biroulet L, Loftus EV Jr, Colombel JF, Sandborn WJ. The natural history of adult Crohn's disease in population-based cohorts. Am J Gastroenterol. 2010 Feb;105(2):289-97. doi: 10.1038/ajg.2009.579. Epub 2009 Oct 27. PMID: 19861953.
45. Jostins L, Ripke S, Weersma RK, et al. Host-microbe interactions have shaped the genetic architecture of inflammatory bowel disease. *Nature.* 2012 Nov 1;491(7422):119–124. https://doi.org/10.1038/nature11582. PMID: 23128233; PMCID: PMC3491803.

46. Danese S, Colombel JF, Reinisch W, Rutgeerts PJ. Review article: infliximab for Crohn's disease treatment—shifting therapeutic strategies after 10 years of clinical experience. *Aliment Pharmacol Ther.* 2011;33:857–869.
47. Danese S, Fiocchi C. Ulcerative colitis. *N Engl J Med.* 2011;365:1713–1725.
48. Neurath MF, Travis SP. Mucosal healing in inflammatory bowel diseases: a systematic review. *Gut.* 2012 Nov;61(11):1619–1635. https://doi.org/10.1136/gutjnl-2012-302830. Epub 2012 Jul 27. PMID: 22842618.
49. Merga Y, Campbell BJ, Rhodes JM. Mucosal barrier, bacteria and inflammatory bowel disease: possibilities for therapy. *Dig Dis.* 2014;32:475–483.
50. Corfield AP, Wagner SA, Clamp JR, Kriaris MS, Hoskins LC. Mucin degradation in the human colon: production of sialidase, sialate O-acetylesterase, N-acetylneuraminate lyase, arylesterase, and glycosulfatase activities by strains of fecal bacteria. *Infect Immun.* 1992;60:3971–3978.
51. Robertson AM, Wright DP. Bacterial glycosulphatases and sulphomucin degradation. *Can J Gastroenterol.* 1997;11:361–366.
52. Macpherson AJ, Uhr T. Induction of protective IgA by intestinal dendritic cells carrying commensal bacteria. *Science.* 2004;303:1662–1665.
53. Schultsz C, Van Den Berg FM, Ten Kate FW, Tytgat GN, Dankert J. The intestinal mucus layer from patients with inflammatory bowel disease harbors high numbers of bacteria compared with controls. *Gastroenterology.* 1999;117:1089–1097.
54. McGuckin MA, Eri R, Simms LA, Florin THJ, Radford-Smith G. Intestinal barrier dysfunction in inflammatory bowel diseases. *Inflamm Bowel Dis.* 2009;15:100–113.
55. Kim JM. Antimicrobial proteins in intestine and inflammatory bowel diseases. *Intest Res.* 2014;12:20–33.
56. Koslowski MJ, Kübler I, Chamaillard M, et al. Genetic variants of Wnt transcription factor TCF-4 (TCF7L2) putative promoter region are associated with small intestinal Crohn's disease. *PLoS ONE.* 2009;4, e4496.
57. Flach C-F, Eriksson A, Jennische E, Lange S, Gunnerek C, Lönnroth I. Detection of elafin as a candidate biomarker for ulcerative colitis by whole-genome microarray screening. *Inflamm Bowel Dis.* 2006;12:837–842.
58. Yuk J-M, Jo E-K. Toll-like receptors and innate immunity. *J Bacteriol Virol.* 2011;41:225.
59. Abraham BP, Kane S. Fecal markers: calprotectin and lactoferrin. *Gastroenterol Clin North Am.* 2012;41:483–495.
60. Sidhu R, Wilson P, Wright A, et al. Faecal lactoferrin—a novel test to differentiate between the irritable and inflamed bowel? *Aliment Pharmacol Ther.* 2010;31:1365–1370.
61. Togawa J-I, Nagase H, Tanaka K, et al. Oral administration of lactoferrin reduces colitis in rats via modulation of the immune system and correction of cytokine imbalance. *J Gastroenterol Hepatol.* 2002;17:1291–1298.
62. Ganz T. Hepcidin, a key regulator of iron metabolism and mediator of anemia of inflammation. *Blood.* 2003;102:783–788.
63. Dalziel TK. Chronic interstitial enteritis. *Dis Colon Rectum.* 1989;32(12):1076–1078. https://doi.org/10.1007/BF02553886.
64. Pierce ES. Where are all the Mycobacterium avium subspecies paratuberculosis in patients with Crohn's disease? *PLoS Pathog.* 2009;5, e1000234.
65. Greenstein RJ. Is Crohn's disease caused by a mycobacterium? Comparisons with leprosy, tuberculosis, and Johne's disease. *Lancet Infect Dis.* 2003;3:507–514.
66. Sartor RB. Does mycobacterium avium subspecies paratuberculosis cause Crohn's disease? *Gut.* 2005;54:896–898.
67. Dalile B, Van Oudenhove L, Vervliet B, Verbeke K. The role of short-chain fatty acids in microbiota-gut-brain communication. *Nat Rev Gastroenterol Hepatol.* 2019;16:461–478.

68. Fernandes J, Su W, Rahat-Rozenbloom S, Wolever TMS, Comelli EM. Adiposity, gut microbiota and faecal short chain fatty acids are linked in adult humans. *Nutr Diabetes*. 2014;4, e121.
69. Khatib M. *The Impact of Food Structure and Promoting Short Chain Fatty Acids Production on Energy Homeostasis and Appetite Regulation*; 2019. https://books.google.com/books/about/The_Impact_of_Food_Structure_and_Promoti.html?hl=&id=gkLizQEACAAJ.
70. Louis P, Young P, Holtrop G, Flint HJ. Diversity of human colonic butyrate-producing bacteria revealed by analysis of the butyryl-CoA:acetate CoA-transferase gene. *Environ Microbiol*. 2010;12:304–314.
71. Nibali L, Henderson B. *The Human Microbiota and Chronic Disease: Dysbiosis as a Cause of Human Pathology*. John Wiley & Sons; 2016. https://books.google.com/books/about/The_Human_Microbiota_and_Chronic_Disease.html?hl=&id=sQ6lDAAAQBAJ.
72. Deleu S, Machiels K, Raes J, Verbeke K, Vermeire S. Short chain fatty acids and its producing organisms: an overlooked therapy for IBD? *EBioMedicine*. 2021;66, 103293.
73. Machiels K, Joossens M, Sabino J, et al. A decrease of the butyrate-producing species Roseburia hominis and Faecalibacterium prausnitzii defines dysbiosis in patients with ulcerative colitis. *Gut*. 2014;63:1275–1283.
74. Brown AJ, Goldsworthy SM, Barnes AA, et al. The Orphan G protein-coupled receptors GPR41 and GPR43 are activated by propionate and other short chain carboxylic acids. *J Biol Chem*. 2003;278:11312–11319.
75. Kussmann M, Stover PJ. *Nutrigenomics and Proteomics in Health and Disease: Towards a Systems-Level Understanding of Gene-Diet Interactions*. John Wiley & Sons; 2017. https://books.google.com/books/about/Nutrigenomics_and_Proteomics_in_Health_a.html?hl=&id=ce5xDgAAQBAJ.
76. Tan J, McKenzie C, Potamitis M, Thorburn AN, Mackay CR, Macia L. The role of short-chain fatty acids in health and disease. *Adv Immunol*. 2014;121:91–119. https://doi.org/10.1016/B978-0-12-800100-4.00003-9.
77. Tan Y, Zou K-F, Qian W, Chen S, Hou X-H. Expression and implication of toll-like receptors TLR2, TLR4 and TLR9 in colonic mucosa of patients with ulcerative colitis. *J Huazhong Univ Sci Technol Med Sci*. 2014;34:785–790.
78. Glauben R, Batra A, Fedke I, et al. Histone hyperacetylation is associated with amelioration of experimental colitis in mice. *J Immunol*. 2006;176:5015–5022.
79. Maslowski KM, Vieira AT, Ng A, et al. Regulation of inflammatory responses by gut microbiota and chemoattractant receptor GPR43. *Nature*. 2009;461:1282–1286.
80. Vatanen T, Kostic AD, d'Hennezel E, et al. Variation in microbiome LPS immunogenicity contributes to autoimmunity in humans. *Cell*. 2016;165:1551.
81. Fuke N, Nagata N, Suganuma H, Ota T. Regulation of gut microbiota and metabolic endotoxemia with dietary factors. *Nutrients*. 2019;11. https://doi.org/10.3390/nu11102277.
82. Sun L, Yu Z, Ye X, et al. A marker of endotoxemia is associated with obesity and related metabolic disorders in apparently healthy Chinese. *Diabetes Care*. 2010;33: 1925–1932.
83. Heinbockel L, Weindl G, Martinez-de-Tejada G, et al. Inhibition of lipopolysaccharide- and lipoprotein-induced inflammation by antitoxin peptide Pep19-2.5. *Front Immunol*. 2018;9:1704.
84. Chow JC, Young DW, Golenbock DT, Christ WJ, Gusovsky F. Toll-like receptor-4 mediates lipopolysaccharide-induced signal transduction. *J Biol Chem*. 1999;274: 10689–10692.
85. Yu M, Wang H, Ding A, et al. HMGB1 signals through toll-like receptor (TLR) 4 and TLR2. *Shock*. 2006;26:174–179.

86. Duchmann R, Kaiser I, Hermann E, Mayet W, Ewe K, Meyer zum Büschenfelde KH. Tolerance exists towards resident intestinal flora but is broken in active inflammatory bowel disease (IBD). *Clin Exp Immunol.* 1995;102:448–455.
87. Simren M, Månsson A, Langkilde AM, et al. Food-related gastrointestinal symptoms in the irritable bowel syndrome. *Digestion.* 2001;63:108–115.
88. Haas SV, Haas MP. The treatment of celiac disease with the specific carbohydrate diet; report on 191 additional cases. *Am J Gastroenterol.* 1955;23:344–360.
89. Suskind DL. *Nutrition in Immune Balance (NIMBAL) Therapy: Using Diet to Treat Inflammatory Bowel Disease.* Nimbal Publishing; 2015. https://books.google.com/books/about/Nutrition_in_Immune_Balance_NIMBAL_Thera.html?hl=&id=H1VtrgEACAAJ.
90. Kakodkar S, Farooqui AJ, Mikolaitis SL, Mutlu EA. The specific carbohydrate diet for inflammatory bowel disease: a case series. *J Acad Nutr Diet.* 2015;115:1226–1232.
91. Obih C, Wahbeh G, Lee D, et al. Specific carbohydrate diet for pediatric inflammatory bowel disease in clinical practice within an academic IBD center. *Nutrition.* 2016;32:418–425.
92. Elsaygh H, Alhafez I, Jaber M, Al-Rasheed M, Zaher A, ElSaygh J. A literature review of major clinical trials that contributed to treatment protocols of irritable bowel syndrome with diarrhea. *Cureus.* 2022;14, e26092.
93. Gibson PR. The evidence base for efficacy of the low FODMAP diet in irritable bowel syndrome: is it ready for prime time as a first-line therapy? *J Gastroenterol Hepatol.* 2017;32(suppl 1):32–35.
94. Barrett JS. How to institute the low-FODMAP diet. *J Gastroenterol Hepatol.* 2017;32:8–10.
95. Corthay A. How do regulatory T cells work? *Scand J Immunol.* 2009;70:326–336.
96. Aguiar SLF, Miranda MCG, Guimarães MAF, et al. High-salt diet induces IL-17-dependent gut inflammation and exacerbates colitis in mice. *Front Immunol.* 2017;8:1969.
97. Tubbs AL, Liu B, Rogers TD, Sartor RB, Miao EA. Dietary salt exacerbates experimental colitis. *J Immunol.* 2017;199:1051–1059.
98. Wei Y, Lu C, Chen J, et al. High salt diet stimulates gut Th17 response and exacerbates TNBS-induced colitis in mice. *Oncotarget.* 2017;8:70–82.
99. Pant A, Maiti TK, Mahajan D, Das B. Human gut microbiota and drug metabolism. *Microb Ecol.* 2022;1–15. https://doi.org/10.1007/s00248-022-02081-x. Epub ahead of print. PMID: 35869999; PMCID: PMC9308113.
100. Keller JJ, Ooijevaar RE, Hvas CL, et al. A standardised model for stool banking for faecal microbiota transplantation: a consensus report from a multidisciplinary UEG working group. *United European Gastroenterol J.* 2021 Mar;9(2):229–247. https://doi.org/10.1177/2050640620967898. Epub 2021 Mar 9. PMID: 33151137; PMCID: PMC8259288.
101. Kedia S, Virmani S, K Vuyyuru S, et al. Faecal microbiota transplantation with anti-inflammatory diet (FMT-AID) followed by anti-inflammatory diet alone is effective in inducing and maintaining remission over 1 year in mild to moderate ulcerative colitis: a randomised controlled trial. *Gut.* 2022 Aug 16. https://doi.org/10.1136/gutjnl-2022-327811. gutjnl-2022-327811. Epub ahead of print. PMID: 35973787.
102. Carretta MD, Quiroga J, López R, Hidalgo MA, Burgos RA. Participation of short-chain fatty acids and their receptors in gut inflammation and colon cancer. *Front Physiol.* 2021 Apr 8;12, 662739. https://doi.org/10.3389/fphys.2021.662739. PMID: 33897470; PMCID: PMC8060628.
103. Das B, Nair GB. Homeostasis and dysbiosis of the gut microbiome in health and disease. *J Biosci.* 2019 Oct;44(5):117. PMID: 31719226.

CHAPTER EIGHT

Gut microbiome dysbiosis in malnutrition

Meenal Chawla[a], Rashi Gupta[b], and Bhabatosh Das[a,*]
[a]Molecular Genetics Laboratory, Infection and Immunology Division, Translational Health Science and Technology Institute, NCR Biotech Science Cluster, Faridabad, India
[b]Department of Microbiology, Institute of Home Economics, University of Delhi, India
*Corresponding author: e-mail address: bhabatosh@thsti.res.in

Contents

Abstract

Malnutrition is a global health issue and the leading cause of childhood morbidity and mortality under 5 years old. Malnutrition comprises undernutrition (stunting, wasting, underweight), overweight, and obesity. Infancy and child malnutrition are substantially influenced by a number of variables, such as insufficient nutrients, early birth, intestinal inflammation, and gastrointestinal tract microbiota. A variety of environmental factors have been identified that modulate the structure and diversity of newborns' gut microbiomes and their long-term health. Significant data demonstrate that the functional potency and compositional diversity of the microbiome differ in different types

Progress in Molecular Biology and Translational Science, Volume 192
ISSN 1877-1173
https://doi.org/10.1016/bs.pmbts.2022.07.011

of malnutrition. The divergence in the gut microbiome composition between malnourished and healthy children can be observed at an age as young as 12 months. This focuses on variations in the gut microbiome that may influence adult obesity/health status, beginning in the early years of life. The therapeutic potential of supporting a healthy microbiome in malnourished children is being studied as a technique to aid in the fight against malnutrition. The goal of this chapter was to determine the makeup of gut microbiota in obese and undernourished children under the age of 5 years.

Abbreviations

DBM	double burden of malnutrition
FMT	fecal microbiota transplantation
HMO	human milk oligosaccharides
IBD	inflammatory bowel disease
SAM	severe acute malnutrition
SCFA	short-chain fatty acids
UNICEF	United Nations International Children's Emergency Fund
WHO	World Health Organization

1. Introduction

Malnutrition, often known as undernutrition, is a state of insufficient nutrition that extends from under- to overnutrition.[1,2] In children under the age of five, malnutrition is the primary cause of mortality, accounting for more than half of the deaths in this age range. The problem is particularly serious in low- and medium-income countries. In 2021, United Nations International Children's Emergency Fund (UNICEF), World Health Organization (WHO) and the World Bank Group, reported 149.2 million children with <5 years of age were stunted, 45.4 million were wasted, and 38.9 million were overweight worldwide, with Africa and Asia being the most impacted. Continents having, respectively, 39% and 55% of global stunting prevalent in such children. Over 45% of all deaths among children under the age of five are caused by undernutrition. Malnutrition has significant and long-term developmental, financial, cultural, and health effects for individuals and governments around the world.[3]

Even though infections are firmly known to contribute to etiology of undernutrition, new research suggests that the microbiome contributes a vital part as well. Microbiome is a distinct microbial community consisting of bacteria, viruses, fungi and other microscopic living things that inhabits in a well defined habitat.[4] The gastrointestinal tract and skin of humans are the predominant sites whereby trillions of microorganisms have been identified. In our body, bacterial cells outweigh the human cells number.[5] From birth to death, the gut microbiota regulates food digestion, the central nervous system, the immune system, and certain other biological functions.[6–9] It is evident in experimental animal models that enteric pathogens have the strongest potential to generate microbial dysbiosis.[10]

Earlier, under- and overnutrition were thought to be two distinct health problems affecting different groups. However, the term "double burden of malnutrition" (DBM) has been coined to describe the cohabitation of undernutrition and obesity, as well as diet-related non-communicable diseases.[11] Malnutrition in children is a global health issue that must be addressed. In order to promote healthy growth and development, the first 1000 days of an infant's life offer a chance to manipulate the microbiota using probiotics, prebiotics, diet, antibiotics, or fecal microbiota transplantation (FMT).[12] Variations in the microbiota can be triggered by a mixture of environmental variables, including nutrition, chemicals, medicines, and pathogens.[13]

Since, child malnutrition is increasingly being considered as a multifaceted problem including a variety of nutritive, infectious, biological, cultural, and psychosocial variables.[14] To lower the overall disease-related mortality and morbidity, it is necessary to address each one of these malnutrition-related issues. In this chapter, we reviewed the constitution and dynamics of the microbiome in healthy subjects, health disorders caused by microbiome perturbations and the factors that trigger them, microbial signatures linked to malnutrition, and therapeutics to cure it in infants and children under the years for five.

2. Microbiome composition and dynamics in children

The microbial composition and functional potential of microbes have a substantial impact on neonatal growth and development. The major composition of gut microbiome in healthy children includes five phyla namely, Firmicutes, Bacteroidetes, Actinobacteria, Proteobacteria, and Verrucomicrobia[15] with Bacteroidetes and Firmicutes being the most dominant ones.[15,16] The diversity of bacteria that are initially colonized in infant gut microbiome depends significantly on the route of delivery and feeding type that determines whether the child is breastfed (BF) or formula-fed (FF). Infants delivered via vaginal delivery are colonized primarily by bacteria that dominate the mother's vagina, such as *Lactobacillus, Prevotella,* and *Sneathia* species, whereas infants delivered via C-section are colonized primarily by bacteria that originated in the environment, dominated by *Staphylococcus,* followed by *Corynebacterium*, and *Propionibacterium* species.[17]

The phylum Actinobacteria is abundant in infant intestines, with the genus *Bifidobacterium* accounting for more than half of the infant microbiota.[18] However, there is a distinction between the BF and FF infants. Higher Actinobacteria and lower Bacteroidetes and Firmicutes have been observed in BF infants in comparison to FF infants.[19] The gut microbiota of BF neonates is dominated by *Bifidobacteriumlongum* subsp. *infantis*, which is specifically adapted to the digestion of human milk oligosaccharides (HMO).[20,21] *Staphylococcus* and *Streptococcus* dominate the human milk microbiome, but it also contains *Bifidobacterium, Lactobacillus, Clostridium,* and *Veillonella*, all of which are resident species in the initial microbiome of infants.[22] Enrichment of the gut with anaerobic bacteria that produce short-chain fatty acids (SCFA) like propionate, butyrate, and acetate from dietary fiber fermentation, is beneficial since SCFA help to reduce inflammatory cytokine production, promote mucus production in the intestine, and induce secretory IgA.[23] *Lactobacillus* and *Bifidobacterium* will help infants establish innate and acquired immunity in the initial stages of life. Gut microbiomes of children after 2–3 years of age start resembling the microbiome of the adult.[24]

3. Early life perturbations of microbiome and associated health disorders

Premature alterations of the infant gut microbiota, might result in long-term metabolic disorders. Inflammatory bowel disease (IBD), autism, mental problems, psoriatic arthritis, and obesity are among the diseases and disorders that have been linked to the gut microbial ecology.[25] Malnutrition is one of the several conditions that can manifest itself in a variety of ways in children and can be divided into three broad groups. First group consists of children including abnormal linear development (low height for age or stunting) or insufficient body mass accumulation (low weight-for-height or wasting). Second group includes the under nutrition caused by a lack of macronutrients like protein, global energy, and micronutrients like zinc, selenium, or vitamins. It has been considered as a form of malnutrition that impacts both children and adults. The third group includes other manifestations of malnutrition such as obesity, overweight, and the consequent diet-related non-communicable diseases.[26] Malnutrition has instant negative health repercussions and has a long-term negative impact on the overall well-being of a child. Malnourished children are more prone to mortality or sickness, and may experience growth and developmental impairments.

3.1 Causes of malnutrition

Variety of reasons have been recognized that give rise to malnutrition and UNICEF classified it into three categories in 1990: (i) immediate causes, (ii) underlying causes, and (iii) basic causes. Inadequate nutritional intake and illness (e.g., diarrhea) are the primary causes. As malnutrition is primarily caused by a lack of proper nutrition, the gut microbiota has been associated to the condition. Inadequate maternity and child care, lack of access to food in the home, lack of access to health, and poor sanitation are the underlying causes of malnutrition. These are variables that contribute significantly to children's malnutrition and illnesses. The resources, economic, political, and cultural systems of the area are the basic causes.[27] Further the risk factors or the causes affecting the gut microbiome leading to malnutrition are discussed in Table 1 and diagrammatic representation of different forms of malnutrition and causes are depicted in Fig. 1.

3.2 Forms of malnutrition

3.2.1 Chronic malnutrition

Chronic or recurring malnutrition, poor prenatal nutrition, and health, frequent diseases, and unsuitable care of a child in the early stage of life give rise to stunting which is defined by lack of height for one's age. Stunting prevents youngsters from attaining their full physical and intellectual potential.

3.2.2 Acute malnutrition

Extreme weight loss owing to a lack of food, as well as the presence of infectious diseases such as diarrhea, acute respiratory tract infections, malaria is the causes of wasting. Wasting is also known as acute malnutrition. Children that are wasting moderately or severely are more likely to die. WHO has defined wasting using anthropometric cutoffs and clinical signs in children (children of 6–59 months).

(i) Moderate acute malnutrition (MAM): Determined as weight-for-height z-score (WHZ) between -2 and -3 or mid-upper arm circumference (MUAC) ranging from 115 and $<$125 mm. MUAC is the circumference of the relaxed left upper arm, determined in mid-position and measures the degree of muscle wasting.

Table 1 Causes that modulate microbiome leading to malnutrition.

S·No.	Causes	Subject age	Study findings/Microbes associated	Health status	Reference
1.	Living condition	Children, 16 months	High relative abundance of *Campylobacter* and other enteric pathogens.	Malnutrition	28
2.	Basic amenities	0–59 months	Severe wasting affects as many as 6% of people in rural regions, 7% of scheduled tribes (STs), and 8% of those in families with illiterate parents.	Wasting	29
3.	Household food insecurity	preschool children	As food insecurity got increasingly profound, the risk of child stunting and underweight rose in a dose–response manner.	Stunting and underweight	30
4.	Income of individual households	0–60 months from 12 countries	Malnutrition is decreased by the same percent whether income is down for individual households or the country as a whole.	Malnutrition	31
5.	Mother's background	0–59 months	Children whose mothers are 15 to 19 years old have the highest rate of severe wasting, whereas children whose mothers do not suffer stillbirth and are connected to mainstream media have a lower rate of severe wasting.	Wasting	29
6.	Order of birth and repeated infections	6–24 months	Later birth order (height-for-age z-score-HAZ −0.38) and repeated diarrhoeal infection (HAZ −0.23).were among the significant risk factor contribute toward stunting.	Stunting	32
7.	Sex of child	0–35 months	Boys are more malnourished as compared to girls	Stunting	33
8.	Infections	Children	Decrease in *Methanobrevibactersmithii* and an enrichment in potentially pathogenic *Proteobacteria, Fusobacteria* and *Streptococcus gallolyticus*	Kwashiorkor	34
9.	Abnormal breastfeeding and weaning	Children, < 2 years	Decreased abundance of *Bifidobacteria, increased abundance of anaerobes*	Malnutrition	35
10.	Nutrient intake	Children, 1–3 years	Increased abundance of Bacteroidales, reduced *Bifidobacteria, Lactobacilli.*	Kwashiorkor	36

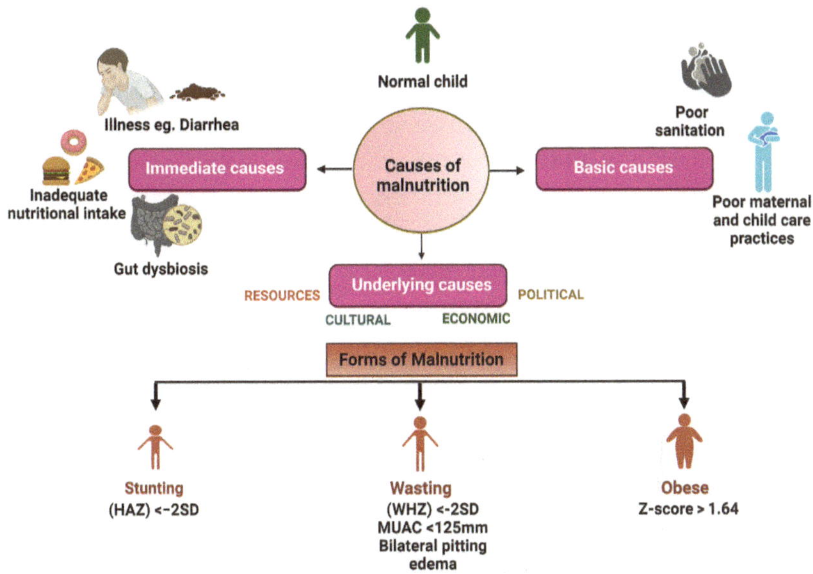

Fig. 1 An overview of three different forms of malnutrition, including stunting, wasting, and obesity and several reasons such as immediate causes, underlying causes, and basic causes that leads to malnutrition are depicted here.

(ii) Severe Acute Malnutrition (SAM): Defined as weight-for-height/length *Z*-score (WHZ) < -3 or MUAC <115 mm and nutritional edema.[37,38] Further, SAM has been differentiated into two types Marasmus and kwashiorkor. However, it is an older classification of malnutrition. Marasmus defines children who are extremely thin in comparison to their height but not suffering from bilateral pitting edema and kwashiorkor also known as edematous malnutrition, a nutritional disorder associated with edema (Fig. 1).[39]

4. Factors influence the composition and diversity of microbiota in infants

4.1 Impact of delivery mode and gestational age

Multiple factors affecting the GI microbiome colonization, establishment, and stabilization have been reported, delivery mode is the most significant one among all. Infants born vaginally have a prominent difference in early gut microbial composition in comparison to those born by cesarean section (C-section, CS).[40–42] The lack of vaginal exposure is largely to blame for the variance in neonatal gut microbiome. According to Mitchell et al., the formation of a microbial community in infants necessitates more than just topical contact to maternal vaginal microbes.[43] The *Bacteroides* species were discovered in substantial numbers during the first week of life in infants born by CS, but they disappeared in the second week and this disparity could be caused by a lack of a supportive element, the introduction of an antagonist, or differences in colonizing strain fitness. Further, the prevalence of rectal transmission from mother to child is revealed by comparing bacterial strain patterns in maternal vaginal and rectal samples with newborns.[43] Perinatal antibiotic use is one of the factors that further differentiate the microbiota makeup of infants delivered by C-section from infants delivered vaginally. Postnatal medicines can severely disrupt the natural succession of the microbiome because a newborn is uncertain to be recolonized with vaginal microorganisms.[44] Antibiotics can be affected over the long run by microbial succession, richness, and resistance.

Premature birth (birth before 33 weeks of pregnancy) has a massive impact on the intestinal microbial community and has a negative impact on the newborns' overall health.[45] Premature infants' intestines are unable to colonize anaerobic symbiotic bacteria like *Bacteroides* and *Bifidobacterium*, and their feces have a comparatively higher prevalence of *Enterococcus* and *Enterobacteriaceae* than term.[46] Premature newborns' immune systems may interact with their immature microbiota, resulting in inflammation and infectious illnesses.[47]

4.2 Breastfeeding and maternal genetics

Breastfeeding also aids in the colonization and stabilization of microbes in the neonate gut. The variety and richness of the fecal microbiota of BF neonates is decreased and is more stable with higher *Bifidobacterium*[48] and depleted in *E. coli, Clostridium difficile, Bacteroides,* and *lactobacilli.*[49] Human milk provides *B. longum* and its prebiotics (galactooligosaccharides) to the infant intestine, encouraging its extremely broad range lantibiotic.[50] This suppresses *Enterobacteriaceae* proliferation, which has been linked to malnutrition-related enteropathy.[51] Recently, studies have shown that *Roseburium* and *Eubacterium*, two butyrate-producing bacteria, have recently shown to flourish on complex HMO.[52] Furthermore, *Bifidobacterium* and *Lactobacillus* spp. are identified in breastmilk, as well as maternal and newborn feces, implying that they could be transmitted from mother to child via breastfeeding.[53] It's now becoming apparent that early feeding methods and their impact on newborns' gut microbiome have both immediate and long-term health repercussions in later childhood.[54] Also, secretory antibodies of the IgA class (sIgA) released from breast milk safeguard neonates from invasive disease, as well as help in maintaining a healthy gut microbiota and modulating gene expression in intestinal cells in the long run.[55]

Genome-wide maternal variation of HMO has been found to be linked with variation in infant fecal microbiota.[56] The composition of HMO differs among breastfeeding women and changes over time. Fucosyltransferase (FUT2 and FUT3) polymorphisms are significantly responsible for inter-individual variance, resulting in different milk groups.[57] Along with this, global variation in breast milk microbiome itself also derives changes in infant microbiome.[58]

4.3 Dietary intake and geographical variation

Diet is another element that affects the creation and function of the microbiome throughout life, and it has a bigger impact than culture, sanitation, cleanliness, geography, and climate. Sugar, animal fat, and calorie-dense meals are anticipated to be rapidly limiting the microbiota's adaptive potential and diminishing its diversity in industrialized countries.[59] The fecal microbiota of European children was compared to that of children from a remote African village in Burkina Faso, where a fiber-rich diet is consumed. Children from Burkina Faso had high levels of bacteria from the genera *Prevotella* and *Xylanibacter,* which are reported to have a group of bacterial genes that encodes for cellulose and xylan hydrolysis, but European children, had none at all.[59] In children below 5 years of age, the impact of dietary intake on fecal microbiota composition has found to be influenced. Dairy and veggie proteins account for 7–10% of the diversity in the microbial community in Australian 2- to 3-year-old children, whereas fruit intake accounts for 8% of the heterogeneity in microbiota composition.[60] The consumption of high-glucose or -sucrose foods alters the gut flora, altering the F/B ratio.[61] In Australian children aged 1–2 years, the intake of meat/fish and fruit is linked with higher Bacteroidetes (*Bacteroidesthetaiotaomicron*) as well as *Ruminococcus, Lachnospiraceae and Bacteroides* species.[62] Another research indicates that the origin of proteins taken, whether plant- or animal-based, may be a critical aspect affecting microbial diversity. Non-whole grain foods, which contain indigestible starches and fibers and are high in vitamin B (e.g., niacin, thiamin, and riboflavin) and minerals (e.g., iron), which appears to be strongly linked to microbial diversity.[63]

Inter-individual and nutrition-derived factors drove variation in fecal microbiota diversity in children around the world. In a recent study conducted in India, children living in Mumbai's urban slums between the ages of 10 and 18 months had microbial diversity that varied by 47% due to age and diets high in iron and polyunsaturated fatty acids. There were differences in β-diversity between children who consumed fats and oils the day before and those who did not.[64]

5. Gut microbiome signatures in malnourished children

The gut microbiome is a composite ecosystem possessing trillions of microbes primarily living in a symbiotic association with the host and largely affecting host metabolism, nutrition absorption, functioning and development of the immune system. Also it may affect several putative processes involved in human pathophysiology, including obesity and malnutrition in children of different age groups.[65–67] Malnutrition can therefore be a consequence of inadequate or excess intake of calories. As discussed, it covers two broad categories namely overnutrition and undernutrition.

5.1 Microbiome signatures in obese children

Obesity is a modern-day health concern worldwide and is a well known threat for various disorders including diabetes, cancer and cardiovascular diseases in both developing and developed countries.[68] Obesity is becoming more prominent among both adults and children due to changed lifestyle habits such as unhealthy dietary intake and reduced physical activity, especially during COVID-19 pandemic time.[69]

Childhood obesity is a complex disorder having varied outcomes. Obese and overweight children are more likely to grow to be obese adults with the higher risk rate for cardiovascular diseases, hypertension, and insulin resistance owing to decreased high-density lipoprotein cholesterol (HDL) and increased blood pressure.[70] However, in addition to genetics and dietary habits, intestinal microbiome also has a strong correlation with obesity and related metabolic disorders.[71,72] The gut microflora is a crucial emerging factor affecting the metabolic homeostasis and promoting metabolic complications linked with obesity in children. When compared to healthy children, obese children's microbiome has altered bacterial diversity and metabolic pathways, which highlights their close relationship to obesity.

Several studies highlight the variations in the gut microbiota composition and energy metabolism between lean and obese children under the age of 5 years. These alterations in fecal microbiota of infants and children have been linked with the chances of being obese at later stage of their lives. Lower abundance of *Bifidobacterial* and higher of *Staphylococcus aureus* at ages 6 and 12 months was observed in obese/overweight infants than those with normal weight.[73] Many studies suggest the inverse relation between the diversity of gut microbiota and overweight/obesity.[71] An inverse association was found between high relative abundance of *Ruminococcus* (*Firmicutes*), *Akkermansia* (*Verrucomicrobia*) *Parabacteroidetes* (*Bacteroidetes*) and *Peptostreptococcae U.* (*Firmicutes*) with overweight/obesity in 3-year-old children in the United States whereas high relative abundance of *Dorea* (*Firmicutes*), *Roseburia* (*Firmicutes)* and *Streptococcus (Firmicutes)* was positively associated with it.[74] Other studies in children above 3 years of age suggest an association between overweight/obesity and higher abundance of *Bacteroides* or *Bacteroides/Prevotella* group[75–77] and lower levels of *Akkermansia muciniphila.*[76,78] These results suggest that changes in intestinal microbiome composition can be used as a signature to predict overweight/obesity in children. A major role played by gut microbes lies in the extraction of energy from food through various mechanisms. Even plant polysaccharides and carbohydrates that the host is unable to digest can be converted by the gut microorganisms to short chain fatty acids (SCFAs) like propionate, butyrate, and acetate. These SCFAs serve as the main energy source for intestinal epithelial cells and are necessary for the liver's processes of lipogenesis and gluconeogenesis.[79]

5.2 Microbiome signatures in undernourished children

SAM is a life-threatening condition for children. Approximately 19 million children with age <5 are affected by SAM annually. Chronic malnutrition culminates in the deaths of about 3 million children under 5 years of age per year.[80] In addition to other factors stated above, the important contribution to poor health status of these children is made by their gut microbiome. Diarrhea, a major contributing factor in the under nutrition of children, is linked with the gut microbiome dysbiosis with increased Proteobacteria and decreased *Bifidobacterium* and *Lactobacillus* species.[25] Such association suggests a close connection between malnutrition and gut dysbiosis, especially in children in low-income countries.

The dysbiosis of the intestinal microbiota in malnourished children is believed to begin with the reduction in *Bifidobacteria*, followed by the invasion by potential pathogenic bacteria including *Fusobacterium mortiferum*,

Streptococcus spp., *Staphylococcus aureus, Escherichia coli* and *Enterococcus faecalis* that cause malabsorption of essential nutrients and diarrhea.[81–83] A similar trend was reported in children of south India under the age of 2 years where malnutrition was associated with depletion in *Bifidobacterium longum* and *Lactobacillus mucosae.*[84] Similar observation in a study conducted in Bangladesh on children aged 2–3 years, malnourished children have a higher bacterial population of the phylum *Proteobacteria,* accounting for 46% compared to 5% in healthy children, and a lower bacterial population of *Bacteroidetes,* accounting for 18% compared to 44% in healthy children. The phylum Proteobacteria comprised pathogenic taxa such as *Escherichia,* and *Klebsiella* that were 9-fold and 174-fold greater, respectively, in malnourished than healthy infants.[85] A rise in the pathogenic bacteria *Shigella, Edwardsiella,* and *Salmonella* in the guts of malnourished children has been observed.[86]

Actinobacteria phylum members are the early invaders of an infant gut.[87] A significant decrease in the abundance of Actinobacteria especially butyrogenic bacteria *Bifidobacterium* and *Faecalibacterium* was found in the gut of children of 2–5 years of age suffering from SAM in suburbs of Chandigarh, India.[88] However in this study, healthy subjects had a significantly higher abundance of *E. coli* than malnourished children, which is contradictory to the previous study. The abundance of the other target phyla such as Bacteroidetes, Firmicutes, and Proteobacteria and genera *Faecalibacterium* and *Roseburia* did not differ significantly. The results obtained using qPCR-based quantification and the 16S rRNA amplicon sequencing methods were nearly identical.[88] Culturomics and metagenomics yielded similar results in a study conducted in Niger and Senegal, where they detected 12 species in controls but none in the feces of children with kwashiorkor. Whereas in the gut microbiota of children with kwashiorkor, there are more potentially pathogenic and oxidative stress-resistant Fusobacteria and Proteobacteria. *Streptococcus gallolyticus* was the sole bacterial species linked to kwashiorkor in this study.[34] This species was also found in abundance in children with SAM in a large metagenomics research conducted in Asia.[89] Commensals like *Bacteroides* and *Ruminococcus* also showed decrease in abundance in these children as compared to healthy guts of similar age group children. Similar depletion of Actinobacteria was found in the gut of malnutritioned children in Mexico, Africa and Kolkata, India.[28,34,90] *Lactobacillus* spp. is a crucial member of the intestinal microbiome, believed to improve the nutritional health and immunity in children.[91] However, a significantly lower abundance (0.10%) of *Lactobacillus* spp. was observed in the gut of healthy children of 2–5 years of age as compared to those of SAM children with higher abundance (0.55%).[28,34,85,88] Another study reported a significant increase in the abundance of Proteobacteria and decreased abundance of Firmicutes in the gut microbiome of malnourished children when compared to their healthy counterparts.[82]

The gut microbiota of healthy children was enriched with bacteria belonging to orders Lactobacillales, Pseudomonadales, Burkholderiales, Bifidobacteriales, Actinomycetales, Desulfovibrionales, Enterobacteriales and Rhizobiales, where as a significant abundance of enteric pathogens under the order Campylobacterales was reported in the gut of malnourished children in Kolkata, India. The presence of these pathogenic bacteria causes intestinal inflammation leading to malabsorption of nutrients, reduced digestion of indigestible polysaccharides and less availability of microbe derived nutrients in their gut.[28] However, the impaired health and nutritional status in malnourished children is not only because of the increase in pathogenic microbes but also due to reduction of several commensal genera including *Faecalibacterium, Roseburia, Eubacterium, Butyrivibrio* and *Phascolarctobacterium*. These genera are recognized for their positive and probiotic benefits, including anti-inflammatory characteristics and intestinal synthesis of SCFA.[92–94] Table 2 summarizes the recent scientific studies that link the microbiome to malnutrition.

Table 2 Recent scientific studies linking the microbiome to malnutrition.

S.No	Age and Location of Children under study	Microbiome signatures in malnourished children	References
1.	Less than or equal to 60 months, West Bengal, India	Increased profusion of Proteobacteria Decreased profusion of Firmicutes	82
2.	16 months children, Kolkata, India	Higher relative abundance of *Campylobacter*	28
3.	7–18 months, children with diarrhea, Dhaka	Increased Proteobacteria; decreased*Bifidobacterium* and *Lactobacillus* species	95
4.	2–5 years, sub-Saharan Africa	Overabundance of *Lactobacillus salivarius* and *Prevotellahisticola*	96

Continued

Table 2 Recent scientific studies linking the microbiome to malnutrition.—cont'd

S.No	Age and Location of Children under study	Microbiome signatures in malnourished children	References
5.	0 to 3 years, Malawi and Bangladesh	Reduced *Bifidobacteriumlongum*, *Lactobacillus mucosae*, and alpha diversity as well as elevated *Acidaminococcus*spp abundance and glutamate fermentation pathways	97
6.	6–59 months, Niger and Senegal	Depletion of obligate anaerobes and the methanogenicarchaeal species *Methanobrevibactersmithii* in SAM	83
7.	10–18 months, Mumbai, India	Higher (80%) relative abundance of *Proteobacteria*, with only 4% *Actinobacteria*	64
8.	2–3 years, Bangladesh	174-fold and 9-fold higher abundance of pathogenic genera, namely *Klebsiella* and *Escherichia*	85
9.	3 to 24 months, South India	Enriched inflammogenic bacteria belonging to the Proteobacteria phylum, with decreased probiotic species Bifidobacterialongum	84
10.	2–5 years, Chandigarh	Decrease in the abundance of Actinobacteria especially butyrogenic bacteria *Bifidobacterium* and *Faecalibacterium*	88

6. Microbiome-based therapeutics for malnourished children

The gut microflora of a malnourished child is generally interpreted as anomalous microflora when compared to the healthy child. Such a dysbiosis causes subclinical conditions characterized by malabsorption, intestinal inflammation, impaired dynamics of commensal gut flora, concurrent infection, and weakened immunity. The reductions in diversity of beneficial bacterial genera in the gut leads to poor gut functioning and further competition for nutrients leading to the degradation of child's health. Therefore, it's important to maintain a healthy microflora in children to ensure their ability to exclude pathogenic bacteria and perform normal functioning of the gut (Fig. 2).

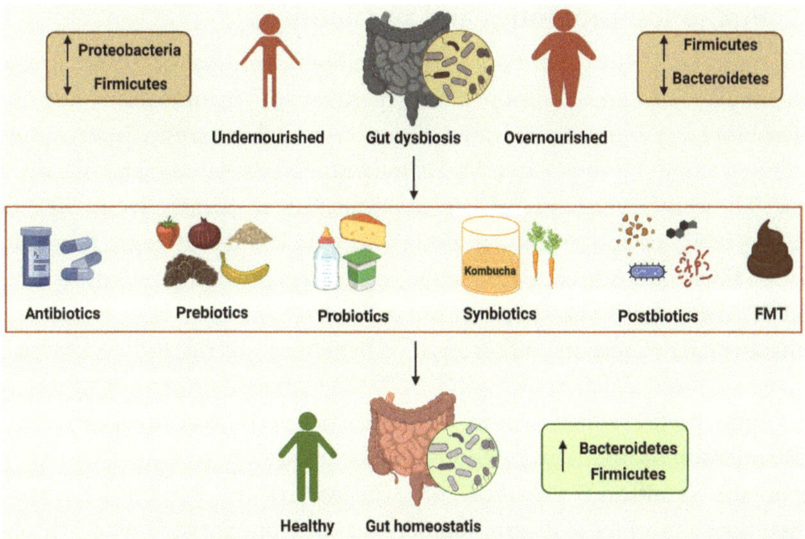

Fig. 2 An overview of microbiome-based therapies for malnourished children including antibiotics, prebiotics, probiotics, symbiotic, postbiotics and fecal microbiota transplantation (FMT).

6.1 Ready-to-use therapeutic foods (RUTF)

One important approach to treat malnutrition in children is the use of RUTF which has revolutionized the treatment of severe malnutrition in children under the age of 5 years.[98,99] During recovery phases, malnourished children receive treatment for electrolyte imbalances, dehydration and bacterial infections in the form of RUTF. RUTF comprises combinations of different foods including a peanut-paste blended with vegetable oil and milk, enriched with minerals and vitamins; partial soybean, which are protein and sorghum flour replacements; oats added to milk-based RUTF and alternative maize, soya and sorghum-based formulas with no milk protein.[100] Such RUTF treatments result in considerable recovery of malnourished children. A considerable reduction in Bacteroidetes and increase in *Ruminococcus, Bifidobacterium*, *Lactobacillus* and *Faecalibacterium* spp. was observed in Malawian twins of <3 years of age suffering with Kwashiorkor after they received treatment with RUTF.[36] Another study found an increased abundance of *Faecalibacterium prausnitzii* and *Clostridioides* spp. in them. Malnourished Bangladeshi children treated with microbiota-directed complementary foods (MDCF) comprising of chickpea flour, soyabean, peanut and green banana.[101]

6.2 Prebiotics, probiotics and postbiotics

Although RUTF treatment gives high-quality nutritional outcomes in malnourished children, additional management of malnutrition targeting microbiota may present a new direction for therapeutic interventions intended at preventing or treating malnutrition and related metabolic disorders. These interventions include manipulation in dietary habits like the use of prebiotics and probiotics along with FMT. Prebiotics are fermented dietary fibers and non digestible oligosaccharides including inulin, lactulose and resistant starch that host cannot assimilate but they help in selective enhancement of activity and growth of beneficial gut microflora especially *Lactobacillus* and *Bifidobacteria* spp.[102,103] The natural sources of inulin are asparagus, garlic, wheat, onion, oats, banana, and soybeans. However, a large quantity of these natural foods would be required to fulfill the requirement for significant elevation of *Bifidobacteria* in the human gut.[104] Therefore, a number of dietary supplements and functional foods containing fructooligosaccharides such as infant foods, biscuits, breads, yogurts and drinks have been developed and are now commercially available.[103]

The WHO defines probiotics as living bacteria that, when fed in sufficient concentrations, offer health benefits to the host. By preventing infections and micronutrient shortages and improving the absorption of critical minerals like zinc, calcium, and vitamin B12, probiotics may help a malnourished children recover.[105,106] A very common method to provide probiotics is in the form of dietary supplements or processed foods. Some common probiotic-carrying foods include yogurt, yakult, cheese, unfermented and fermented milks, smoothies, juices, nutrition bars, and infant formula. The primary microbial constituents of these commercially available probiotic supplements are *Lactobacillus, Streptococcus* and *Bifidobacterium* spp., and *Saccharomyces boulardii*, which are normal inhabitants of the human gut microflora. Therefore, intervention with prebiotics and probiotics causes a modification in gut microbiota composition in both healthy and malnourished children. A combination of four recognized probiotic bacteria including *Leuconostocmesenteroides, Pediococcuspentosaceus, Lactobacillus plantarum,* and *L. paracasei* subsp. *Paracasei* and four recognized prebiotic substrates (pectin, inulin, resistant starch and oat bran), when added to a nutrient dense and high energy RUTF, improved the health status of severely malnourished Malawian children.[107] Synbiotics, on the other hand, are a mix of prebiotics and probiotics, which can be used to address dysbiosis in the gut microbiota. In vitro research suggests that synbiotics are more potent at regulating the intestinal flora than prebiotics and probiotics.[108]

Postbiotics are being used in pediatric settings since they have several perks over probiotics, such as avoiding intestinal translocation or exacerbating local inflammation. A novel class of chemicals called postbiotics has the capacity to affect the microbiota. They can include nonviable bacteria as well as a variety of compounds created through the fermentation and metabolization of various substances by specific bacterial strains. The best technique for promoting microbiome health in this group appears to be using them in vulnerable population groups like infants.[109] The effects of postbiotics in infants have been reviewed before.[110]

6.3 Antibiotics

As studies have revealed that children hospitalized for severe malnutrition had a high frequency of clinically serious infections. Because of this finding, treatment recommendations now suggest the daily use of antibiotics even for children treated as outpatients, despite the fact that outpatients are theoretically far less likely to develop a systemic infection in comparison to patients with difficult illnesses who demand inpatient care.[111] Growth promoting effects of antibiotics have been related with increase in weight in high-income settings children.[112] The addition of routine antibiotics to nutritional therapy has been studied to see if it improves recovery rates and reduces death in children with severe acute malnutrition. A total of 2767 children suffering from SAM were included in the study. 88.7%, 90.9%, and 85.1% of the children in the amoxicillin, cefdinir, and placebo groups, respectively, improved. The rate of weight gain among children who recovered was higher among those who got antibiotics.[113] Antibiotic use raised height by 0.04 cm/month and weight by 23.8 g/month in children in large and middle-income countries, according to a comprehensive meta-analysis.[114] Antibiotics are given in large doses to children in low-income countries where malnutrition is common and this reduces child mortality.[115] Gentamicin plus ampicillin or penicillin is currently suggested as first-line antibiotics for treating SAM.[116] Antibiotics not only eliminate pathogenic bacteria from the gut but also alter its microbial composition. However, conflicting roles of antibiotics in recovering growth of severely malnourished children in Niger, Malawi and Kenya have been reported by some groups.[113,117] Also, antibiotic use in the absence of clinical infections raises the risk of spreading antimicrobial resistance.

6.4 Fecal microbiota transplantation (FMT)

Current approaches in treatment of SAM have only mild effects with ~30% of children not responding to standard nutritional regimen. For such children who fail to respond there is a need to find a new mode of treatment. This recent strategy that has gained recognition in the treatment of malnourishment is FMT. FMT is the term for the transmission of healthy donors' fecal microbiota to recipients. It's a new strategy that's been used to treat infections, immunological illnesses, liver ailments, intestinal encephalopathy, and cancer in recent years.[118,119] Unlike probiotic therapy or other approaches, FMT focuses on the gut microbiota as a separate entity and offers a secure and reliable strategy for gut microbiota regeneration.[120] In the process of FMT, stool collected from a highly screened, healthy donor is homogenized, filtered and introduced into the bowel of an unhealthy recipient. It aims to re-establish and change the recipient's gut dysbiosis to regularize the composition, gaining therapeutic benefits.[121] FMT is frequently employed for the treatment of *Clostridium difficile* infections for children[122] and has not been much tested in children who are malnourished. A small pilot study of FMT is being conducted on children with SAM who are not responding to therapy, and it will give important information on the viability and efficacy of this treatment for severe malnutrition.[123]

7. Conclusion: Challenges and perspectives

Undernutrition due to inadequate nutrient uptake or inefficient energy assimilation from diets may cause low weight-for-height (wasting), low height-for-age (stunting) or low weight-for-age (underweight) in children. The human genome has limited metabolic capacity for digestion of complex dietary components. Microbial communities living in the gastrointestinal tract are the major source of metabolic enzymes, vitamins, and other essential micronutrients. Such functions are not uniformly distributed in the genomes of all the resident microbes. Subsets of bacterial species are enriched with such metabolic capacity. In addition, absorption of digested nutrients is also vital. An inflamed intestinal milieu due to the dysbiosis in the gut microbiome reduces the absorption capacity of the gut, which can also lead to undernutrition. Thus, for normal development, both adequate nutrient uptake and indigenous gut microbiome diversity are equally important. In the event of a change, microbiome-based therapeutics such as the use of prebiotics, probiotics, and postbiotics may aid in the restoration of native community composition.

The comparison of the change in microbiota composition of healthy children with that of malnourished or obese children using phylogenetic analysis is the most frequent method to determine the modifications in the intestinal microbiota composition. However, drawing definitive conclusions about the relationship between malnourished states and the prevalence and relative abundance of specific microbial communities is more difficult because, in addition to geographical and inter- and intraindividual variations, a variety of factors such as mode of delivery, irregular breastfeeding, food intake, infections, and antibiotic use are evident. It's also difficult to say if the observed alterations in microbiome makeup are the cause or result of malnutrition. In the future, it may be possible to pinpoint particular compositional or even functional diversity changes that can be utilized as a biomarker for the disease and have therapeutic potential despite such large variance. Multiomics techniques, combining epigenetic and transcriptomic investigations in large, prospective human birth cohorts, might help us better understand how the microbiome interacts and influences aspects of child development.

Acknowledgments

M.C. received research fellowship from Council of Scientific and Industrial Research (CSIR), Govt. of India. Images are created using BioRender.

References

1. (Geneva). WHO, UNICEF (New York, NY). *WHO Child Growth Standards and the Identification of Severe Acute Malnutrition in Infants and Children: A Joint Statement by the World Health Organization and the United Nations Children's Fund*; 2009.
2. Williams AM, Suchdev PS. Assessing and improving childhood nutrition and growth globally. *Pediatr Clin North Am.* 2017;64(4):755–768. https://doi.org/10.1016/j.pcl.2017.03.001.
3. Malnutrition. Accessed February 15, 2022. https://www.who.int/news-room/fact-sheets/detail/malnutrition.
4. Sweet MJ, Bulling MT. On the importance of the microbiome and Pathobiome in coral health and disease. *Front Mar Sci.* 2017. https://doi.org/10.3389/fmars.2017.00009.
5. Sender R, Fuchs S, Milo R. Revised estimates for the number of human and bacteria cells in the body. *PLoS Biol.* 2016;14(8), e1002533.
6. Cryan JF, Dinan TG. Mind-altering microorganisms: The impact of the gut microbiota on brain and behaviour. *Nat Rev Neurosci.* 2012;13(10):701–712.

7. Brown L, Rosner B, Willett WW, Sacks FM. Cholesterol-lowering effects of dietary fiber: A meta-analysis. *Am J Clin Nutr.* 1999;69(1):30–42.
8. Levy M, Kolodziejczyk AA, Thaiss CA, Elinav E. Dysbiosis and the immune system. *Nat Rev Immunol.* 2017;17(4):219–232. https://doi.org/10.1038/nri.2017.7.
9. Xu X, Ying J. Gut microbiota and immunotherapy. *Front Microbiol.* 2022;13, 945887.
10. Wen L, Ley RE, Volchkov PY, et al. Innate immunity and intestinal microbiota in the development of type 1 diabetes. *Nature.* 2008;455(7216):1109–1113.
11. Popkin BM, Corvalan C, Grummer-Strawn LM. Dynamics of the double burden of malnutrition and the changing nutrition reality. *Lancet.* 2020;395(10217):65–74.
12. Robertson RC, Manges AR, Finlay BB, Prendergast AJ. The human microbiome and child growth - first 1000 days and beyond. *Trends Microbiol.* 2019;27(2):131–147.
13. Stiemsma LT, Michels KB. The role of the microbiome in the developmental origins of health and disease. *Pediatrics.* 2018;141(4). https://doi.org/10.1542/peds.2017-2437.
14. Seferidi P, Hone T, Duran AC, Bernabe-Ortiz A, Millett C. Global inequalities in the double burden of malnutrition and associations with globalisation: A multilevel analysis of demographic and health surveys from 55 low-income and middle-income countries, 1992-2018. *Lancet Glob Health.* 2022;10(4):e482–e490.
15. Qin J, Li R, Raes J, et al. A human gut microbial gene catalogue established by metagenomic sequencing. *Nature.* 2010;464(7285):59–65.
16. Tang WHW, Wilson Tang WH, Kitai T, Hazen SL. Gut microbiota in cardiovascular health and disease. *Circ Res.* 2017;120(7):1183–1196. https://doi.org/10.1161/circresaha.117.309715.
17. Dominguez-Bello MG, Costello EK, Contreras M, et al. Delivery mode shapes the acquisition and structure of the initial microbiota across multiple body habitats in newborns. *Proc Natl Acad Sci U S A.* 2010;107(26):11971–11975.
18. Turroni F, Peano C, Pass DA, et al. Diversity of bifidobacteria within the infant gut microbiota. *PLoS One.* 2012;7(5):e36957.
19. Derrien M, Alvarez AS, de Vos WM. The gut microbiota in the first decade of life. *Trends Microbiol.* 2019;27(12):997–1010.
20. Sela DA, Chapman J, Adeuya A, et al. The genome sequence of Bifidobacterium longum subsp. infantis reveals adaptations for milk utilization within the infant microbiome. *Proc Natl Acad Sci U S A.* 2008;105(48):18964–18969.
21. Forbes JD, Azad MB, Vehling L, et al. Association of Exposure to formula in the hospital and subsequent infant feeding practices with gut microbiota and risk of overweight in the first year of life. *JAMA Pediatr.* 2018;172(7):e181161.
22. Moossavi S, Sepehri S, Robertson B, et al. Composition and variation of the human Milk microbiota are influenced by maternal and early-life factors. *Cell Host Microbe.* 2019;25(2):324–335.e4.
23. van den Elsen LW, Poyntz HC, Weyrich LS, Young W, Forbes-Blom EE. Embracing the gut microbiota: The new frontier for inflammatory and infectious diseases. *Clin Transl Immunol.* 2017;6(1):e125.
24. Yatsunenko T, Rey FE, Manary MJ, et al. Human gut microbiome viewed across age and geography. *Nature.* 2012;486(7402):222–227.
25. Iddrisu I, Monteagudo-Mera A, Poveda C, et al. Malnutrition and gut microbiota in children. *Nutrients.* 2021;13(8). https://doi.org/10.3390/nu13082727.
26. International Food Policy Research Institute (IFPRI). *Global Nutrition Report 2016: From Promise to Impact: Ending Malnutrition by 2030.* Intl Food Policy Res Inst; 2016.
27. Fund UNC. United Nations Children's fund. Strategy for improved nutrition of children and women in developing countries. *Indian J Pediatr.* 1991;58(1):13–24. https://doi.org/10.1007/bf02810402.
28. Gupta SS, Mohammed MH, Ghosh TS, Kanungo S, Nair GB, Mande SS. Metagenome of the gut of a malnourished child. *Gut Pathog.* 2011;3:7.

29. Dinachandra Singh K, Alagarajan M, Ladusingh L. What explains child malnutrition of indigenous people of Northeast India? *PLoS One.* 2015;10(6):e0130567.
30. Hackett M, Melgar-Quiñonez H, Alvarez MC. Household food insecurity associated with stunting and underweight among preschool children in Antioquia, Colombia. *Rev Panam Salud Publica.* 2009;25(6):506–510.
31. World Bank Group. Reducing child malnutrition: How far does income growth take us? *World Bank Econ Rev.* 2001.
32. Saxton J, Rath S, Nair N, et al. Handwashing, sanitation and family planning practices are the strongest underlying determinants of child stunting in rural indigenous communities of Jharkhand and Odisha, eastern India: A cross-sectional study. *Matern Child Nutr.* 2016;12(4):869–884.
33. Tumilowicz A, Habicht JP, Pelto G, Pelletier DL. Gender perceptions predict sex differences in growth patterns of indigenous Guatemalan infants and young children. *Am J Clin Nutr.* 2015;102(5):1249–1258.
34. Tidjani Alou M, Million M, Traore SI, et al. Gut bacteria missing in severe acute malnutrition, can we identify potential probiotics by Culturomics? *Front Microbiol.* 2017;8:899.
35. Motarjemi Y, Käferstein F, Moy G, Quevedo F. Contaminated weaning food: A major risk factor for diarrhoea and associated malnutrition. *Bull World Health Organ.* 1993;71 (1):79–92.
36. Smith MI, Yatsunenko T, Manary MJ, et al. Gut microbiomes of Malawian twin pairs discordant for kwashiorkor. *Science.* 2013;339(6119):548–554.
37. Schwinger C, Golden MH, Grellety E, Roberfroid D, Guesdon B. Severe acute malnutrition and mortality in children in the community: Comparison of indicators in a multi-country pooled analysis. *PLoS One.* 2019;14(8):e0219745. https://doi.org/10.1371/journal.pone.0219745.
38. Ghimire U, Aryal BK, Gupta AK, Sapkota S. Severe acute malnutrition and its associated factors among children under-five years: A facility-based cross-sectional study. *BMC Pediatr.* 2020;20(1). https://doi.org/10.1186/s12887-020-02154-1.
39. Dipasquale V, Cucinotta U, Romano C. Acute malnutrition in children: Pathophysiology, clinical effects and treatment. *Nutrients.* 2020;12(8). https://doi.org/10.3390/nu12082413.
40. Bokulich NA, Chung J, Battaglia T, et al. Antibiotics, birth mode, and diet shape microbiome maturation during early life. *Sci Transl Med.* 2016;8(343) [343ra82].
41. Reyman M, van Houten MA, van Baarle D, et al. Author correction: Impact of delivery mode-associated gut microbiota dynamics on health in the first year of life. *Nat Commun.* 2019;10(1):5352.
42. Shao Y, Forster SC, Tsaliki E, et al. Stunted microbiota and opportunistic pathogen colonization in caesarean-section birth. *Nature.* 2019;574(7776):117–121.
43. Mitchell CM, Mazzoni C, Hogstrom L, et al. Delivery mode affects stability of early infant gut microbiota. *Cell Rep Med.* 2020;1(9):100156.
44. Langdon A, Crook N, Dantas G. The effects of antibiotics on the microbiome throughout development and alternative approaches for therapeutic modulation. *Genome Med.* 2016;8(1):39.
45. La Rosa PS, Warner BB, Zhou Y, et al. Patterned progression of bacterial populations in the premature infant gut. *Proc Natl Acad Sci U S A.* 2014;111 (34):12522–12527.
46. Cong X, Xu W, Janton S, et al. Gut microbiome developmental patterns in early life of preterm infants: Impacts of feeding and gender. *PLoS One.* 2016;11(4):e0152751.
47. Collado MC, Cernada M, Neu J, Pérez-Martínez G, Gormaz M, Vento M. Factors influencing gastrointestinal tract and microbiota immune interaction in preterm infants. *Pediatr Res.* 2015;77(6):726–731.

48. Alam N, Monira S. Effect of formula composition on the development of infant gut microbiota. *J Pediatr Gastroenterol Nutr.* 2011. https://doi.org/10.3410/f.12196956.13359054. Faculty Opinions – Post-Publication Peer Review of the Biomedical Literature Published online.
49. Penders J, Thijs C, Vink C, et al. Factors influencing the composition of the intestinal microbiota in early infancy. *Pediatrics.* 2006;118(2):511–521.
50. Lee JH, Li X, O'Sullivan DJ. Transcription analysis of a lantibiotic gene cluster from Bifidobacterium longum DJO10A. *Appl Environ Microbiol.* 2011;77(17): 5879–5887.
51. Kau AL, Planer JD, Liu J, et al. Functional characterization of IgA-targeted bacterial taxa from undernourished Malawian children that produce diet-dependent enteropathy. *Sci Transl Med.* 2015;7(276) [276ra24].
52. Pichler MJ, Yamada C, Shuoker B, et al. Butyrate producing colonic Clostridiales metabolise human milk oligosaccharides and cross feed on mucin via conserved pathways. *Nat Commun.* 2020;11(1):3285.
53. Martín V, Maldonado-Barragán A, Moles L, et al. Sharing of bacterial strains between breast milk and infant feces. *J Hum Lact.* 2012;28(1):36–44.
54. Exam CME. 2: Events within the first year of life, but not the neonatal period, affect risk for later development of inflammatory bowel diseases. *Gastroenterology.* 2019;156(8): e18. https://doi.org/10.1053/j.gastro.2019.04.036.
55. Rogier EW, Frantz AL, Bruno MEC, et al. Secretory antibodies in breast milk promote long-term intestinal homeostasis by regulating the gut microbiota and host gene expression. *Proc Natl Acad Sci U S A.* 2014;111(8):3074–3079.
56. Lackey KA, Williams JE, Meehan CL, et al. Corrigendum: What's Normal? Microbiomes in human Milk and infant feces are related to each other but vary geographically: The INSPIRE study. *Front Nutr.* 2020;7:12.
57. Samuel TM, Binia A, de Castro CA, et al. Impact of maternal characteristics on human milk oligosaccharide composition over the first 4 months of lactation in a cohort of healthy European mothers. *Sci Rep.* 2019;9(1):11767.
58. Pace RM, Williams JE, Robertson B, et al. Variation in human Milk composition is related to differences in Milk and infant fecal microbial communities. *Microorganisms.* 2021;9(6). https://doi.org/10.3390/microorganisms9061153.
59. De Filippo C, Cavalieri D, Di Paola M, et al. Impact of diet in shaping gut microbiota revealed by a comparative study in children from Europe and rural Africa. *Proc Natl Acad Sci U S A.* 2010;107(33):14691–14696.
60. Smith-Brown P, Morrison M, Krause L, Davies PSW. Dairy and plant based food intakes are associated with altered faecal microbiota in 2 to 3 year old Australian children. *Sci Rep.* 2016;6:32385.
61. Do MH, Lee E, Oh MJ, Kim Y, Park HY. High-glucose or -fructose diet cause changes of the gut microbiota and metabolic disorders in mice without body weight change. *Nutrients.* 2018;10(6). https://doi.org/10.3390/nu10060761.
62. Matsuyama M, Morrison M, Cao KAL, et al. Dietary intake influences gut microbiota development of healthy Australian children from the age of one to two years. *Sci Rep.* 2019;9(1):12476.
63. Herman DR, Rhoades N, Mercado J, Argueta P, Lopez U, Flores GE. Dietary habits of 2- to 9-year-old American children are associated with gut microbiome composition. *J Acad Nutr Diet.* 2020;120(4):517–534.
64. Huey SL, Jiang L, Fedarko MW, et al. Nutrition and the gut microbiota in 10- to 18-month-old children living in urban slums of Mumbai, India. *mSphere.* 2020;5(5). https://doi.org/10.1128/mSphere.00731-20.
65. Ley RE. Obesity and the human microbiome. *Curr Opin Gastroenterol.* 2010; 26(1):5–11.

66. Caesar R, Tremaroli V, Kovatcheva-Datchary P, Cani PD, Bäckhed F. Crosstalk between gut microbiota and dietary lipids aggravates WAT inflammation through TLR signaling. *Cell Metab.* 2015;22(4):658–668.
67. Chassaing B, Gewirtz AT. Gut microbiota, low-grade inflammation, and metabolic syndrome. *Toxicol Pathol.* 2014;42(1):49–53.
68. Ng M, Fleming T, Robinson M, et al. Global, regional, and national prevalence of overweight and obesity in children and adults during 1980-2013: A systematic analysis for the global burden of disease study 2013. *Lancet.* 2014;384(9945):766–781.
69. Bhattacharya S, Aggarwal P, Bera OP, et al. COVID-19 and childhood obesity (CO-BESITY) in the era of new normal life: A need for a policy research. *J Public Health Res.* 2021;10(s2). https://doi.org/10.4081/jphr.2021.2673.
70. Umer A, Kelley GA, Cottrell LE, Giacobbi Jr P, Innes KE, Lilly CL. Childhood obesity and adult cardiovascular disease risk factors: A systematic review with meta-analysis. *BMC Public Health.* 2017;17(1):683.
71. Sonnenburg JL, Bäckhed F. Diet-microbiota interactions as moderators of human metabolism. *Nature.* 2016;535(7610):56–64.
72. Davis CD. The gut microbiome and its role in obesity. *Nutr Today.* 2016;51 (4):167–174. https://doi.org/10.1097/nt.0000000000000167.
73. Kalliomäki M, Collado MC, Salminen S, Isolauri E. Early differences in fecal microbiota composition in children may predict overweight. *Am J Clin Nutr.* 2008;87(3):534–538.
74. Karvonen AM, Sordillo JE, Gold DR, et al. Gut microbiota and overweight in 3-year old children. *Int J Obes (Lond).* 2019;43(4):713–723.
75. Riva A, Borgo F, Lassandro C, et al. Pediatric obesity is associated with an altered gut microbiota and discordant shifts in F irmicutes populations. *Environ Microbiol.* 2017;19(1):95–105. https://doi.org/10.1111/1462-2920.13463.
76. Borgo F, Verduci E, Riva A, et al. Relative abundance in bacterial and fungal gut microbes in obese children: A case control study. *Child Obes.* 2017;13(1):78–84.
77. Bervoets L, Van Hoorenbeeck K, Kortleven I, et al. Differences in gut microbiota composition between obese and lean children: A cross-sectional study. *Gut Pathog.* 2013;5(1):10.
78. Karlsson CLJ, Önnerfält J, Xu J, Molin G, Ahrné S, Thorngren-Jerneck K. The microbiota of the gut in preschool children with Normal and excessive body weight. *Obesity.* 2012;20(11):2257–2261. https://doi.org/10.1038/oby.2012.110.
79. Turnbaugh PJ, Ley RE, Mahowald MA, Magrini V, Mardis ER, Gordon JI. An obesity-associated gut microbiome with increased capacity for energy harvest. *Nature.* 2006;444(7122):1027–1031.
80. Severe childhood malnutrition. *Nat Rev Dis Primers.* 2017;3:1. https://doi.org/10.1038/nrdp.2017.68.
81. Becker-Dreps S, Allali I, Monteagudo A, et al. Gut microbiome composition in Young Nicaraguan children during diarrhea episodes and recovery. *Am J Trop Med Hyg.* 2015;93(6):1187–1193.
82. Ghosh TS, Gupta SS, Bhattacharya T, et al. Gut microbiomes of Indian children of varying nutritional status. *PLoS One.* 2014;9(4):e95547.
83. Million M, Tidjani Alou M, Khelaifia S, et al. Increased gut redox and depletion of anaerobic and methanogenic prokaryotes in severe acute malnutrition. *Sci Rep.* 2016;6:26051.
84. Dinh DM, Ramadass B, Kattula D, et al. Longitudinal analysis of the intestinal microbiota in persistently stunted Young children in South India. *PLoS One.* 2016;11(5): e0155405.
85. Monira S, Nakamura S, Gotoh K, et al. Gut microbiota of healthy and malnourished children in Bangladesh. *Front Microbiol.* 2011;2:228.

86. Million M, Diallo A, Raoult D. Gut microbiota and malnutrition. *Microb Pathog.* 2017;106:127–138. https://doi.org/10.1016/j.micpath.2016.02.003.
87. Hill CJ, Lynch DB, Murphy K, et al. Erratum to: Evolution of gut microbiota composition from birth to 24 weeks in the INFANTMET cohort. *Microbiome.* 2017;5(1):21.
88. Balasubramaniam C, Mallappa RH, Singh DK, et al. Gut bacterial profile in Indian children of varying nutritional status: A comparative pilot study. *Eur J Nutr.* 2021;60 (7):3971–3985.
89. Subramanian S, Huq S, Yatsunenko T, et al. Persistent gut microbiota immaturity in malnourished Bangladeshi children. *Nature.* 2014;510(7505):417–421.
90. Méndez-Salazar EO, Ortiz-López MG, de los Ángeles Granados-Silvestre M, Palacios-González B, Menjivar M. Corrigendum: Altered gut microbiota and compositional changes in firmicutes and proteobacteria in Mexican undernourished and obese children. *Front Microbiol.* 2018;9. https://doi.org/10.3389/fmicb.2018.02693.
91. Xiao Y, Zhao J, Zhang H, Zhai Q, Chen W. Mining lactobacillus and Bifidobacterium for organisms with long-term gut colonization potential. *Clin Nutr.* 2020;39 (5):1315–1323.
92. Gevers D, Knight R, Petrosino JF, et al. The human microbiome project: A community resource for the healthy human microbiome. *PLoS Biol.* 2012;10(8), e1001377. https://doi.org/10.1371/journal.pbio.1001377.
93. Turnbaugh PJ, Hamady M, Yatsunenko T, et al. A core gut microbiome in obese and lean twins. *Nature.* 2009;457(7228):480–484.
94. Peuhkuri K, Vapaatalo H, Korpela R. Even low-grade inflammation impacts on small intestinal function. *World J Gastroenterol.* 2010;16(9):1057–1062.
95. Monira S, Shabnam SA, Alam NH, Endtz HP, Cravioto A, Alam M. 16S rRNA gene-targeted TTGE in determining diversity of gut microbiota during acute diarrhoea and convalescence. *J Health Popul Nutr.* 2012;30(3):250–256.
96. Vonaesch P, Morien E, Andrianonimiadana L, et al. Stunted childhood growth is associated with decompartmentalization of the gastrointestinal tract and overgrowth of oropharyngeal taxa. *Proc Natl Acad Sci U S A.* 2018;115(36):E8489–E8498.
97. Gough EK, Stephens DA, Moodie EEM, et al. Erratum to: Linear growth faltering in infants is associated with Acidaminococcus sp. and community-level changes in the gut microbiota. *Microbiome.* 2016;4:5.
98. Ciliberto MA, Sandige H, Ndekha MJ, et al. Comparison of home-based therapy with ready-to-use therapeutic food with standard therapy in the treatment of malnourished Malawian children: A controlled, clinical effectiveness trial. *Am J Clin Nutr.* 2005;81 (4):864–870.
99. Linneman Z, Matilsky D, Ndekha M, Manary MJ, Maleta K, Manary MJ. A large-scale operational study of home-based therapy with ready-to-use therapeutic food in childhood malnutrition in Malawi. *Matern Child Nutr.* 2007;3(3):206–215. https://doi.org/10.1111/j.1740-8709.2007.00095.x.
100. Schoonees A, Lombard MJ, Musekiwa A, Nel E, Volmink J. Ready-to-use therapeutic food (RUTF) for home-based nutritional rehabilitation of severe acute malnutrition in children from six months to five years of age. *Cochrane Database Syst Rev.* 2019;5 [CD009000].
101. Gehrig JL, Venkatesh S, Chang HW, et al. Effects of microbiota-directed foods in gnotobiotic animals and undernourished children. *Science.* 2019;365(6449). https://doi.org/10.1126/science.aau4732.
102. Lim CC, Ferguson LR, Tannock GW. Dietary fibres as "prebiotics": Implications for colorectal cancer. *Mol Nutr Food Res.* 2005;49(6):609–619.
103. Kolida S, Gibson GR. Prebiotic capacity of inulin-type fructans. *J Nutr.* 2007;137 (11 Suppl):2503S–2506S.

104. Manning TS, Gibson GR. Microbial-gut interactions in health and disease. Prebiotics. *Best Pract Res Clin Gastroenterol.* 2004;18(2):287–298.
105. He M, Yang YX, Han H, Men JH, Bian LH, Wang GD. Effects of yogurt supplementation on the growth of preschool children in Beijing suburbs. *Biomed Environ Sci.* 2005;18(3):192–197.
106. Scholz-Ahrens KE, Ade P, Marten B, et al. Prebiotics, probiotics, and synbiotics affect mineral absorption, bone mineral content, and bone structure. *J Nutr.* 2007;137 (3 Suppl 2):838S–846S.
107. No authors. *The American Journal of Clinical Nutrition.*; 2003.
108. Perturbations associated with hungry gut microbiome and postbiotic perspectives to strengthen the microbiome health. *Future Foods.* 2021;4 [100043].
109. Mojgani N, Dadar M. *Probiotic Bacteria and Postbiotic Metabolites: Role in Animal and Human Health.* Springer Nature; 2021.
110. Wegh CAM, Geerlings SY, Knol J, Roeselers G, Belzer C. Postbiotics and their potential applications in early life nutrition and beyond. *Int J Mol Sci.* 2019;20(19). https://doi.org/10.3390/ijms20194673.
111. Lazzerini M, Tickell D. Antibiotics in severely malnourished children: Systematic review of efficacy, safety and pharmacokinetics. *Bull World Health Organ.* 2011;89 (8):594–607.
112. Bailey LC, Forrest CB, Zhang P, Richards TM, Livshits A, DeRusso PA. Association of antibiotics in infancy with early childhood obesity. *JAMA Pediatr.* 2014;168(11):1063–1069.
113. Trehan I, Goldbach HS, LaGrone LN, et al. Antibiotics as part of the management of severe acute malnutrition. *Malawi Med J.* 2016;28(3):123–130.
114. Gough EK, Moodie EEM, Prendergast AJ, et al. The impact of antibiotics on growth in children in low and middle income countries: Systematic review and meta-analysis of randomised controlled trials. *BMJ.* 2014;348:g2267.
115. Keenan JD, Bailey RL, West SK, et al. Azithromycin to reduce childhood mortality in sub-Saharan Africa. *N Engl J Med.* 2018;378(17):1583–1592. https://doi.org/10.1056/nejmoa1715474.
116. Clinicaltrials.gov. *First Line Antimicrobials in Children with Complicated Severe Acute Malnutrition - Full Text View.* ClinicalTrials.gov; 2022 [Accessed February 28, 2022. https://clinicaltrials.gov/ct2/show/NCT03174236].
117. Isanaka S, Langendorf C, Berthé F, et al. Routine amoxicillin for uncomplicated severe acute malnutrition in children. *N Engl J Med.* 2016;374(5):444–453.
118. Mullish BH, McDonald JAK, Thursz MR, Marchesi JR. Fecal microbiota transplant from a rational stool donor improves hepatic encephalopathy: A randomized clinical trial. *Hepatology.* 2017;66(4):1354–1355.
119. Wortelboer K, Nieuwdorp M, Herrema H. Fecal microbiota transplantation beyond Clostridioides difficile infections. *EBioMedicine.* 2019;44:716–729.
120. Ianiro G, Masucci L, Quaranta G, et al. Randomised clinical trial: Faecal microbiota transplantation by colonoscopy plus vancomycin for the treatment of severe refractory Clostridium difficile infection-single versus multiple infusions. *Aliment Pharmacol Ther.* 2018;48(2):152–159.
121. Gupta S, Allen-Vercoe E, Petrof EO. Fecal microbiota transplantation: In perspective. *Therap Adv Gastroenterol.* 2016;9(2):229–239.
122. Gough E, Shaikh H, Manges AR. Systematic review of intestinal microbiota transplantation (fecal bacteriotherapy) for recurrent Clostridium difficile infection. *Clin Infect Dis.* 2011;53(10):994–1002. https://doi.org/10.1093/cid/cir632.
123. ClinicalTrials.gov. *Transfer of Healthy Gut Flora for Restoration of Intestinal Microbiota Via Enema for Patients in the Rehabilitative Phase of Malnutrition - Full Text View.* ClinicalTrials.gov; 2022 [Accessed February 28, 2022. https://clinicaltrials.gov/ct2/show/NCT03087097].

CHAPTER NINE

Human microbiome and cardiovascular diseases

Md Jahangir Alam[a], Vaishnavi Puppala[a], Shravan K. Uppulapu[a], Bhabatosh Das[b], and Sanjay K. Banerjee[a,*]

[a]Department of Biotechnology, National Institute of Pharmaceutical Education and Research (NIPER), Guwahati, Assam, India

[b]Molecular Genetics Laboratory, Infection and Immunology Division, Translational Health Science and Technology Institute, NCR Biotech Science Cluster, Faridabad, India

*Corresponding author e-mail address: sanjay@niperguwahati.in

Contents

Abstract

A number of microorganisms are co-evolved with the host, among which bacteria are the predominant organisms in the colonic site. The human microbiota contributes to various physiological functions, including the digestion and degradation of food components, harvesting of inaccessible nutrients, immune system regulation, maintenance of gut barrier function, and regulation of brain function and behavior. Microbes in the gut produce a wealth of metabolites from the exogenous dietary substances or endogenous metabolic compounds produced by the host and the resident microorganisms. These microbial-derived metabolites are the major factors in the host-microbiota

Progress in Molecular Biology and Translational Science, Volume 192
ISSN 1877-1173
https://doi.org/10.1016/bs.pmbts.2022.07.012

cross-talk and influence the host's cardiometabolic health directly or indirectly depending on the structure and function of the microbial community. Evidence suggests that the perturbation in the composition and function of gut microbiota (referred to as gut dysbiosis) is associated with the development of several diseased conditions such as that of the gastrointestinal tract or colorectal cancer, metabolic diseases such as obesity, diabetes, immune disorders e.g. asthma, allergies, depression, anxiety and cardiometabolic disease. Several pathological conditions in the gastrointestinal tract may impair the intestinal barrier that allows translocation of bacteria and their metabolites to a remote organ such as the heart, which may ultimately be associated with systemic inflammation and the development of CVDs.

In this chapter, we will discuss various gut microbiota-dependent metabolites, which have a significant role in cardiovascular diseases' pathologic processes and their risk factors. Finally, we will discuss the therapeutic potential of the gut-metabolite-heart axis as a novel target for the treatment of CVD and highlight the current updates and exciting directions for future research.

Abbreviations

AhR aryl-hydrocarbon receptor
BA bile acid
BCAA branched-chain amino acid
BP blood pressure
CA cholic acid
CAD coronary artery disease
CAR constitutive androstane receptor
CDCA chenodeoxycholic acid
CKD chronic kidney disease
CMD cardiometabolic disease
CVD cardiovascular disease
DCA deoxycholic acid
DMB 3,3-dimethyl-1-butanol
FMAO3 flavin monooxygenases 3
FMT fecal microbiota transplantation
FXR farnesoid X receptor
GF germfree
GM gut microbiota
GPCR G-protein-coupled receptors
HF heart failure
IAA indole-3-acid-acetic
IAAld indole-3-acetaldehyde
IAld indole-3-aldehyde
IDO1 indoleamine 2,3-dioxygenase
ImP imidazole propionate
IPA indole-3-propionic acid
IRS insulin receptor substrate
IS indoxyl sulfate
LCA lithocholic acid

LPS	lipopolysaccharide
LVEF	left ventricular ejection fraction
LXR	liver X receptor
MACE	major adverse cardiac event
MI	myocardial infarction
NOX	NADPH oxidase
PAA	phenylacetic acid
PAG	phenylacetylglutamine
PAMP	pattern-associated molecular patterns
pCS	p-cresyl sulfate
PXR	pregnane X receptor
RIG-I	tetinoic acid-inducible gene I
RLR	RIG-I-like receptors
ROS	reactive oxygen species
SCFA	short chain fatty acid
T2DM	type 2 diabetes
TF	tissue factor
TG	triglycerides
TGR5	Takeda G-protein–coupled receptor 5
TLR	toll-like receptors
TMA	trimethylamine
TMAO	trimethylamine N-oxide
VDR	vitamin D receptor

1. Introduction

The human body is a resident of various organs and co-ordinating all the organs to perform physiological activities for human well-being. The coordination of organs in the human body is also influenced by the presence of microbial communities in humans, especially in the gastrointestinal tract. The human microbiota, a pseudo-organ, can be defined as a dynamic community of various organisms such as prokaryotes, eukaryotes, archaea, and fungi.[1] The microbes are co-evolved with the host of which bacteria are the predominant organisms in the colonic site.[2] Bacteria are the dominant species which majorly include *Bacteriodates, Firmicutes, Actinobacteria, Proteobacteria,* and *Verrucomicrobia.*[3] The human microbiome is estimated to comprise $\sim 10^{13}$–10^{14} genome of the microbiota (equivalent to >45 million non-redundant genes) termed as the microbiome and represents ~50–100 fold more genes than that of the human genome.[4,5] Since the 1880s, the role of gut microbiota, including *Escherichia coli* and other fecal bacteria, on human health such as digestion and intestinal diseases has been hypothesized.[6] With the recent advancement in the omics approaches, such as 16s RNA gene sequencing, transcriptomics, metagenomics and metabolomics, we have started to understand the complex gut microbiome and deciphered the potential molecular mechanism of its interaction with the host.[7] The human microbiome project (HMP) and the metagenomics of the human intestinal tract (MetaHIT) have led to a foundation to define the composition and properties of the gut microbiome.[8–10]

The human microbiota, sometimes termed as "second brain," contributes to a variety of physiological functions, including digestion of food components,[11] synthesis of vitamins,[12,13] degradation and harvesting of inaccessible nutrients,[14] maturation of the immune system and regulation of gut barrier function[15–17] and regulation of brain function and behaviour.[18] The microbes in the gut produce a range of reservoirs of metabolic substances from the exogenous dietary materials or endogenous compounds generated by the host and the microorganisms.[19] Gut microbiota is considered an essential organ. These microbial metabolites are the key actors in the host-microbiota cross-talk and play a crucial role in the aforementioned activities. For example, *Firmicutes* are enriched with CAZymes, a class of enzymes critical for the digestion of complex dietary fiber.[20] As evident from the Human Microbiome Project, because of variation in taxonomic composition and unique microbiome between individuals, total metabolites (metabolome) produced by microbes are of a diverse range, reflecting microbial functions.[7,8] Although gut microbiota, considered the largest endocrine organ, generally have necessary and beneficial biological processes, they may play a role in the development of harmful effects on the body depending on the microbial community structure and function like diverse reservoirs of metabolite production and the host state such as health, gender,[21] types of diet,[22,23] medication use, housing conditions, and social network characteristics.[24] Depending on various internal and external factors, the microbiota grows, colonizes, and communicates with the host. The change in microbiota composition occurs in different stages of life, starting from birth and up to adulthood.[25,26]

A number of evidence suggest that the alteration in the composition and function of gut microbiota during both early life and adulthood is associated with the development of several diseased conditions such as that of the gastrointestinal tract[27] or colorectal cancer,[28] metabolic diseases such as obesity, diabetes, immune disorders e.g. asthma, allergies, depression, anxiety and cardiometabolic disease.[24,29,30] Moreover, the metabolic diseases resulting from an altered gut microbiome may adversely affect the gestational environment and may affect offspring in the next generation.[31,32] Alteration in the microbiome composition is also associated with mortality risk. One of the recent long-term prospective studies involving a population in Finland ($n = 7211$) established an association between gut microbiome variation and cause-specific mortality risk, and potentially also disease risk. This vigorous association study suggested a crucial involvement of the Enterobacteriaceae family with the mortality risk.[33]

This chapter will focus on the recent findings from human, animal, and *in vitro* studies investigating how the gut microbiome or microbiota-derived metabolites can affect host pathophysiology, particularly the cardiovascular system. We will also discuss ways to improve cardiovascular outcomes, such as developing inhibitors for specific microbial pathways. Finally, we'll go over the current situation in developing therapeutic approaches for cardiometabolic diseases that target the gut microbiome.

2. The gut metabolome and the host pathophysiology

Gut microbiota has shown a crucial impact on the physiological state and onset of several diseases. Few of the reviews extensively focused on gut microbiota involved in regulating several chronic disorders such as neurodegenerative diseases, metabolic diseases, cardiovascular diseases, and gastrointestinal diseases. In future, thus modulating gut microbiota may provide a new dimension for the prevention and treatment of human disease.[34] The studies have seen a significant association of the gut microbiota with the brain and liver, forming axes called the gut-brain axis[35] and the gut-liver axis, respectively.[36–38] The production or release of the metabolites/gut-derived products innervate the axes mentioned above, leading to various diseases and ailments in the host body. The imbalance in the regulation of the gut microbial activities, composition, and homeostasis leads to leakage of the gut epithelium due to disruption of adhesion proteins and translocation of the pathogenic bacteria-derived products. These lead to the activation of various immune responses in the systemic circulation. An imbalance in the gut microbiota causes infections due to the entry of opportunistic pathogens, resulting in different infectious diseases such as *H. pylori* infection.[39] It also results in the increased growth of other bacteria such as Fusobacterium and *H. pylori* which contributes to the risk of causing gastric cancer.[39] Inappropriate immune responses and activation of carcinogenesis in the gut may cause colorectal cancer and extra-intestinal cancer such as hepatocellular carcinoma. It is reported that disturbances in the gut epithelial barrier cause immune impairment, causing diseases such as Inflammatory bowel syndrome.[39] Decrease in the abundance of gut Firmicutes leads to increased levels of pro-inflammatory cytokines (IL12, IFN-γ) and reduced levels of anti-inflammatory cytokine (IL-10). In addition, studies have also reported an indirect link between gut microbiota and the risk of cardiovascular diseases and associated them with a decrease in the Bacteriodates and Firmicutes ratio.[40,41]

Translocation of bacteria and bacterial metabolites may cause the production of uremic toxins. Further, the synthesis of nitrogenous compounds activates the systemic pro-inflammatory cytokines.[42] Neurological disorders are caused due to the improper release of neuroactive chemicals and neurotransmitters.[43] On investigation, it was found that the presence or outgrowth of a few bacterial genera such as *Akkermanisia, Desulfovibrio, Lactococcus, Oscillopsia, and Sutterella* was linked with social behavior and autism disorders.[44,45] Streptococcus being the commensal bacteria, can also become opportunistic and are more prevalent in patients with depression.[46] An exciting role played by the gut microbiome is that it shapes the social contacts of humans and shapes the gut microbiome on social connections.[35] It has also been described that there is a connection between the microbiome and systemic and fecal metabolic environments.[47]

Metabolic disorders or poor metabolic health have been a major concern worldwide due to their increasing prevalence, which arises primarily due to various factors such as a sedentary lifestyle, intake of high-energy food, and smoking. Both the gut microbiota and endogenously formed host metabolites from dietary sources (small molecules, <1500 Da) interact in the gut and produce a diverse range of new metabolites which not only alter the host metabolome but also the microbiome composition.[48] These changes are also represented in blood and may be responsible for cardiovascular disorders. Depending upon the nature of microbiota-derived metabolites such as short-chain fatty acids, branched-chain amino acids, bile acids, tryptophan, indole derivatives and trimethylamine N-oxide, different kinds of cardiovascular disorders may arise.[49–53] The first evidence of the mechanistic association between microbiota and metabolism comes from a study on germ-free (GF) mice, which revealed that the gut microbiota increased the body fat content by 60% and induced insulin resistance by suppressing the Fiaf protein (a lipoprotein lipase inhibitor).[54] With the evolution of metabolomics and metagenomics, several potential causality links have been established in the last decades.[55] For instance, an inverse correlation has been established between the abundance of *Bacteroides thetaiotaomicron* and serum glutamate level in obese subjects.[56] Similarly, a positive correlation has been found between insulin resistance and a few microbial species, such as *Bacteroides vulgatus* and *Prevotella copri*.[57] In metabolic disorders, alteration in the levels of metabolites such as triglycerides, phospholipids, imidazole propionate (ImP), aromatic and branched-chain amino acid (BCAA), short-chain fatty acid (SCFA), bile acid (BA) and dysregulation of metabolic pathways such as lipolysis, aminogenesis and ketogenesis have been observed in human samples.[58–63] Recently, hippurate, a hepatic phase 2 conjugation product of microbial benzoate, has been shown to be involved in regulating metabolic health.[64]

A plethora of studies has shown the alteration of the microbiome due to environmental and dietary perturbations which include dietary changes, including fiber deficient diet,[9,65] antibiotic use,[66] neurological conditions,[67–69] geography,[70] immigration,[71] and immune response that impact allergy.[9] Experiments on germ-free and colonized mice proved that the gut microbiome not only affects the colonized site, i.e., the gut but also distant body sites, including the kidney, liver,[59,72] heart [73] and plasma.[74]

3. Mechanism of interaction between the gut microbiome and the host

The gut microbiota has been involved in the incidence, development, and progression of various diseases. The microbiota or the metabolites derived from them coordinate with the host in a bidirectional way and modulates the underlying host process and physiology (Figure 1). Sensing and coordination start with the mammalian intestine which is a physical barrier to separate the microbiota from other parts of the body. Similarly, the host system also senses the microbial metabolites to regulate the barrier function of the intestinal epithelial layer.[75] The microbiota also regulates the host immunity by stimulating the innate immune cells and activity-specific receptors located on the membrane surface of the epithelium site.[76] These receptors include toll-like receptors (TLR), NOD-like receptors (NLRs), Retinoic acid-inducible gene I (RIG-I)-like receptors (RLRs)[77,78] and pattern recognition receptors (PRR) which can be found in general and might be cell type-specific.[79] The molecular structure present on foreign organisms or pathogens called pattern-associated molecular patterns (PAMPs) and are recognized through the pattern recognition receptors such as TLRs. Few of the receptors get activated by bacteria-specific ligands such as TLR activation by the endotoxin lipopolysaccharide (LPS) from gram-negative bacteria.[80] Activation of a range of immune cells such as macrophages, NK cells,

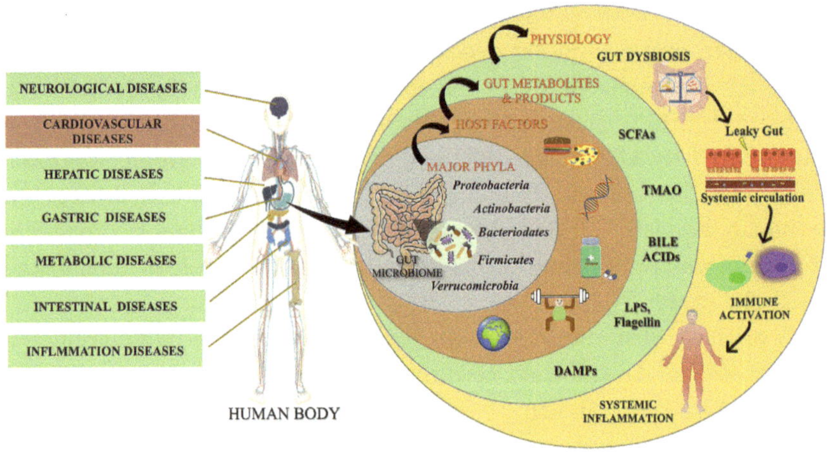

Fig. 1 Roles of the gut microbiome on host health. Gut microbiota and metabolites derived from them influence human physiology in a variety of ways by acting on the distal organs. Dysbiosis of the host gut microbiome as well as an alteration in the metabolome may result in the perturbation of host physiology characterized by leaky gut and systemic inflammation.

dendritic cells, and the cytokines released by them cause immune compromise and may lead to several autoimmunity disorders.[81] Many physiological factors such as the host's health and other environmental factors adversely influence or disturb the host-microbiota homeostasis and its regulation system toward the host, which may finally cause gut dysbiosis. It leads to microbial translocation, small intestinal bacterial overgrowth (SIBO), and an increase in intestinal permeability. Intestinal permeability disrupts the intestinal epithelial barrier causing a "leaky gut," characterized by the leakage of gut-derived products and bacterial products/metabolites such as LPS, lipoteichoic acid, flagellin, and bacterial DNA.[82–84] The gut microbial components not only affect the intestine but also damage the remotely located cell, tissue, or organs by entering through the systemic circulation and finally alter various facets of host's physiological changes.[51]

Recent studies have highlighted several non-olfactory GPCRs (G-protein-coupled receptors) that can sense small molecules derived from microbiota that impact host physiology.[7] For instance, a recent study using a forward chemical genetics screen identified dozens of bacteria that activated GPCRs. These include histamine and phenethylamine,[85] tryptamine (stimulate $5HT_4R$)[86] and *N*-acyl amides.[87] Microbial metabolites also interact with other receptors and activate signaling pathways for example indole 3-aldehyde from *Lactobacillus reuteri* bind to the aryl-hydrocarbon receptor (AhR).[88]

4. Gut microbiota, metabolome, and CVDs

There is a complex relationship between microbiome composition, intestinal wall integrity, proinflammatory state, and the development of CVDs.[89] It has been often observed that barrier function in heart failure patients, as compared to that of healthy people, was impaired and leaky, leading to migration of bacterial products from the intestinal lumen into the circulation, which can lead to a pro-inflammatory state.[90,91] In the state of compromised intestinal integrity, endotoxins, i.e., LPS derived from several gram-negative bacteria can travel to the circulation, where they are primarily recognized by TLR (toll-like receptor) on the sentinel cells of the immune system.[92] Upon binding of bacterial ligands to the sentinel cells of the host, the TLR signaling pathway is triggered and initiates the release of proinflammatory cytokines, which facilitate a proinflammatory state in the host.

Cardiometabolic diseases are generally characterized by a high disequilibrium in the composition of microbial taxa or gut bacterial richness, which has been shown in patients with myriad CVD risk factors, including metabolic diseases.[93,94] For example, patients with cardiometabolic disease (CMD) have an enriched abundance of gut microbiome species of Enterobacteriaceae, including *Escherichia coli*, *Enterobacter aerogenes* and *Klebsiella* spp. as compared with healthy control subjects.[95] Similarly, the gut microbiome of the patients with ischemic heart failure has an increased abundance of *Acinetobacter, Ruminococcus*, and *Veillonella* spp. and a decreased abundance of *Oscillibacter* spp., *Alistipes* and *Faecalibacterium*.[96] It is evident that the microbiome or its metabolites cause various metabolic disorders, which together represent a high-risk factor for cardiovascular diseases and mortality.[49] With the advent of multi-omics studies, researchers tried to identify an association between the gut microbiome and CMD. The first study linking the microbiome and CVD identified trimethylamine N-oxide (TMAO), a microbial metabolite, as a causing agent of CVD.[52,97,98] Fig. 1 shows several important pathways connecting gut microbiota to CVD, and a number of related phenotypes and known molecular factors, such as identified host receptors.

4.1 Bile acids: Predominant metabolites of the gut microbiota

Bile acids (BAs) are host-produced small molecules synthesized from cholesterol through oxidation of cholesterol in hepatocytes and play a major role in the emulsification and absorption of fat-soluble dietary nutrients.

Recent evidence has shown that apart from significant roles, BAs also play additional roles in the modulation of lipid metabolism, glucose/insulin metabolism, and inflammation.[99] The primary BAs like cholic acid (CA) and chenodeoxycholic acid (CDCA) are produced by the mitochondrial CYP27A1 and conjugated to amino acids taurine or glycine, termed bile salts.[100,101] To maintain the bile acid pool, bile salts that arrive at the ileum can be recycled through the enterohepatic circulatory system or excreted in the faeces.[102,103] The gut microbiota plays a vital role in increasing the BA diversity through directly sensing and transforming primary BA into major secondary BA such as deoxycholic acid (DCA) and lithocholic acid (LCA) from CA and CDCA, respectively. Over twenty different secondary bile acids have been reported in human adult feces produced by several bacterial genera through deconjugation, oxidation, epimerization, 7-dehydroxylation, esterification, and desulfatation.[104] These secondary BA produced from the gut microbiome are identified by the host to regulate numerous aspects of host health through interaction with several nuclear and membrane-bound receptors, including FXR (farnesoid X receptor; expressed in most enterocytes), LXR (liver X receptor), PXR (pregnane X receptor), and specific GPCRs like TGR5 (takeda G-protein–coupled receptor 5), vitamin D receptor (VDR), and constitutive androstane receptor (CAR).[105–110] The recognition and activation of the nuclear receptor by bile acids leads to a variety of biological effects in the host, including lipid and glucose metabolism and host immunity.[111–113]

Perturbation in the composition of the microbiome and bile acids and alteration of interaction between them have been reported to be associated with several cardiometabolic diseases and could be used as potential targets for the treatment of CVDs.[102,114] Such perturbation also leads to primary BA accumulation since microbiota cannot metabolize primary bile acids during metabolic disorders. For example, BA can hamper deconjugation reaction in the normal gut microbiota and thus starts to accumulate in the colon of patients.[115] Changes in the level of primary BA like CDCA cause a decrease in low-density lipoprotein and plasma triglycerides (TG), which ultimately lead to a state of low-grade inflammation.[116] Previous studies showed that levels of BA in the serum of hypertensive patients are directly proportional to the disease severity.[117] A number of studies proved that FXR is expressed in the aorta and cardiomyocytes and involved in the progression of metabolic dysfunction, myocardial ischemia/reperfusion injury, and cardiomyopathy through its modulation by BAs.[118–120] Generally, primary bile acids have pro-inflammatory effects, and recently it has been shown that gut microbiota can regulate CVDs by inhibiting the synthesis of BAs through activation of GPR5 expression.[113]

4.2 Short-chain fatty acids

Short-chain fatty acids (SCFAs), are one of the most studied microbial metabolites, are produced through anaerobic fermentation of dietary fibers by anaerobic bacteria (such as *Bifidobacteria* and *Lactobacillus*) and are made up of fewer than six carbons such as acetate (two-carbon); propionate (three-carbon) and butyrate (four-carbon).[121–123] The dietary polysaccharides ingested by mammals cannot be digested due to a lack of the specific enzyme, and eventually, the undigested polysaccharides are metabolized into SCFAs by intestinal microbes.[124] These SCFAs are reported to be implicated in a myriad of physiological functions and are beneficial in most circumstances such as maintenance of mucosal barrier function,[125] glucose and lipid metabolism,[126] control of energy expenditure,[124,127,128] immunity and inflammation[62] and inhibition of the growth of harmful bacteria.[129]

Accumulating evidence using untargeted metabolomics indicates that in addition to the above-mentioned function of intestinal microbiota and SCFAs derived from them can impact host blood pressure homeostasis, myocardial repair, and other cardiometabolic diseases.[124,130] Moreover, early antibiotic exposure has also been linked to alteration in the microbiota diversity[131] and cardiometabolic diseases, such as diabetes.[132,133]

Recent evidence suggests that alteration in dietary behavior, such as a diet rich in fermentable fibers can lead to dysbiosis of the intestinal microbiota, which can disturb the microbial production of SCFAs that is ultimately associated with blood pressure instability.[134,135] For example, an initial clinical intervention study reported that fiber consumption has been linked to lower blood pressure.[136] A cross-sectional study using 16s rRNA gene sequencing, transcriptomics and metagenomics showed that plasma and faces were abundant in SCFAs and were enriched in *Ruminococcaceae NK4A214, Faecalibacterium prausnitzii, Ruminococcaceae_UCG-010, Roseburia hominis and Christensenellaceae_R-7* in normotensive groups, whereas, *Bacteroides plebeius, Bacteroides coprocola* and genera of Lachnospiraceae were of increased abundance in hypertensive patients.[134]

Line of evidence indicates that SCFAs regulate blood pressure through binding to their cognate GPCRs. These extracellular receptors, including

Olfr78 (olfactory receptor 78) and Gpr41 (G-protein receptor 41) are present in intestinal epithelial cells, vascular smooth muscle cells and kidneys,[41,137] and SCFA binding to the receptors modulate blood pressure (BP) by releasing renin and changing vascular resistance.[138,139] The role of microbial-derived SCFA (acetate) in regulating blood pressure and cardiac contractility has further been investigated in mice. The results showed that an increase in acetate level diminished the cardiac output and reduced the vascular resistance[140] and depletion of the SCFA pool increased BP. Moreover, another SCFA (propionate) protected the mice with angiotensin-II-induced hypertension and Apo $E^{-/-}$ atherosclerotic mice by attenuating cardiac hypertrophy, fibrosis, and vascular dysfunction.[141] Recent studies provided further evidence that cecal microbiota transplantation of SCFAs altered cardiovascular pathology such as arterial stiffness and myocardial infarction (MI) and improved repair capacity after MI.[142,143]

The studies on fecal transfer between hypertensive and normal counterparts provided inconsistent data. The transmission of the disease showed mixed results may be due to host genetic or epigenetic factors. One of the studies showed that the transfer of feces from human donors with hypertension into germ-free (GF) mice showed the transfer of heightened BP.[41] Contrary to this finding, transplantation of feces from normotensive donors into hypertensive recipients exacerbated hypertension in the recipient rats.[144]

Individual SCFAs derived from a diverse range of microbial flora may have distinct effects on host organs. Deciphering the role of each SCFA and their receptors, and signaling pathways may be helpful in understanding the biological significance and finding new therapeutic strategies.

4.3 Branched-chain amino acids

The essential and most abundant branched-chain amino acids (BCAAs) synthesized from gut microbiota, including leucine, isoleucine, and valine, play crucial roles in protein synthesis, glucose and lipid metabolism and biological processes such as hepatocyte proliferation, and immunity.[145] Although moderate intake of BCAAs has beneficial roles and promotes a healthy microbiota,[146] the excessive accumulation of it may exert various adverse effects such as insulin resistance, diabetes, neurological distress and cardiomyopathy[147] by activation of mTOR, JNK, and insulin receptor substrate (IRS)-1 signaling pathways.[148,149] Increased levels of BCAA in genetically obese mice (ob/ob mice) have been linked to an increased risk of metabolic disorder and insulin resistance in future.[150] In addition, patients with insulin resistance have a reduced abundance of bacteria that can take up BCAAs, such as *Butyrivibrio crossotus* and *Eubacterium siraeum*.[57] The results of a randomized, crossover study suggested that an animal-based high-fat diet increased the plasma levels of BCAAs like valine and leucine compared to a plant-based low-fat diet and was associated with heightened risk for cardiometabolic abnormalities.[151]

There is an independent association between the rise in circulatory BCAAs and ambulatory BP in individuals with high hyperglycemic condition.[152] The high level of BCAAs is also correlated independently with several coronary artery disease (CAD) risk factors and increased carotid intima-media thickness in subclinical atherosclerosis.[153] Recently, a study on patients with CAD diagnosed by coronary angiography showed that BCAAs are significantly related to the development of CAD independent of confounding factors i.e. diabetes, hypertension and dyslipidemia.[154] Increased accumulation of BCAAs augments the excessive generation of mitochondrial reactive oxygen species (ROS),[155] leading to mitochondrial dysfunction and associated cardiometabolic diseases such as diabetes, hypertension, atherosclerosis, and CAD.[156] Recently, Yang *et al.*, demonstrated that diabetic cardiomyopathy is characterized by decreased gut microbial diversity and increased abundance of the BCAA-producing microbes in mice. These changes were associated with intestinal barrier impairment, which eventually lead to heart rate variability and cardiac damage. Pyridostigmine, a reversible cholinesterase inhibitor, restored the above-mentioned parameters as well as reduced circulating BCAA levels with concomitant attenuation in cardiac dysfunction in mice having diabetic cardiomyopathy.[157]

4.4 Trimethylamine N-oxide

In recent years, one of the microbial metabolites, trimethylamine N-oxide (TMAO), drew considerable attention from cardiovascular biologists for having adverse effects on the heart. Western diets including cheese, egg yolk, shellfish and meat and are abundant in the dietary components viz. choline, phosphatidylcholine, and L-carnitine, which serve as the precursors for trimethylamine (TMA; an odorous gas) production in the gut. Subsequently, other nutrient components such as betaine,[98] γ-butyrobetaine[158] and trimethyllysine[159] are found to be attributed as the precursors for TMA formation. In the gut, the microbiota is involved in the conversion of the dietary source to TMA by utilizing microbial-specific enzyme choline-TMA lyases (encoded by the *cutC/D* genes), carnitine oxygenase (CntA) and betaine reductase (GrdH).[160,161] After being absorbed from the gut, it reaches the liver through the circulation, where it subsequently metabolized to TMAO[162] through the enzymatic reaction of hepatic flavin monooxygenases 3 (FMO3)[52,98,163] (Fig. 2).

In a randomized-controlled dietary intervention and heavy isotope tracer studies, Wang *et al.* have shown recently that consumption of red meat, but not white meat or non-meat, may induce the formation of microbial TMAO from carnitine but not from choline. Red meat intake also increased two other gut microbiota-produced metabolites of carnitine i.e. γ-butyrobetaine, and crotonobetaine.[164]

The first incidence of the link between gut microbiome-derived metabolite, TMAO, and CVD risk comes from the study in a large clinical cohort. Authors have generated unbiased small-molecule metabolic profiles in plasma using a metabolomics approach that predicts risk for CVD. They found that diet supplemented with choline, TMAO, or betaine elevated atherosclerosis through upregulation of several macrophage scavenger receptors which are dependent on the gut flora.[98] Later, a large number of studies including meta-analyses[165–167] have shown that increased level of TMAO is directly related to various CVDs such as accelerated atherosclerosis,[52] acute coronary syndrome,[168–170] coronary artery disease,[169] thrombosis,[171] peripheral artery disease[172] and heart failure (HF)[173–175] and is linked to a higher risk of cardiovascular disease and all-cause mortality.[97] Studies have revealed that TMAO not only have a direct role in the pathogenesis of atherosclerosis but also it can aggravate atherogenesis by promoting platelet reactivity,[171,176–179] increasing vascular inflammation,[180–183] accelerating cholesterol metabolism,[52] inducing TF (tissue factor) expression[184] and heightened heart failure.[185–187]

A number of studies indicate that meat consumption dose-dependently increases the risk of CVD and mortality.[188] It has been shown that choline and carnitine from red meat or TMAO can be metabolized by intestinal microbiota to increase the production of TMAO and accelerated atherogenesis.[52] Microbial transplantation studies have shown that when cecal microbes producing high TMA (choline-induced) were transferred to atherosclerosis-resistant NZW/LacJ mice, the recipient mice became more prone to atherosclerosis.[189] This transmissibility to dietary choline-induced TMA production and formation of thrombus was due to the direct role of

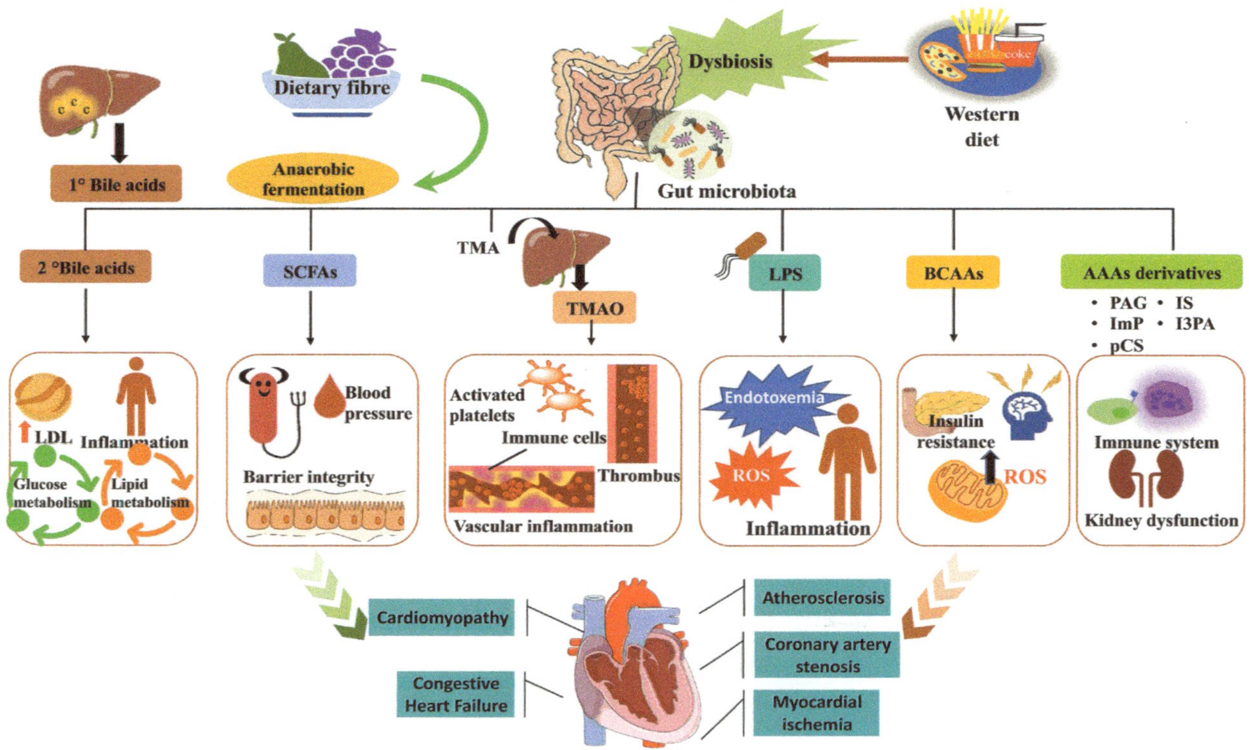

Fig. 2 See figure legend on next page.

gut microbial *cutC* gene as evidenced by gain-of-function in *C. sporogenes* mutant.[160,177,190] Several *in vitro* and animal studies showed the involvement of TMAO to increase atherosclerosis.[191–195]

Platelet function and its reactivity play an important role in thrombus formation and atherosclerosis. In this context, a large number of studies have shown that microbe-derived TMAO can induce platelet hyperactivity and thereby can promote thrombotic events. For example, Zhu *et al.* have shown that TMAO from gut microbes or choline supplementation contributes to platelet hyperreactivity and enhances thrombosis potential by altering calcium signaling.[171,196] Using a murine $FeCl_3$-induced carotid artery injury model, the effect of TMAO was further investigated by FMO3 knockdown or overexpression. Manipulating the expression of FMO3 directly alters systemic TMAO levels, platelet hyperactivity, thrombosis rate,[178,197] and also impacts cholesterol balance by upregulating macrophages scavenger receptors, namely CD36 and scavenger receptor A.[52,198,199] Studies on *in vitro* and rodents have demonstrated that TMAO can cause cardiac fibrosis, cardiac dysfunction, and accelerated endothelial dysfunction by activation of signaling cascades such as TLR4-NLRP3-TGF-β[200] and PKC/NF-κB/VCAM-1[201] pathway accompanied by

Fig. 2 Direct and indirect roles of gut microbiota-derived metabolites on the pathogenesis of CVDs. The difference in the abundance of microbial flora inside the gut and the kind of nutritional intake affects the metabolome composition. Western diet influence the host physiology by getting absorbed into the circulation and/or liver and then converted into secondary metabolites. The circulating metabolites may be the indicators of myocardial ischemia, peripheral artery disease, congestive heart failure etc. Primary bile acids (1° BAs) are converted to secondary bile acids (2° BAs) by the microbiome. Converted 2° BAs are involved in anti-inflammatory pathways. BA metabolism is altered in patients with metabolic syndrome and is associated with dyslipidemia and insulin resistance. SCFAs from dietary fibres can regulate blood pressure by combining with Olfr78 and GPR41. The microbiome also metabolises histidine to ImP, tyrosine to pCS, and choline to TMA. ImP impairs glucose tolerance and insulin signaling, and increased pCS level is associated with increased cardiovascular events in patients with CKD. Increased TMAO level is directly related to inflammation, T2DM, atherosclerosis and cardiovascular diseases. Increased level of BCAA has been linked to a high risk of cardiometabolic abnormalities. LPS cause the release of pro-inflammatory cytokines mediated by TLR4 stimulation and their immune responses are ultimately associated with the development of CVDs. CVD: cardiovascular disease, SCFA: short chain fatty acid, ImP: imidazole propionate, pCS: p-cresyl sulfate, TMA: trimethylamine, TMAO: trimethylamine oxide, BCAA: branched-chain amino acids, T2DM: Type 2 diabetes, LPS: lipopolysaccharide, AAA: aromatic amino acids, PAG: phenylacetylglutamine, IS: indoxyl sulfate, I3PA: indole-3-propionic acid.

ROS generation and a surge in inflammatory cytokines which ultimately lead to atherosclerosis. Recently, Wang *et al.* showed that the high-choline or L-carnitine diet-induced microbial production of TMAO in apolipoprotein E null mice was substantially inhibited by 3,3-dimethyl-1-butanol (DMB; a TMA lyases inhibitor). Their study suggests that the inhibitor may serve as a potential therapeutic approach for the treatment of cardiometabolic diseases.[194]

Extensive experimental evidence supports the involvement of vascular inflammation as a critical risk factor for the progression of atherosclerosis mediated by immune cells and inflammatory cytokines.[202] Recent studies on animal models have shown that gut microbiota-derived TMAO or its direct administration heightened vascular inflammation which was accompanied by the activation of endothelial cells, increased oxidative stress and impaired cell proliferation and migration,[203] increased expression of vascular adhesion molecules,[181] activation of the NLRP3 inflammasome, mitochondrial reactive oxygen species generation, altered levels of TXNIP (thioredoxin-interacting protein), and a decreased lysosomal stability.[180,204,205] To investigate whether the gut microbiome plays important role in age-related arterial dysfunction, the aged mice were treated with broad-spectrum and poorly-absorbed antibiotics to suppress the gut microbiome and measured plasma levels of TMAO. In aged mice, antibiotics treatment recovered endothelial dysfunction characterized by lower oxidative stress and reduced arterial stiffness.[206]

Being a crucial factor for blood coagulation, TF (tissue factor) forms a thermostatic barrier for organ surface, which perturbation leads to arterial and venous thrombosis.[207] As evidence suggests, TMAO can induce the expression of tissue factors in endothelial cells *in vitro*[208] and promote vascular inflammation and thrombosis in patients with Type 2 diabetes (T2DM).[184] A recent study on germ-free and TLR2-deficient mice (having reduced thrombus growth) revealed that colonization with microbiota increased platelet integrin-dependent thrombus formation and decreased hepatic VWF (von Willebrand factor) synthesis.[209]

Patients with heart failure (HF) have already been shown to have changes in gut microbial composition, particularly those with low microbial diversity and depleted core gut microbiota.[96,210,211] However, the link between gut microbiome composition and the pathophysiology of HF remains elusive. In a prospective study of 22 patients with decompensated or compensated HF, there is a significant increase in the plasma concentration of TMAO.[212] Moreover, bacterial 16S rRNA sequencing revealed that levels of Actinobacteria and Bifidobacterium at the phylum level and genus level, respectively, were increased in HF patients compared to those in the control patients.[212] In a prospective study of 955 patients with reduced ejection fraction, it has been shown that increased TMAO levels in the plasma are associated with a heightened risk of the composite outcome of cardiovascular death. Moreover, the lower plasma TMAO levels are associated with the FMO3 AA-genotype (homozygous normal) in FMO3 E158K polymorphisms.[213] In patients with HFpEF who have been examined before or after discharge have elevated plasma levels of TMAO (>20.37 μmol/L) and were associated with an increased risk of HFpEF or cardiac events later.[214,215] In contrary to the beneficial effects of TMAO, a recent finding suggests that administration of TMAO (120 mg/kg) improves mitochondrial energy metabolism and heart function right ventricle heart failure in rats.[216] Although few of the research depicted the beneficial effect of diet enriched in TMAO-producing components (l-carnitine) on cardiovascular outcomes i.e. reduced aortic lesion sizes,[217,218] gut microbiota-derived TMAO is majorly involved in various cardiometabolic diseases especially aggravation of atherogenesis.

4.5 Aromatic amino acids (AAAs) and their derivatives

AAAs include benzene ring-containing amino acids viz. tyrosine (Tyr), tryptophan (Trp), and phenylalanine (Phe). Trp and Phe are essential amino acids obtained from the diet with high protein content such as meats, fish, nuts, eggs, oats, milk, and cheese.[219,220] Besides having a role in protein synthesis, Tyr is a precursor to a variety of bioactive molecules, including epinephrine, norepinephrine, dopamine, and thyroid hormone, and Trp is a precursor of serotonin, melatonin, and niacin.[221,222]

Three main pathways can metabolize dietary Trp in the gastrointestinal tract are serotonin (5-hydroxytryptamine [5-HT]) route via Trp hydroxylase 1 (TpH1),[43] kynurenine pathway via indoleamine 2,3-dioxygenase (IDO) 1[223] and direct transformation into several molecules, such as indole and its derivatives by the action of the gut microbiota. These indole derivatives act on a nuclear receptor aryl hydrocarbon receptor (AhR), also called the dioxin receptor.[88] Some of the indole derivatives of Trp include indole-3-acid-acetic (IAA), indole-3-aldehyde (IAld), indole-3-acetaldehyde (IAAld), indole-3-propionic acid (IPA), and indoleacrylic acid, and are ligands for AhR.[224] These indole derivatives interact with the AhR and have a variety of consequences on the host, including immune response at the intestinal barrier, metabolism, behavior, and lifespan.[225]

4.5.1 Indoxyl sulfate

The indoles derived through microbial metabolism may consequently be absorbed and transformed into a uremic toxin i.e. indoxyl sulfate (IS) in the liver, then excreted in urines and accumulated during chronic kidney disease (CKD).[226] During accumulation, these metabolites were suggested to exert pro-inflammatory and oxidative stress and to be involved in the pathogenesis of cardiorenal diseases such as atherosclerosis, arteriosclerosis, atrial fibrillation, and congestive heart failure.[227–229] More recently, the effects of indole and indoxyl sulfate have been explored in the context of gut microbiota and associated CVDs. For example, gut bacteria-derived metabolites indole and indoxyl sulfate influence arterial blood pressure via central and peripheral mechanisms via serotonin-dependent signaling.[230] The evidence establishing a direct link between gut microbiome-derived Trp metabolites and CVDs is limited. Kinenurinic acid produced in the host by IDO1 (a rate-limiting enzyme of the kynurenine pathway) has been found to be associated with unstable atheromatous plaques in patients with atherosclerotic lesions. Administration of Kynurenic acid in the Ido1-deficient mice reduced IL10 (a potent anti-inflammatory cytokine) production and promoted atherosclerosis in mice.[231]

Accumulating evidence showed that the increased activity of IDO1 and levels of IDO1-generated kynurenine in humans are associated with risk factors for atherosclerosis, acute myocardial infarction and coronary heart disease in the suspected stable angina patients.[232–235] Together, these results show a potential association of IDO1 and its product with CVD and warns the future study on exploring the relation between Trp metabolites and cardiometabolic diseases.

4.5.2 Indole-3-propionic acid

One of the indole derivatives from the direct transformation of Trp by gut microbiota (mainly by *Clostridium sporogenes*) is indole-3-propionic acid (IPA)[224,236] which upon binding to its receptor AhR affects intestinal permeability and host immunity.[237–240] The reports showed that IPA has beneficial effects *in vivo*, and its level was found to be negatively correlated with several metabolic diseases. Elevated levels of IPA were clinically associated with a lower risk of chronic kidney disease.[241] In the rodent models of steatohepatitis, IPA treatment reduced weight gain[242], microbial dysbiosis and inflammation,[243] and inflammation through PXR in a colitis model.[244] Moreover, low levels of IPA were linked to the occurrence of T2DM.[245,246]

Similarly, an elevated level of IPA is inversely associated with the incidence of cardiometabolic diseases in humans and mice. For instance, IPA levels were negatively correlated with the burden of atherosclerosis.[247,248] In this study on the advanced atherosclerosis cohort, plasma microbiome-modulated IPA concentration was negatively associated with advanced atherosclerosis.[247] Recently, Gesper *et al.* evaluated the effect of IPA on mitochondrial function *ex vivo* on cardiomyocytes as well as in the hearts of C57BL/6J mice. The authors found that acute treatment enhanced maximal mitochondrial respiration in cardiomyocytes and improved cardiac contractility in a dose-dependent manner in isolated perfused mouse hearts. The mechanistic study indicated that microbial-derived IPA has a significant impact on mitochondrial function in cardiomyocytes and cardiac physiology.[249]

4.5.3 Phenylacetylglutamine

Another well-known gut microbial metabolite[250] derived from an aromatic amino acid, phenylalanine, is the phenylacetylglutamine (PAG). PAG is the result of the microbial conversion of phenylalanine into phenylacetic acid (PAA) followed by glutamine conjugation.[251] PAA is synthesized in the large intestine from dietary phenylalanine, which has not been absorbed in the small intestine. The PAA may consequently be absorbed into the portal system and transformed to PAG in the liver then excreted in the urines. A line of evidence demonstrated that there is a dense interconnection between the elevated circulating PAG levels with increased abundance of microbial taxa such as Bifidobacteriaceae,[250] genera in the Lachnospiraceae, Christensenellaceae, and Ruminococcaceae[252] and Coriobacteriaceae[253] families. The recent finding showed a positive association of PAG with gut microbial diversity,[254] early kidney dysfunction[252,255], obesity,[256] and diabetes.[257,258] In addition to metabolic diseases mentioned above, PAG has been found to be a key risk factor for CVD and mortality after adjusting for comorbid conditions.[255]

The mechanistic association between PAG and CVDs has recently been studied in a cohort of 4000 patients undergoing cardiac diagnosis and has been described by Stanley Hazen and colleagues.[259] The elevated level of gut microbiota-derived PAG increased platelet hyper-reactivity and thrombotic potential via adrenoreceptors thereby increasing the risk of a major adverse cardiac event (MACE) in patients with T2DM. Through microbial engineering and transplantation into germ-free mice, the authors confirmed that gut microbial generation of PAG in the host is due to the relative function of the gut microbial genes viz. *fldH* gene cluster and the *porA* gene. They established several mechanistic connections between PAG and PAG-induced CVD risk through genetic, biophysical, and pharmacological studies. Using *ex vivo and in vivo* studies, PAG was found to enhance platelet function in whole blood, PRPs, and isolated platelets and facilitate the promotion of thrombosis potential in arterial injury models. Various biophysical, loss-of-function and gain-of-function, and pharmacological studies further elucidate that PAG-mediated cellular response and CVD-relevant phenotypes occur through interaction with adrenoreceptors, especially via α_{2A}, α_{2B}, and β_2 subtypes on human platelets.[259] Moreover, PAG-mediated prothrombotic effects and platelet hyperactivity were reversed by adrenoreceptor siRNA knockdown or administrating a β-adrenergic receptor antagonist (or β-blocker, carvedilol or propranolol) in a mice model of arterial injury. More recently, a study further showed that microbiota-derived PAG increased the risk of future CAD independently of other cardiovascular risk factors.[260] These studies suggest that several phenotypes, especially of CVDs exhibited by microbial metabolite interaction with adrenoreceptors, may be taken care of by using pharmacological inhibitors such as β-blockers and may be helpful for future therapeutic intervention.

4.5.4 Histidine derivative: Imidazole propionate

The role of a histidine derivative, imidazole propionate (ImP), has recently been explored. It has been identified as a gut microbial metabolite derived from dietary histidine and found to be enhanced in the systemic circulation of patients with T2D. Recently, an experiment on mice showed that a high concentration of ImP impairs insulin signaling and glucose tolerance through the activation of p38 MAPK and subsequent activation of mTORC1 when administered to the subjects.[59] Moreover, in a large MetaCardis cohort, serum ImP level was increased in patients with prediabetes and diabetes with low bacterial gene richness and Bacteroides 2 enterotype. Intake of a histidine-rich diet is also associated with the Bacteroides 2 enterotype, abundant in the genes involved in imidazole propionate biosynthesis.[261] Interestingly, a sub-analysis for the presence of CVD identified that subjects with CVD had significantly increased levels of ImP in the patients as compared to the control.[261]

4.5.5 p-Cresyl sulfate

In the distal end of the colon, phenylalanine and tyrosine are metabolized by bacteria into several phenolic compounds, of which p-cresol is one of the components. In the mucosa of the colon and the liver, detoxification of p-cresol produces p-cresyl sulfate (pCS). Following the entry into the circulatory system, pCS binds irreversibly to plasma albumin and is finally secreted from renal tubules.[262] Similar to indoxyl sulfate and PAG, pCS is one of the uremic toxins. Due to impaired excretion and accumulation during CKD, elevated serum/plasma pCS levels have been implicated in several toxic biological actions and worsening outcomes.[241,263,264]

In addition to adverse effects of pCS in general, elevated serum total level of pCS is significantly associated with coronary artery stenosis[265] and CV mortality among patients having CKD and diabetes,[266–269] and occurrence of MACEs,[270] Moreover, in several prospective studies, the increased serum or urine level of p-cresol (pCS before sulfation) was also correlated with increased cardiovascular events in the patients with CKD or non-diabetic hemodialysis.[271–273] Recent evidence from a study further suggests that pCS promoted cardiac apoptosis and diastolic dysfunction in mice that underwent nephrectomy. pCS also induced apoptosis in H9c2 cells by upregulating the expression of NADPH oxidase (NOX) subunits and the generation of ROS, which were suppressed by apocynin (an inhibitor of NOX) and N-acetylcysteine.[274] Taken together, these findings suggest that serum levels of p-cresol or pCS may be valuable markers in predicting CVD risk in patients with renal dysfunction.

4.6 Lipopolysaccharides (LPS)

The outer layer of gram-negative bacteria is composed of a glycolipid called lipopolysaccharide (LPS; endotoxin), which is the most potent inflammatory toxin and plays a crucial role in bacteria-host interaction modulating the immune response.[275] Most of the biological effects of LPS are due to lipid A, one of the three components of LPS structure.[276] LPS acts as a specific ligand for TLR4 receptors, stimulating the release of the potent pro-inflammatory cytokines such as TNFα, IL-1β, and IL-18 which may be responsible for various organ failure including cardiac dysfunction.[277–279] For several decades, higher endotoxin levels in the blood and endotoxin-induced immune activation have been associated with decompensated heart failure.[280] Later, several clinical studies have shown that the release of pro-inflammatory cytokines by the LPS is responsible for the development of CVDs.[281–283] It is now clear that the endotoxins derived from gut microbiota have been mechanistically associated with systemic inflammation, immune disorder, and cardiac dysfunctions, presumably through oxidative stress, calcium overload, and myocardial loss.[284–286]

It has also been observed that there was an increased translocation of endotoxin from the bowel into the circulation during acute heart failure as evidenced by elevated levels of endotoxin, soluble CD14, TNFα and IL6 in the hepatic veins of the patients compared to the left ventricle.[287] In a recent prospective study of patients with atrial fibrillation, the circulating plasma LPS levels were negatively associated with high adherence to a Mediterranean diet and predictive of MACE by enhancing platelet activation.[288] In the FINRISK97 cohort, the higher serum endotoxin level was associated with an increased risk of clinically incident diabetes after excluding other risk factors and was independent of metabolic syndrome.[289] In a study on peritoneal dialysis patients, plasma LPS level was significantly higher than control and correlated with serum CRP and albumin levels indicating systemic inflammation and features of atherosclerosis in the PD patients. The LPS level was also higher in the patients with pre-existing CVD.[290] A recent study suggests that administration of LPS to rats can induce systemic inflammation and cardiovascular oxidative stress as indicated by an increased level of white blood cells, malondialdehyde, and decreased level of the anti-oxidants the in heart, aorta, and serum. Treatment with curcumin improved oxidative stress and systemic inflammation induced by LPS.[291]

In the experimental setup with obese subjects, increased intestinal permeability to LPS to the circulation results in a heightened level of serum LPS termed metabolic endotoxemia, which elicits TLR4-mediated chronic low-grade inflammation, insulin resistance, and cardiac damage.[292,293] Large epidemiological studies suggest that endotoxemia is closely related to the pathogenesis of atherosclerotic heart disease. TLR4-mediated inflammatory processes by LPS are implicated in its progression to clinical heart disease.[294] Another study showed the association of endotoxemia in end-stage kidney disease (ESRD) patients and acute coronary syndrome as evidenced by a significant increase in the endotoxin level in ESRD patients with ACS.[295] Moreover, Zhang *et al.* showed that disturbance in the gut microbiota may induce increased LPS levels in mice and endotoxin exposure increased the incidence of cardia bifiBifidaing gastrulation in the chick embryo. The adverse effects of LPS on the embryo may be explained by the altered expressions of Nkx2.5 and GATA5 and increased production of excess ROS.[296]

TLR4 response elicited by LPS of different bacterial species characterized by structural variation in lipid A moieties. For instance, TLR4 response by tetra- and penta-acylated lipid A moieties of Bacteroides is milder than the hexa-acylated lipid A moiety of Escherichia.[297,298] In addition, the atypical LPS structure of *Porphyromonas gingivalis* contains heterogeneous lipid A species. One of the variants of lipid A exhibits TLR4 antagonist activity at the local site but it helps the pathogen avoid detection by the host innate immune system and contributes to inflammation at the remote site which leads to vascular inflammation and atherosclerosis progression in the ApoE$^{-/-}$ mice.[299]

LPS structural difference and associated TLR4 response may have resulted from the microbial diversity in the gut and thus may have harmful or beneficial effects on the host. Several human studies have shown that altered gut microbial composition is associated with atrial fibrillation.[300] More recently, a prospective study with atrial fibrillation patients showed that *Parabacteroides*, *Lachnoclostridium*, *Streptococcus*, and *Alistipes* were enriched while *Enterobacter* was depleted.[301] Despite the adverse effects of LPS on the cardiovascular system, several studies indicate its beneficial effects. For example, increasing bifidobacteria in the gut microbiota of mice improved high-fat diet-induced diabetes[292] and the LPS from Bacteroides inhibited cardiovascular diseases like atherosclerosis.[302,303] As evidenced recently, administration of a western diet with live *Akkermansia muciniphila* in ApoE deficient mice reduced intestinal permeability and decreased fecal and circulatory LPS levels, which was linked to lower aortic atherosclerosis without affecting lipid metabolism.[304] In a recent proof-of-concept study in humans, it was discovered that giving pasteurized (but not live) *Akkermansia muciniphila* in obese people with metabolic syndrome for 3 months reduced plasma LPS levels.[305]

5. Therapeutic uses of gut microbe/probiotics

The gut microbiota (GM) forms a symbiotic relationship with the host and plays a critical role in the host's health.[306] GM has been reported to have a critical role in the pathogenesis of TMAO-induced atherosclerosis,[307] and the determining role of GM in CVD has been documented previously.[308] Numerous microflora in humans such as *Desulfovibrio desulfuricans, Clostridium asparagiforme, Anaerococcus hydrogenalis, Clostridium sporogenes, Clostridium hathewayi, Escherichia fergusonii, Providencia rettgeri, Proteus penneri and Edwardsiella tarda* have shown to display choline to TMA-converting property.[309] Given the fact that gut microbiota interacts with the human host directly or indirectly thereby influencing the cardiometabolic health of the host by producing primary or secondary bioactive microbial metabolites, knowledge of the abundance of the gut microbiota, their genetic constituent and metabolic activity may provide a great understanding on the potential therapeutic strategies. Based on the microbiome composition and metabolome, a number of individual-specific interventions may be proposed to modulate disease-specific microbial diversity and activity such as dietary source, probiotic, prebiotic and antibiotics supplementation, fecal microbiota transplantation and inhibitors of microbial enzymes.

5.1 Probiotic supplementation

A recent study established a connection between the gut microbiome and the risk of cardiovascular disease. Modulating the composition of gut microbiota and/or microbial-derived metabolites might influence CVD risks and pathological changes in the host. As a result, novel therapeutic strategies may be developed for preventing and treating cardiometabolic diseases and CVD utilizing gut microbiota's potential.

Supplementation of Lactobacillus probiotic increased the survival rate of post-infarction mice pre-treated with antibiotics.[139] Mayerhofer *et al.* conducted the GutHeart trial (Phase II, randomized, open-label, controlled trial) and found that manipulating the gut microbiota, by providing probiotic *Saccharomyces boulardii,* improved LVEF (left ventricular ejection fraction) in patients with HF.[310] A recently published study has evaluated the effect of probiotic administration (*Saccharomyces boulardii*) in patients with chronic systolic HF and showed a reduction in left atrial diameter and an improvement in LVEF.[311] Gan *et al.* have shown that administration of the probiotics *Lactobacillus rhamnosus* GR-1 reduces left ventricular hypertrophy and strengthens left ventricular function in a rat model after inducing a myocardial infarction by occluding a coronary artery for 6 weeks.[312]

5.1.1 Hypertension

Probiotics containing other Lactobacillus spp. have also been shown to reduce the risk of cardiovascular disease. Treatment with *Lactobacillus murinus* made the mice resistant to high salt-induced hypertension by decreasing the T_H17 cell abundance.[313] In spontaneously hypertensive rat (SHR) hearts, oral administration of probiotic-fermented purple sweet potato yoghurt (PSPY) showed a beneficial role in reducing the cardiomyocyte apoptosis.[314]

Pre-biotics such as dietary fiber have the potential to change the microbiota composition of the host. Feeding the mice with a high-fiber diet and treating them with excess mineralocorticoids showed an increase in the acetate-producing bacterium *Bacteroides acidifaciens*. The high-fiber diet as well as the acetate supplementation significantly decreased the systolic and diastolic blood pressures, left ventricular hypertrophy and cardiac fibrosis. All the changes enhancing cardiovascular health and function were accompanied by the down-regulation of the cardiac and renal early growth response protein 1 (Egr1) gene.[315]

5.1.2 Atherosclerosis

Human fecal 16S ribosomal RNA gene sequencing analysis of gut microbial profile in CAD patients showed a significantly lower abundance of *Bacteroides dorei* and *Bacteroides vulgatus* compared to non-CAD controls. Gavage with live *B. dorei* and *B. vulgatus* decreased the plasma LPS levels and lessened atherosclerotic lesion formation in Apoe-deficient mice.[302] Similarly, increasing plasma TMAO levels in mice by feeding choline and carnitine-rich diets promote atherosclerotic plaque formation while depleting dietary supplements or suppressing the intestinal microbiota inhibit TMAO production thereby reducing atherosclerosis.[98,316]

Treatment with *Lactobacillus rhamnosus* GG can reduce atherosclerosis by decreasing the elevated level of atherosclerosis-related biomarkers.[317] Some microbial species help to prevent the formation of atherosclerotic plaques. The abundance of butyrate-producing bacterial genus Eubacterium and Roseburia has been found to be inversely correlated to the development of atherosclerotic lesions in patients and genetically modified mice models.[318,319]

Lactobacillus plantarum is a probiotic supplement that comes from fermented plant foods.[320] Wang *et al.* reported that intragastrical administration of *Lactobacillus plantarum* L2 to rats prevented ischemia/reperfusion (I/R)-induced bacterial translocation to extraintestinal sites viz. mesenteric lymph nodes, spleen, liver and kidney by maintaining mucosal integrity.[321] Lactobacillus plantarum consumption reduced serum levels of leptin and fibrinogen, resulting in the reduction of atherosclerosis in patients with a smoking habit.[322] VSL#3 is an anti-inflammatory probiotic blend made up of eight live bacterial strains that have been shown to reduce the risk of atherosclerotic plaque rupture and vascular inflammation in overweight adults and improve dyslipidemia profiles.[323,324]

5.1.3 Diabetes-associated obesity and hyperlipidemia

In people and rodents with obesity and diabetes, the abundance of *Akkermansia muciniphila* is associated with lower levels of mucin that resides in the mucus layer when compared to healthy people.[325–327] Daily oral supplementation of *A. muciniphila* slightly decreased body weight, reduced insulinemia, improved insulin sensitivity and total plasma cholesterol.[305]

5.2 Microbial enzyme inhibition

Inhibition of microbial choline TMA lyase provided a novel strategy to prevent gut microbes from converting choline to TMA.[176] As a result, circulating TMAO could be a helpful indicator for diagnosing CVD. Organ *et al.* induced LV hypertrophy and HF in mice, which they kept for 15 weeks on their respective diets. Compared to the control group, the choline-diet group had worse cardiac remodeling and dysfunction and higher levels of TGF-B, MMP2, collagen 1a1, collagen 3a1, and fibronectin-1 after 15 weeks. These effects were normalized with TMA lyase inhibitor iodomethylcholine treatment.[328]

5.3 Fecal microbiota transplantation

The transfer of fecal samples from a donor into the gastrointestinal tract of a patient with depleted microbiota in patients to restore normal gut microbiota is known as fecal microbiota transplantation (FMT). On the basis of the source of the fecal microbiota for transplantation, FMT can be categorized into allogenic FMT (fecal sample from a healthy donor) and autologous FMT (transfer of own fecal sample). The most commonly used FMT method is the Allogenic FMT.[329] Fecal microbiota transfer aims to introduce a new gut microbiota to a patient's gastrointestinal system.[330] FMT also plays an important role in combating antibiotic resistance as induces an 81% remission rate. For example, FMT effectively and safely treated *Clostridium difficile*-associated diarrhea in humans.[331,332]

5.3.1 Diabetes/obesity

Few studies have suggested that FMT can also be considered a novel therapeutic option for a whole range of other diseases, like obesity and T2DM.[333] FMT boosts the number of SCFA-producing bacteria like *Roseburia intestinalis* and the protective strain *Akkermansia muciniphila* in the body[325] and may play a role in improving insulin sensitivity by regulating GLP-1.[334] Furthermore, a higher amount of *A. muciniphila* was linked to a healthier metabolic status, particularly in the terms of fasting plasma glucose, body fat distribution, plasma triglycerides.[325] Furthermore, findings from a separate animal study revealed that plasma TMAO levels were linked to the relative abundance of Akkermansia in the jejunum and cecum of γ-butyrobetaine supplemented mice.[316] However, there is a harmful effect of the bacteria, it is enigmatic to conclude the actual outcome of Akkermansia abundance in humans. Interestingly, 6 weeks of small intestinal microbiota infusion from lean donors to obese patients showed enhanced insulin sensitivity.[335]

6. Conclusion

Over the past few years, the evidence suggests that CVD's prevalence and risk factors have increased tremendously worldwide. From various studies on humans, animals, and *in vitro*, a direct link between gut microbiota and CVDs has been established up to a great extent. Despite a large number of studies, the link between the two is not well-established; however, the advent of new technologies and research tools may provide insight into the new aspects of the gut-metabolite-CVD axis. Gut dysbiosis and its metabolites have adverse effects on the cardiovascular system directly or indirectly, and pre-existing comorbid risk factors such as diabetes, insulin resistance, atherosclerosis, and hypertension have been reported to worsen cardiovascular outcomes. A line of evidence indicates that the pro-inflammatory state, oxidative stress and increased thrombotic events play a central role in promoting the associated risk factors that finally progress to various cardiovascular complications.

Further exploration of the molecular and genetic mechanism that leads to the production of diverse metabolites and elucidating the signaling pathways related to associated CVD may provide a clue for therapeutic strategies. Considering the fact that alteration in dietary behavior, such as a diet rich in fat, branched-chain amino acids, and a western diet, can lead to dysbiosis of the gut microbiota, various animal and human dietary interventions, probiotics, or pre-biotics may be used to restore microbial diversity for the betterment of the host. Furthermore, a new therapeutic approach based on microbiome composition, food habits, health condition, and genetic background of the host can be suggested, and personalized medicine will be a hope in the future. The finding from microbiota transplantation studies indicates that FMT may serve as an excellent opportunity for intervention in the treatment of CVDs. Alteration of gut microbiota by phytochemicals represents exciting therapeutic potential in treating metabolic syndrome and various cardiovascular ailments.[336,337]

Moreover, few phytochemicals such as 3,3-dimethyl-1-butanol (DMB; a TMA lyases inhibitor) have been reported to block microbial TMA production.[194] Thus similar approach using natural and synthetic molecules as inhibitors of the microbial enzyme(s) of the metabolite production may be helpful in the development of novel drugs. As the plasma levels of several metabolites have been increased in various clinical conditions, these metabolites may serve as early biomarkers to identify the risk of CVD progression. Although technological advancements such as multi-omics hold the promise for deciphering host-microbiome interactions and biomarker discovery, further studies, including oversized, prospective interventional, will be needed to assess causality in clinical settings.

Acknowledgments

The authors are thankful to the Department of Biotechnology (DBT) for providing financial support of this work (Project no. BT/PR22881/BRB/10/1654/2018) and the National Institute of Pharmaceutical Education and Research, Guwahati core fund. Thanks are due to the Indian Council of Medical Research for supporting Research Associate position to MJA (Grant number 3/1/2(19)/OBS/2019-NCD–II).

Conflict of interest

The authors have no conflicts of interest, financial or otherwise.

References

1. Baquero F, Nombela C. The microbiome as a human organ. *Clin Microbiol Infect.* 2012;18(Suppl 4):2–4. https://doi.org/10.1111/j.1469-0691.2012.03916.x.
2. Peroni DG, Nuzzi G, Trambusti I, Di Cicco ME, Comberiati P. Microbiome composition and its impact on the development of allergic diseases. *Front Immunol.* 2020;11:700. https://doi.org/10.3389/fimmu.2020.00700.
3. Hillman ET, Lu H, Yao T, Nakatsu CH. Microbial ecology along the gastrointestinal tract. *Microbes Environ.* 2017;32(4):300–313. https://doi.org/10.1264/jsme2.ME17017.
4. Kho ZY, Lal SK. The human gut microbiome—a potential controller of wellness and disease. *Front Microbiol.* 2018;9:1835. https://doi.org/10.3389/fmicb.2018.01835.
5. Porter NT, Martens EC. The critical roles of polysaccharides in gut microbial ecology and physiology. *Annu Rev Microbiol.* 2017;71:349–369. https://doi.org/10.1146/annurev-micro-102215-095316.
6. Escherich T. The intestinal bacteria of the neonate and breast-fed infant. 1884. *Rev Infect Dis.* 1988;10(6):1220–1225. https://doi.org/10.1093/clinids/10.6.1220.
7. Glowacki RWP, Martens EC. In sickness and health: effects of gut microbial metabolites on human physiology. *PLoS Pathog.* 2020;16(4):e1008370. https://doi.org/10.1371/journal.ppat.1008370.
8. Human Microbiome Project Consortium. Structure, function and diversity of the healthy human microbiome. *Nature.* 2012;486(7402):207–214. https://doi.org/10.1038/nature11234.
9. Li J, Jia H, Cai X, et al. An integrated catalog of reference genes in the human gut microbiome. *Nat Biotechnol.* 2014;32(8):834–841. https://doi.org/10.1038/nbt.2942.
10. Qin J, Li R, Raes J, et al. A human gut microbial gene catalogue established by metagenomic sequencing. *Nature.* 2010;464(7285):59–65. https://doi.org/10.1038/nature08821.

11. Cantarel BL, Lombard V, Henrissat B. Complex carbohydrate utilization by the healthy human microbiome. *PLoS One*. 2012;7(6):e28742. https://doi.org/10.1371/journal.pone.0028742.
12. Rowland I, Gibson G, Heinken A, et al. Gut microbiota functions: metabolism of nutrients and other food components. *Eur J Nutr*. 2018;57(1):1–24. https://doi.org/10.1007/s00394-017-1445-8.
13. Yoshii K, Hosomi K, Sawane K, Kunisawa J. Metabolism of dietary and microbial vitamin B family in the regulation of host immunity. *Front Nutr*. 2019;6:48. https://doi.org/10.3389/fnut.2019.00048.
14. Nash MJ, Frank DN, Friedman JE. Early microbes modify immune system development and metabolic homeostasis—the "restaurant" hypothesis revisited. *Front Endocrinol (Lausanne)*. 2017;8:349. https://doi.org/10.3389/fendo.2017.00349.
15. Alam A, Neish A. Role of gut microbiota in intestinal wound healing and barrier function. *Tissue Barriers*. 2018;6(3):1539595. https://doi.org/10.1080/21688370.2018.1539595.
16. Cheng HY, Ning MX, Chen DK, Ma WT. Interactions between the gut microbiota and the host innate immune response against pathogens. *Front Immunol*. 2019;10:607. https://doi.org/10.3389/fimmu.2019.00607.
17. Chung H, Pamp SJ, Hill JA, et al. Gut immune maturation depends on colonization with a host-specific microbiota. *Cell*. 2012;149(7):1578–1593. https://doi.org/10.1016/j.cell.2012.04.037.
18. Ochoa-Repáraz J, Kasper LH. The second brain: is the gut microbiota a link between obesity and central nervous system disorders? *Curr Obes Rep*. 2016;5(1):51–64. https://doi.org/10.1007/s13679-016-0191-1.
19. Sittipo P, Shim J, won, Lee YK. Microbial metabolites determine host health and the status of some diseases. *Int J Mol Sci*. 2019;20(21):5296. https://doi.org/10.3390/ijms20215296.
20. El Kaoutari A, Armougom F, Gordon JI, Raoult D, Henrissat B. The abundance and variety of carbohydrate-active enzymes in the human gut microbiota. *Nat Rev Microbiol*. 2013;11(7):497–504. https://doi.org/10.1038/nrmicro3050.
21. Fransen F, van Beek AA, Borghuis T, et al. The impact of gut microbiota on gender-specific differences in immunity. *Front Immunol*. 2017;8:754. https://doi.org/10.3389/fimmu.2017.00754.
22. Sittipo P, Shim JW, Lee YK. Microbial metabolites determine host health and the status of some diseases. *Int J Mol Sci*. 2019;20(21):E5296. https://doi.org/10.3390/ijms20215296.
23. David LA, Maurice CF, Carmody RN, et al. Diet rapidly and reproducibly alters the human gut microbiome. *Nature*. 2014;505(7484):559–563. https://doi.org/10.1038/nature12820.
24. Amato KR, Arrieta MC, Azad MB, et al. The human gut microbiome and health inequities. *PNAS*. 2021;118(25). https://doi.org/10.1073/pnas.2017947118.
25. Donaldson GP, Lee SM, Mazmanian SK. Gut biogeography of the bacterial microbiota. *Nat Rev Microbiol*. 2016;14(1):20–32. https://doi.org/10.1038/nrmicro3552.
26. Mueller NT, Bakacs E, Combellick J, Grigoryan Z, Dominguez-Bello MG. The infant microbiome development: mom matters. *Trends Mol Med*. 2015;21(2):109–117. https://doi.org/10.1016/j.molmed.2014.12.002.
27. Ni J, Wu GD, Albenberg L, Tomov VT. Gut microbiota and IBD: causation or correlation? *Nat Rev Gastroenterol Hepatol*. 2017;14(10):573–584. https://doi.org/10.1038/nrgastro.2017.88.
28. Alhinai EA, Walton GE, Commane DM. The role of the gut microbiota in colorectal cancer causation. *Int J Mol Sci*. 2019;20(21):E5295. https://doi.org/10.3390/ijms20215295.

29. Ding R-X, Goh W-R, Wu R-N, et al. Revisit gut microbiota and its impact on human health and disease. *J Food Drug Anal.* 2019;27(3):623–631. https://doi.org/10.1016/j.jfda.2018.12.012.
30. Brown JM, Hazen SL. The gut microbial endocrine organ: bacterially derived signals driving cardiometabolic diseases. *Annu Rev Med.* 2015;66:343–359. https://doi.org/10.1146/annurev-med-060513-093205.
31. Ferretti P, Pasolli E, Tett A, et al. Mother-to-infant microbial transmission from different body sites shapes the developing infant gut microbiome. *Cell Host Microbe.* 2018;24(1):133–145.e5. https://doi.org/10.1016/j.chom.2018.06.005.
32. Reynolds RM, Labad J, Buss C, Ghaemmaghami P, Räikkönen K. Transmitting biological effects of stress in utero: implications for mother and offspring. *Psychoneuroendocrinology.* 2013;38(9):1843–1849. https://doi.org/10.1016/j.psyneuen.2013.05.018.
33. Salosensaari A, Laitinen V, Havulinna AS, et al. Taxonomic signatures of cause-specific mortality risk in human gut microbiome. *Nat Commun.* 2021;12(1):2671. https://doi.org/10.1038/s41467-021-22962-y.
34. Chen Y, Zhou J, Wang L. Role and mechanism of gut microbiota in human disease. *Front Cell Infect Microbiol.* 2021;11:625913. https://doi.org/10.3389/fcimb.2021.625913.
35. Mussell M, Kroenke K, Spitzer RL, Williams JBW, Herzog W, Löwe B. Gastrointestinal symptoms in primary care: prevalence and association with depression and anxiety. *J Psychosom Res.* 2008;64(6):605–612. https://doi.org/10.1016/j.jpsychores.2008.02.019.
36. Albillos A, de Gottardi A, Rescigno M. The gut-liver axis in liver disease: pathophysiological basis for therapy. *J Hepatol.* 2020;72(3):558–577. https://doi.org/10.1016/j.jhep.2019.10.003.
37. Martín-Mateos R, Albillos A. The role of the gut-liver axis in metabolic dysfunction-associated fatty liver disease. *Front Immunol.* 2021;12:660179. https://doi.org/10.3389/fimmu.2021.660179.
38. Tripathi A, Debelius J, Brenner DA, et al. The gut-liver axis and the intersection with the microbiome. *Nat Rev Gastroenterol Hepatol.* 2018;15(7):397–411. https://doi.org/10.1038/s41575-018-0011-z.
39. Schulz C, Schütte K, Mayerle J, Malfertheiner P. The role of the gastric bacterial microbiome in gastric cancer: Helicobacter pylori and beyond. *Ther Adv Gastroenterol.* 2019;12. https://doi.org/10.1177/1756284819894062.
40. Mariat D, Firmesse O, Florence L, et al. The Firmicutes/Bacteroides ratio of the human microbiota changes with age. *BMC Microbiol.* 2009;9:123. https://doi.org/10.1186/1471-2180-9-123.
41. Li J, Zhao F, Wang Y, et al. Gut microbiota dysbiosis contributes to the development of hypertension. *Microbiome.* 2017;5(1):14. https://doi.org/10.1186/s40168-016-0222-x.
42. Claro LM, Moreno-Amaral AN, Gadotti AC, et al. The impact of uremic toxicity induced inflammatory response on the cardiovascular burden in chronic kidney disease. *Toxins (Basel).* 2018;10(10):384. https://doi.org/10.3390/toxins10100384.
43. Yano JM, Yu K, Donaldson GP, et al. Indigenous bacteria from the gut microbiota regulate host serotonin biosynthesis. *Cell.* 2015;161(2):264–276. https://doi.org/10.1016/j.cell.2015.02.047.
44. Davidson GL, Cooke AC, Johnson CN, Quinn JL. The gut microbiome as a driver of individual variation in cognition and functional behaviour. *Philos Trans R Soc Lond Ser B Biol Sci.* 2018;373(1756):20170286. https://doi.org/10.1098/rstb.2017.0286.
45. Johnson KVA. Gut microbiome composition and diversity are related to human personality traits. *Human Microb J.* 2020;15:100069. https://doi.org/10.1016/j.humic.2019.100069.

46. Lin P, Ding B, Feng C, et al. Prevotella and Klebsiella proportions in fecal microbial communities are potential characteristic parameters for patients with major depressive disorder. *J Affect Disord.* 2017;207:300–304. https://doi.org/10.1016/j.jad.2016.09.051.
47. Visconti A, Le Roy CI, Rosa F, et al. Interplay between the human gut microbiome and host metabolism. *Nat Commun.* 2019;10(1):4505. https://doi.org/10.1038/s41467-019-12476-z.
48. Lamichhane S, Sen P, Dickens AM, Orešič M, Bertram HC. Gut metabolome meets microbiome: a methodological perspective to understand the relationship between host and microbe. *Methods.* 2018;149:3–12. https://doi.org/10.1016/j.ymeth.2018.04.029.
49. Agus A, Clément K, Sokol H. Gut microbiota-derived metabolites as central regulators in metabolic disorders. *Gut.* 2020;3. Published online December. https://doi.org/10.1136/gutjnl-2020-323071.
50. Allin KH, Tremaroli V, Caesar R, et al. Aberrant intestinal microbiota in individuals with prediabetes. *Diabetologia.* 2018;61(4):810–820. https://doi.org/10.1007/s00125-018-4550-1.
51. Fan Y, Pedersen O. Gut microbiota in human metabolic health and disease. *Nat Rev Microbiol.* 2021;19(1):55–71. https://doi.org/10.1038/s41579-020-0433-9.
52. Koeth RA, Wang Z, Levison BS, et al. Intestinal microbiota metabolism of L-carnitine, a nutrient in red meat, promotes atherosclerosis. *Nat Med.* 2013;19(5):576–585. https://doi.org/10.1038/nm.3145.
53. Qin J, Li Y, Cai Z, et al. A metagenome-wide association study of gut microbiota in type 2 diabetes. *Nature.* 2012;490(7418):55–60. https://doi.org/10.1038/nature11450.
54. Bäckhed F, Ding H, Wang T, et al. The gut microbiota as an environmental factor that regulates fat storage. *Proc Natl Acad Sci U S A.* 2004;101(44):15718–15723. https://doi.org/10.1073/pnas.0407076101.
55. Tilg H, Zmora N, Adolph TE, Elinav E. The intestinal microbiota fuelling metabolic inflammation. *Nat Rev Immunol.* 2020;20(1):40–54. https://doi.org/10.1038/s41577-019-0198-4.
56. Liu R, Hong J, Xu X, et al. Gut microbiome and serum metabolome alterations in obesity and after weight-loss intervention. *Nat Med.* 2017;23(7):859–868. https://doi.org/10.1038/nm.4358.
57. Pedersen HK, Gudmundsdottir V, Nielsen HB, et al. Human gut microbes impact host serum metabolome and insulin sensitivity. *Nature.* 2016;535(7612):376–381. https://doi.org/10.1038/nature18646.
58. Fiamoncini J, Rundle M, Gibbons H, et al. Plasma metabolome analysis identifies distinct human metabotypes in the postprandial state with different susceptibility to weight loss-mediated metabolic improvements. *FASEB J.* 2018;32(10):5447–5458. https://doi.org/10.1096/fj.201800330R.
59. Koh A, Molinaro A, Ståhlman M, et al. Microbially produced imidazole propionate impairs insulin signaling through mTORC1. *Cell.* 2018;175(4):947–961.e17. https://doi.org/10.1016/j.cell.2018.09.055.
60. McGlone ER, Bloom SR. Bile acids and the metabolic syndrome. *Ann Clin Biochem.* 2019;56(3):326–337. https://doi.org/10.1177/0004563218817798.
61. Org E, Blum Y, Kasela S, et al. Relationships between gut microbiota, plasma metabolites, and metabolic syndrome traits in the METSIM cohort. *Genome Biol.* 2017;18(1):70. https://doi.org/10.1186/s13059-017-1194-2.
62. Ratajczak W, Rył A, Mizerski A, Walczakiewicz K, Sipak O, Laszczyńska M. Immunomodulatory potential of gut microbiome-derived short-chain fatty acids (SCFAs). *Acta Biochim Pol.* 2019;66(1):1–12. https://doi.org/10.18388/abp.2018_2648.

63. Surowiec I, Noordam R, Bennett K, et al. Metabolomic and lipidomic assessment of the metabolic syndrome in Dutch middle-aged individuals reveals novel biological signatures separating health and disease. *Metabolomics*. 2019;15(2):23. https://doi.org/10.1007/s11306-019-1484-7.
64. Brial F, Chilloux J, Nielsen T, et al. Human and preclinical studies of the host-gut microbiome co-metabolite hippurate as a marker and mediator of metabolic health. *Gut*. 2021. Published online May 11. https://doi.org/10.1136/gutjnl-2020-323314.
65. Kumar MS, Slud EV, Okrah K, Hicks SC, Hannenhalli S, Corrada BH. Analysis and correction of compositional bias in sparse sequencing count data. *BMC Genomics*. 2018;19(1):799. https://doi.org/10.1186/s12864-018-5160-5.
66. Rochfort S. Metabolomics reviewed: a new "omics" platform technology for systems biology and implications for natural products research. *J Nat Prod*. 2005;68(12):1813–1820. https://doi.org/10.1021/np050255w.
67. Chong J, Xia J. Computational approaches for integrative analysis of the metabolome and microbiome. *Meta*. 2017;7(4):E62. https://doi.org/10.3390/metabo7040062.
68. Franzosa EA, Hsu T, Sirota-Madi A, et al. Sequencing and beyond: integrating molecular "omics" for microbial community profiling. *Nat Rev Microbiol*. 2015;13(6):360–372. https://doi.org/10.1038/nrmicro3451.
69. Pasolli E, Asnicar F, Manara S, et al. Extensive unexplored human microbiome diversity revealed by over 150,000 genomes from metagenomes spanning age, geography, and lifestyle. *Cell*. 2019;176(3):649–662.e20. https://doi.org/10.1016/j.cell.2019.01.001.
70. Kaddurah-Daouk R, Kristal BS, Weinshilboum RM. Metabolomics: a global biochemical approach to drug response and disease. *Annu Rev Pharmacol Toxicol*. 2008;48:653–683. https://doi.org/10.1146/annurev.pharmtox.48.113006.094715.
71. Quinn TP, Erb I, Gloor G, Notredame C, Richardson MF, Crowley TM. A field guide for the compositional analysis of any-omics data. *Gigascience*. 2019;8(9):giz107. https://doi.org/10.1093/gigascience/giz107.
72. Zhang C, Yin A, Li H, et al. Dietary modulation of gut microbiota contributes to alleviation of both genetic and simple obesity in children. *EBioMedicine*. 2015;2(8):968–984. https://doi.org/10.1016/j.ebiom.2015.07.007.
73. Suganya K, Son T, Kim KW, Koo BS. Impact of gut microbiota: how it could play roles beyond the digestive system on development of cardiovascular and renal diseases. *Microb Pathog*. 2021;152:104583. https://doi.org/10.1016/j.micpath.2020.104583.
74. Hoyles L, Fernández-Real JM, Federici M, et al. Molecular phenomics and metagenomics of hepatic steatosis in non-diabetic obese women. *Nat Med*. 2018;24(7):1070–1080. https://doi.org/10.1038/s41591-018-0061-3.
75. Ghosh S, Whitley CS, Haribabu B, Jala VR. Regulation of intestinal barrier function by microbial metabolites. *Cell Mol Gastroenterol Hepatol*. 2021;11(5):1463–1482. https://doi.org/10.1016/j.jcmgh.2021.02.007.
76. Zheng D, Liwinski T, Elinav E. Interaction between microbiota and immunity in health and disease. *Cell Res*. 2020;30(6):492–506. https://doi.org/10.1038/s41422-020-0332-7.
77. Rakoff-Nahoum S, Paglino J, Eslami-Varzaneh F, Edberg S, Medzhitov R. Recognition of commensal microflora by toll-like receptors is required for intestinal homeostasis. *Cell*. 2004;118(2):229–241. https://doi.org/10.1016/j.cell.2004.07.002.
78. Rehwinkel J, Gack MU. RIG-I-like receptors: their regulation and roles in RNA sensing. *Nat Rev Immunol*. 2020;20(9):537–551. https://doi.org/10.1038/s41577-020-0288-3.
79. Price AE, Shamardani K, Lugo KA, et al. A map of toll-like receptor expression in the intestinal epithelium reveals distinct spatial, cell type-specific, and temporal patterns. *Immunity*. 2018;49(3):560–575.e6. https://doi.org/10.1016/j.immuni.2018.07.016.

80. Park BS, Lee JO. Recognition of lipopolysaccharide pattern by TLR4 complexes. *Exp Mol Med*. 2013;45:e66. https://doi.org/10.1038/emm.2013.97.
81. Wu HJ, Wu E. The role of gut microbiota in immune homeostasis and autoimmunity. *Gut Microbes*. 2012;3(1):4–14. https://doi.org/10.4161/gmic.19320.
82. Belizário JE, Faintuch J, Garay-Malpartida M. Gut microbiome dysbiosis and immunometabolism: new frontiers for treatment of metabolic diseases. *Mediat Inflamm*. 2018;2018:e2037838. https://doi.org/10.1155/2018/2037838.
83. de Cayres LCF, de Salis LVV, Rodrigues GSP, et al. Detection of alterations in the gut microbiota and intestinal permeability in patients with Hashimoto thyroiditis. *Front Immunol*. 2021. https://doi.org/10.3389/fimmu.2021.579140.
84. Kieser KJ, Kagan JC. Multi-receptor detection of individual bacterial products by the innate immune system. *Nat Rev Immunol*. 2017;17(6):376–390. https://doi.org/10.1038/nri.2017.25.
85. Chen H, Nwe PK, Yang Y, et al. A forward chemical genetic screen reveals gut microbiota metabolites that modulate host physiology. *Cell*. 2019;177(5):1217–1231.e18. https://doi.org/10.1016/j.cell.2019.03.036.
86. Bhattarai Y, Williams BB, Battaglioli EJ, et al. Gut microbiota-produced tryptamine activates an epithelial g-protein-coupled receptor to increase colonic secretion. *Cell Host Microbe*. 2018;23(6):775–785.e5. https://doi.org/10.1016/j.chom.2018.05.004.
87. Cohen LJ, Esterhazy D, Kim SH, et al. Commensal bacteria make GPCR ligands that mimic human signalling molecules. *Nature*. 2017;549(7670):48–53. https://doi.org/10.1038/nature23874.
88. Zelante T, Iannitti RG, Cunha C, et al. Tryptophan catabolites from microbiota engage aryl hydrocarbon receptor and balance mucosal reactivity via interleukin-22. *Immunity*. 2013;39(2):372–385. https://doi.org/10.1016/j.immuni.2013.08.003.
89. Witkowski M, Weeks TL, Hazen SL. Gut microbiota and cardiovascular disease. *Circ Res*. 2020;127(4):553–570. https://doi.org/10.1161/CIRCRESAHA.120.316242.
90. Polsinelli VB, Sinha A, Shah SJ. Visceral congestion in heart failure: right ventricular dysfunction, splanchnic hemodynamics, and the intestinal microenvironment. *Curr Heart Fail Rep*. 2017;14(6):519–528. https://doi.org/10.1007/s11897-017-0370-8.
91. Tang WHW, Li DY, Hazen SL. Dietary metabolism, the gut microbiome, and heart failure. *Nat Rev Cardiol*. 2019;16(3):137–154. https://doi.org/10.1038/s41569-018-0108-7.
92. Hug H, Mohajeri MH, La Fata G. Toll-like receptors: regulators of the immune response in the human gut. *Nutrients*. 2018;10(2):E203. https://doi.org/10.3390/nu10020203.
93. Le Chatelier E, Nielsen T, Qin J, et al. Richness of human gut microbiome correlates with metabolic markers. *Nature*. 2013;500(7464):541–546. https://doi.org/10.1038/nature12506.
94. Yang T, Santisteban MM, Rodriguez V, et al. Gut dysbiosis is linked to hypertension. *Hypertension*. 2015;65(6):1331–1340. https://doi.org/10.1161/HYPERTENSIONAHA.115.05315.
95. Jie Z, Xia H, Zhong SL, et al. The gut microbiome in atherosclerotic cardiovascular disease. *Nat Commun*. 2017;8(1):845. https://doi.org/10.1038/s41467-017-00900-1.
96. Cui X, Ye L, Li J, et al. Metagenomic and metabolomic analyses unveil dysbiosis of gut microbiota in chronic heart failure patients. *Sci Rep*. 2018;8(1):635. https://doi.org/10.1038/s41598-017-18756-2.
97. Tang WHW, Wang Z, Levison BS, et al. Intestinal microbial metabolism of phosphatidylcholine and cardiovascular risk. *N Engl J Med*. 2013;368(17):1575–1584. https://doi.org/10.1056/NEJMoa1109400.
98. Wang Z, Klipfell E, Bennett BJ, et al. Gut flora metabolism of phosphatidylcholine promotes cardiovascular disease. *Nature*. 2011;472(7341):57–63. https://doi.org/10.1038/nature09922.

99. Di Ciaula A, Garruti G, Lunardi Baccetto R, et al. Bile acid physiology. *Ann Hepatol.* 2017;16(Suppl 1):S4–S14. https://doi.org/10.5604/01.3001.0010.5493.
100. Boyer JL. Bile formation and secretion. *Compr Physiol.* 2013;3(3):1035–1078. https://doi.org/10.1002/cphy.c120027.
101. Shiffka SJ, Kane MA, Swaan PW. Planar bile acids in health and disease. *Biochim Biophys Acta Biomembr.* 2017;1859(11):2269–2276. https://doi.org/10.1016/j.bbamem.2017.08.019.
102. Chiang JYL. Bile acid regulation of gene expression: roles of nuclear hormone receptors. *Endocr Rev.* 2002;23(4):443–463. https://doi.org/10.1210/er.2000-0035.
103. Thomas C, Pellicciari R, Pruzanski M, Auwerx J, Schoonjans K. Targeting bile-acid signalling for metabolic diseases. *Nat Rev Drug Discov.* 2008;7(8):678–693. https://doi.org/10.1038/nrd2619.
104. Gérard P. Metabolism of cholesterol and bile acids by the gut microbiota. *Pathogens.* 2013;3(1):14–24. https://doi.org/10.3390/pathogens3010014.
105. Chen ML, Takeda K, Sundrud MS. Emerging roles of bile acids in mucosal immunity and inflammation. *Mucosal Immunol.* 2019;12(4):851–861. https://doi.org/10.1038/s41385-019-0162-4.
106. Hylemon PB, Zhou H, Pandak WM, Ren S, Gil G, Dent P. Bile acids as regulatory molecules. *J Lipid Res.* 2009;50(8):1509–1520. https://doi.org/10.1194/jlr.R900007-JLR200.
107. Jia ET, Liu ZY, Pan M, Lu JF, Ge QY. Regulation of bile acid metabolism-related signaling pathways by gut microbiota in diseases. *J Zhejiang Univ Sci B.* 2019;20(10):781–792. https://doi.org/10.1631/jzus.B1900073.
108. Keitel V, Stindt J, Häussinger D. Bile acid-activated receptors: GPBAR1 (TGR5) and other G protein-coupled receptors. *Handb Exp Pharmacol.* 2019;256:19–49. https://doi.org/10.1007/164_2019_230.
109. Wang B, Tontonoz P. Liver X receptors in lipid signalling and membrane homeostasis. *Nat Rev Endocrinol.* 2018;14(8):452–463. https://doi.org/10.1038/s41574-018-0037-x.
110. Zhou H, Hylemon PB. Bile acids are nutrient signaling hormones. *Steroids.* 2014;86:62–68. https://doi.org/10.1016/j.steroids.2014.04.016.
111. Chiang JYL. Bile acid metabolism and signaling. *Compr Physiol.* 2013;3(3):1191–1212. https://doi.org/10.1002/cphy.c120023.
112. McCarville JL, Chen GY, Cuevas VD, Troha K, Ayres JS. Microbiota metabolites in health and disease. *Annu Rev Immunol.* 2020;38:147–170. https://doi.org/10.1146/annurev-immunol-071219-125715.
113. Zhu Y, Shui X, Liang Z, et al. Gut microbiota metabolites as integral mediators in cardiovascular diseases (Review). *Int J Mol Med.* 2020;46(3):936–948. https://doi.org/10.3892/ijmm.2020.4674.
114. Cho CE, Aardema NDJ, Bunnell ML, et al. Effect of choline forms and gut microbiota composition on trimethylamine-N-oxide response in healthy men. *Nutrients.* 2020;12(8):E2220. https://doi.org/10.3390/nu12082220.
115. Grüner N, Mattner J. Bile acids and microbiota: multifaceted and versatile regulators of the liver-gut axis. *Int J Mol Sci.* 2021;22(3):1397. https://doi.org/10.3390/ijms22031397.
116. Duboc H, Rajca S, Rainteau D, et al. Connecting dysbiosis, bile-acid dysmetabolism and gut inflammation in inflammatory bowel diseases. *Gut.* 2013;62(4):531–539. https://doi.org/10.1136/gutjnl-2012-302578.
117. Wu P, Zhang Y, Liu Y, et al. Effects of cholic acid on blood pressure and production of vascular aldosterone and corticosterone. *Steroids.* 1999;64(4):291–295. https://doi.org/10.1016/s0039-128x(99)00005-7.
118. Desai MS, Mathur B, Eblimit Z, et al. Bile acid excess induces cardiomyopathy and metabolic dysfunctions in the heart. *Hepatology.* 2017;65(1):189–201. https://doi.org/10.1002/hep.28890.

119. Khurana S, Raufman J, Pallone TL. Bile acids regulate cardiovascular function. *Clin Transl Sci*. 2011;4(3):210–218. https://doi.org/10.1111/j.1752-8062.2011.00272.x.
120. Pu J, Yuan A, Shan P, et al. Cardiomyocyte-expressed farnesoid-X-receptor is a novel apoptosis mediator and contributes to myocardial ischaemia/reperfusion injury. *Eur Heart J*. 2013;34(24):1834–1845. https://doi.org/10.1093/eurheartj/ehs011.
121. Bose S, Ramesh V, Locasale JW. Acetate metabolism in physiology, cancer, and beyond. *Trends Cell Biol*. 2019;29(9):695–703. https://doi.org/10.1016/j.tcb.2019.05.005.
122. Deehan EC, Yang C, Perez-Muñoz ME, et al. Precision microbiome modulation with discrete dietary fiber structures directs short-chain fatty acid production. *Cell Host Microbe*. 2020;27(3):389–404.e6. https://doi.org/10.1016/j.chom.2020.01.006.
123. Morrison DJ, Preston T. Formation of short chain fatty acids by the gut microbiota and their impact on human metabolism. *Gut Microbes*. 2016;7(3):189–200. https://doi.org/10.1080/19490976.2015.1134082.
124. Koh A, De Vadder F, Kovatcheva-Datchary P, Bäckhed F. From dietary fiber to host physiology: short-chain fatty acids as key bacterial metabolites. *Cell*. 2016;165(6):1332–1345. https://doi.org/10.1016/j.cell.2016.05.041.
125. Kim CH. Microbiota or short-chain fatty acids: which regulates diabetes? *Cell Mol Immunol*. 2018;15(2):88–91. https://doi.org/10.1038/cmi.2017.57.
126. Chambers ES, Morrison DJ, Frost G. Control of appetite and energy intake by SCFA: what are the potential underlying mechanisms? *Proc Nutr Soc*. 2015;74(3):328–336. https://doi.org/10.1017/S0029665114001657.
127. Hu J, Lin S, Zheng B, Cheung PCK. Short-chain fatty acids in control of energy metabolism. *Crit Rev Food Sci Nutr*. 2018;58(8):1243–1249. https://doi.org/10.1080/10408398.2016.1245650.
128. Litvak Y, Byndloss MX, Bäumler AJ. Colonocyte metabolism shapes the gut microbiota. *Science*. 2018;362(6418). eaat9076. https://doi.org/10.1126/science.aat9076.
129. Sun Y, O'Riordan MXD. Regulation of bacterial pathogenesis by intestinal short-chain fatty acids. *Adv Appl Microbiol*. 2013;85:93–118. https://doi.org/10.1016/B978-0-12-407672-3.00003-4.
130. Cheema MU, Pluznick JL. Gut microbiota plays a central role to modulate the plasma and fecal metabolomes in response to angiotensin II. *Hypertension*. 2019;74(1):184–193. https://doi.org/10.1161/HYPERTENSIONAHA.119.13155.
131. Nobel YR, Cox LM, Kirigin FF, et al. Metabolic and metagenomic outcomes from early-life pulsed antibiotic treatment. *Nat Commun*. 2015;6:7486. https://doi.org/10.1038/ncomms8486.
132. Livanos AE, Greiner TU, Vangay P, et al. Antibiotic-mediated gut microbiome perturbation accelerates development of type 1 diabetes in mice. *Nat Microbiol*. 2016;1(11):16140. https://doi.org/10.1038/nmicrobiol.2016.140.
133. Zhang XS, Li J, Krautkramer KA, et al. Antibiotic-induced acceleration of type 1 diabetes alters maturation of innate intestinal immunity. *eLife*. 2018;7:e37816. https://doi.org/10.7554/eLife.37816.
134. Calderón-Pérez L, Gosalbes MJ, Yuste S, et al. Gut metagenomic and short chain fatty acids signature in hypertension: a cross-sectional study. *Sci Rep*. 2020;10(1):6436. https://doi.org/10.1038/s41598-020-63475-w.
135. Sonnenburg ED, Smits SA, Tikhonov M, Higginbottom SK, Wingreen NS, Sonnenburg JL. Diet-induced extinctions in the gut microbiota compound over generations. *Nature*. 2016;529(7585):212–215. https://doi.org/10.1038/nature16504.
136. Whelton SP, Hyre AD, Pedersen B, Yi Y, Whelton PK, He J. Effect of dietary fiber intake on blood pressure: a meta-analysis of randomized, controlled clinical trials. *J Hypertens*. 2005;23(3):475–481. https://doi.org/10.1097/01.hjh.0000160199.51158.cf.

137. Sivaprakasam S, Prasad PD, Singh N. Benefits of short-chain fatty acids and their receptors in inflammation and carcinogenesis. *Pharmacol Ther.* 2016;164:144–151. https://doi.org/10.1016/j.pharmthera.2016.04.007.
138. Natarajan N, Hori D, Flavahan S, et al. Microbial short chain fatty acid metabolites lower blood pressure via endothelial G protein-coupled receptor 41. *Physiol Genomics.* 2016;48(11):826–834. https://doi.org/10.1152/physiolgenomics.00089.2016.
139. Pluznick JL, Protzko RJ, Gevorgyan H, et al. Olfactory receptor responding to gut microbiota-derived signals plays a role in renin secretion and blood pressure regulation. *Proc Natl Acad Sci U S A.* 2013;110(11):4410–4415. https://doi.org/10.1073/pnas.1215927110.
140. Poll B, Steppan J, Lester L, Berkowitz D, Pluznick J. A short chain fatty acid produced by the gut microbiota plays a role in blood pressure regulation and cardiac contractility. *FASEB J.* 2019;33(S1):569.19. https://doi.org/10.1096/fasebj.2019.33.1_supplement.569.19.
141. Bartolomaeus H, Balogh A, Yakoub M, et al. Short-chain fatty acid propionate protects from hypertensive cardiovascular damage. *Circulation.* 2019;139(11):1407–1421. https://doi.org/10.1161/CIRCULATIONAHA.118.036652.
142. Battson ML, Lee DM, Li Puma LC, et al. Gut microbiota regulates cardiac ischemic tolerance and aortic stiffness in obesity. *Am J Physiol Heart Circ Physiol.* 2019;317(6):H1210–H1220. https://doi.org/10.1152/ajpheart.00346.2019.
143. Tang TWH, Chen HC, Chen CY, et al. Loss of gut microbiota alters immune system composition and cripples postinfarction cardiac repair. *Circulation.* 2019;139(5):647–659. https://doi.org/10.1161/CIRCULATIONAHA.118.035235.
144. Mell B, Jala VR, Mathew AV, et al. Evidence for a link between gut microbiota and hypertension in the Dahl rat. *Physiol Genomics.* 2015;47(6):187–197. https://doi.org/10.1152/physiolgenomics.00136.2014.
145. Tajiri K, Shimizu Y. Branched-chain amino acids in liver diseases. *Transl Gastroenterol Hepatol.* 2018;3:47. https://doi.org/10.21037/tgh.2018.07.06.
146. Li J, Zuo K, Zhang J, et al. Shifts in gut microbiome and metabolome are associated with risk of recurrent atrial fibrillation. *J Cell Mol Med.* 2020;24(22):13356–13369. https://doi.org/10.1111/jcmm.15959.
147. Ogier de Baulny H, Saudubray JM. Branched-chain organic acidurias. *Semin Neonatol.* 2002;7(1):65–74. https://doi.org/10.1053/siny.2001.0087.
148. Newgard CB, An J, Bain JR, et al. A branched-chain amino acid-related metabolic signature that differentiates obese and lean humans and contributes to insulin resistance. *Cell Metab.* 2009;9(4):311–326. https://doi.org/10.1016/j.cmet.2009.02.002.
149. Wang TJ, Larson MG, Vasan RS, et al. Metabolite profiles and the risk of developing diabetes. *Nat Med.* 2011;17(4):448–453. https://doi.org/10.1038/nm.2307.
150. Zhou M, Shao J, Wu CY, et al. Targeting BCAA catabolism to treat obesity-associated insulin resistance. *Diabetes.* 2019;68(9):1730–1746. https://doi.org/10.2337/db18-0927.
151. Park JE, Miller M, Rhyne J, Wang Z, Hazen SL. Differential effect of short-term popular diets on TMAO and other cardio-metabolic risk markers. *Nutr Metab Cardiovasc Dis.* 2019;29(5):513–517. https://doi.org/10.1016/j.numecd.2019.02.003.
152. Mels CM, Schutte AE, Schutte R, et al. The link between vascular deterioration and branched chain amino acids in a population with high glycated haemoglobin: the SABPA study. *Amino Acids.* 2013;45(6):1405–1413. https://doi.org/10.1007/s00726-013-1611-0.
153. Yang R, Dong J, Zhao H, et al. Association of branched-chain amino acids with carotid intima-media thickness and coronary artery disease risk factors. *PLoS One.* 2014;9(6):e99598. https://doi.org/10.1371/journal.pone.0099598.

154. Yang RY, Wang SM, Sun L, et al. Association of branched-chain amino acids with coronary artery disease: a matched-pair case-control study. *Nutr Metab Cardiovasc Dis.* 2015;25(10):937–942. https://doi.org/10.1016/j.numecd.2015.06.003.
155. Jiang YJ, Sun SJ, Cao WX, et al. Excessive ROS production and enhanced autophagy contribute to myocardial injury induced by branched-chain amino acids: roles for the AMPK-ULK1 signaling pathway and α7nAChR. *Biochim Biophys Acta Mol basis Dis.* 2021;1867(1):165980. https://doi.org/10.1016/j.bbadis.2020.165980.
156. Zhenyukh O, Civantos E, Ruiz-Ortega M, et al. High concentration of branched-chain amino acids promotes oxidative stress, inflammation and migration of human peripheral blood mononuclear cells via mTORC1 activation. *Free Radic Biol Med.* 2017;104:165–177. https://doi.org/10.1016/j.freeradbiomed.2017.01.009.
157. Yang Y, Zhao M, He X, Wu Q, Li D, Zang W. Pyridostigmine protects against diabetic cardiomyopathy by regulating vagal activity, gut microbiota, and branched-chain amino acid catabolism in diabetic mice. *Front Pharmacol.* 2021;12:647481. https://doi.org/10.3389/fphar.2021.647481.
158. Koeth RA, Levison BS, Culley MK, et al. γ-Butyrobetaine is a proatherogenic intermediate in gut microbial metabolism of L-carnitine to TMAO. *Cell Metab.* 2014;20 (5):799–812. https://doi.org/10.1016/j.cmet.2014.10.006.
159. Li XS, Wang Z, Cajka T, et al. Untargeted metabolomics identifies trimethyllysine, a TMAO-producing nutrient precursor, as a predictor of incident cardiovascular disease risk. *JCI Insight.* 2018;3(6):99096. https://doi.org/10.1172/jci.insight.99096.
160. Craciun S, Balskus EP. Microbial conversion of choline to trimethylamine requires a glycyl radical enzyme. *Proc Natl Acad Sci U S A.* 2012;109(52):21307–21312. https://doi.org/10.1073/pnas.1215689109.
161. Rath S, Rud T, Pieper DH, Vital M. Potential TMA-producing bacteria are ubiquitously found in mammalia. *Front Microbiol.* 2019;10:2966. https://doi.org/10.3389/fmicb.2019.02966.
162. Dehghan P, Farhangi MA, Nikniaz L, Nikniaz Z, Asghari-Jafarabadi M. Gut microbiota-derived metabolite trimethylamine N-oxide (TMAO) potentially increases the risk of obesity in adults: an exploratory systematic review and dose-response meta- analysis. *Obes Rev.* 2020;21(5):e12993. https://doi.org/10.1111/obr.12993.
163. Fennema D, Phillips IR, Shephard EA. Trimethylamine and trimethylamine N-oxide, a flavin-containing monooxygenase 3 (FMO3)-mediated host-microbiome metabolic axis implicated in health and disease. *Drug Metab Dispos.* 2016;44(11):1839–1850. https://doi.org/10.1124/dmd.116.070615.
164. Wang Z, Bergeron N, Levison BS, et al. Impact of chronic dietary red meat, white meat, or non-meat protein on trimethylamine N-oxide metabolism and renal excretion in healthy men and women. *Eur Heart J.* 2019;40(7):583–594. https://doi.org/10.1093/eurheartj/ehy799.
165. Heianza Y, Ma W, Manson JE, Rexrode KM, Qi L. Gut microbiota metabolites and risk of major adverse cardiovascular disease events and death: a systematic review and meta-analysis of prospective studies. *J Am Heart Assoc.* 2017;6(7):e004947. https://doi.org/10.1161/JAHA.116.004947.
166. Qi J, You T, Li J, et al. Circulating trimethylamine N-oxide and the risk of cardiovascular diseases: a systematic review and meta-analysis of 11 prospective cohort studies. *J Cell Mol Med.* 2018;22(1):185–194. https://doi.org/10.1111/jcmm.13307.
167. Schiattarella GG, Sannino A, Toscano E, et al. Gut microbe-generated metabolite trimethylamine-N-oxide as cardiovascular risk biomarker: a systematic review and dose-response meta-analysis. *Eur Heart J.* 2017;38(39):2948–2956. https://doi.org/10.1093/eurheartj/ehx342.

168. Haghikia A, Li XS, Liman TG, et al. Gut microbiota-dependent trimethylamine N-oxide predicts risk of cardiovascular events in patients with stroke and is related to proinflammatory monocytes. *Arterioscler Thromb Vasc Biol.* 2018;38(9):2225–2235. https://doi.org/10.1161/ATVBAHA.118.311023.
169. Senthong V, Li XS, Hudec T, et al. Plasma trimethylamine N-oxide, a gut microbe-generated phosphatidylcholine metabolite, is associated with atherosclerotic burden. *J Am Coll Cardiol.* 2016;67(22):2620–2628. https://doi.org/10.1016/j.jacc.2016.03.546.
170. Tan Y, Sheng Z, Zhou P, et al. Plasma trimethylamine N-oxide as a novel biomarker for plaque rupture in patients with ST-segment-elevation myocardial infarction. *Circ Cardiovasc Interv.* 2019;12(1):e007281. https://doi.org/10.1161/CIRCINTERVENTIONS.118.007281.
171. Zhu W, Gregory JC, Org E, et al. Gut microbial metabolite TMAO enhances platelet hyperreactivity and thrombosis risk. *Cell.* 2016;165(1):111–124. https://doi.org/10.1016/j.cell.2016.02.011.
172. Senthong V, Wang Z, Fan Y, Wu Y, Hazen SL, Tang WHW. Trimethylamine N-oxide and mortality risk in patients with peripheral artery disease. *J Am Heart Assoc.* 2016;5(10):e004237. https://doi.org/10.1161/JAHA.116.004237.
173. Suzuki T, Heaney LM, Bhandari SS, Jones DJL, Ng LL. Trimethylamine N-oxide and prognosis in acute heart failure. *Heart.* 2016;102(11):841–848. https://doi.org/10.1136/heartjnl-2015-308826.
174. Tang WHW, Wang Z, Fan Y, et al. Prognostic value of elevated levels of intestinal microbe-generated metabolite trimethylamine-N-oxide in patients with heart failure: refining the gut hypothesis. *J Am Coll Cardiol.* 2014;64(18):1908–1914. https://doi.org/10.1016/j.jacc.2014.02.617.
175. Tang WHW, Wang Z, Shrestha K, et al. Intestinal microbiota-dependent phosphatidylcholine metabolites, diastolic dysfunction, and adverse clinical outcomes in chronic systolic heart failure. *J Card Fail.* 2015;21(2):91–96. https://doi.org/10.1016/j.cardfail.2014.11.006.
176. Roberts AB, Gu X, Buffa JA, et al. Development of a gut microbe-targeted nonlethal therapeutic to inhibit thrombosis potential. *Nat Med.* 2018;24(9):1407–1417. https://doi.org/10.1038/s41591-018-0128-1.
177. Skye SM, Zhu W, Romano KA, et al. Microbial transplantation with human gut commensals containing CutC is sufficient to transmit enhanced platelet reactivity and thrombosis potential. *Circ Res.* 2018;123(10):1164–1176. https://doi.org/10.1161/CIRCRESAHA.118.313142.
178. Zhu W, Buffa JA, Wang Z, et al. Flavin monooxygenase 3, the host hepatic enzyme in the metaorganismal trimethylamine N-oxide-generating pathway, modulates platelet responsiveness and thrombosis risk. *J Thromb Haemost.* 2018;16(9):1857–1872. https://doi.org/10.1111/jth.14234.
179. Duttaroy AK. Role of gut microbiota and their metabolites on atherosclerosis, hypertension and human blood platelet function: a review. *Nutrients.* 2021;13(1):E144. https://doi.org/10.3390/nu13010144.
180. Chen ML, Zhu XH, Ran L, Lang HD, Yi L, Mi MT. Trimethylamine-N-oxide induces vascular inflammation by activating the NLRP3 inflammasome through the SIRT3-SOD2-mtROS signaling pathway. *J Am Heart Assoc.* 2017;6(9):e006347. https://doi.org/10.1161/JAHA.117.006347.
181. Seldin MM, Meng Y, Qi H, et al. Trimethylamine N-oxide promotes vascular inflammation through signaling of mitogen-activated protein kinase and nuclear factor-κB. *J Am Heart Assoc.* 2016;5(2):e002767. https://doi.org/10.1161/JAHA.115.002767.

182. Zhang X, Li Y, Yang P, et al. Trimethylamine-N-oxide promotes vascular calcification through activation of NLRP3 (nucleotide-binding domain, leucine-rich-containing family, pyrin domain-containing-3) inflammasome and NF-κB (nuclear factor κB) signals. *Arterioscler Thromb Vasc Biol.* 2020;40(3):751–765. https://doi.org/10.1161/ATVBAHA.119.313414.
183. Fang Q, Zheng B, Liu N, et al. Trimethylamine N-oxide exacerbates renal inflammation and fibrosis in rats with diabetic kidney disease. *Front Physiol.* 2021;12:682482. https://doi.org/10.3389/fphys.2021.682482.
184. Dambrova M, Latkovskis G, Kuka J, et al. Diabetes is associated with higher trimethylamine N-oxide plasma levels. *Exp Clin Endocrinol Diabetes.* 2016;124(4):251–256. https://doi.org/10.1055/s-0035-1569330.
185. Jin B, Ji F, Zuo A, et al. Destructive role of TMAO in T-tubule and excitation-contraction coupling in the adult cardiomyocytes. *Int Heart J.* 2020;61(2):355–363. https://doi.org/10.1536/ihj.19-372.
186. Organ CL, Otsuka H, Bhushan S, et al. Choline diet and its gut microbe-derived metabolite, trimethylamine N-oxide, exacerbate pressure overload-induced heart failure. *Circ Heart Fail.* 2016;9(1):e002314. https://doi.org/10.1161/CIRCHEARTFAILURE.115.002314.
187. Wang G, Kong B, Shuai W, Fu H, Jiang X, Huang H. 3,3-Dimethyl-1-butanol attenuates cardiac remodeling in pressure-overload-induced heart failure mice. *J Nutr Biochem.* 2020;78:108341. https://doi.org/10.1016/j.jnutbio.2020.108341.
188. Abete I, Romaguera D, Vieira AR, Lopez de Munain A, Norat T. Association between total, processed, red and white meat consumption and all-cause, CVD and IHD mortality: a meta-analysis of cohort studies. *Br J Nutr.* 2014;112(5):762–775. https://doi.org/10.1017/S000711451400124X.
189. Gregory JC, Buffa JA, Org E, et al. Transmission of atherosclerosis susceptibility with gut microbial transplantation. *J Biol Chem.* 2015;290(9):5647–5660. https://doi.org/10.1074/jbc.M114.618249.
190. Craciun S, Marks JA, Balskus EP. Characterization of choline trimethylamine-lyase expands the chemistry of glycyl radical enzymes. *ACS Chem Biol.* 2014;9(7):1408–1413. https://doi.org/10.1021/cb500113p.
191. Aldana-Hernández P, Leonard KA, Zhao YY, Curtis JM, Field CJ, Jacobs RL. Dietary choline or trimethylamine n-oxide supplementation does not influence atherosclerosis development in Ldlr-/- and Apoe-/- male mice. *J Nutr.* 2020;150(2):249–255. https://doi.org/10.1093/jn/nxz214.
192. Aldana-Hernández P, Azarcoya-Barrera J, van der Veen JN, et al. Dietary phosphatidylcholine supplementation reduces atherosclerosis in Ldlr-/- male mice2. *J Nutr Biochem.* 2021;92:108617. https://doi.org/10.1016/j.jnutbio.2021.108617.
193. Ding L, Chang M, Guo Y, et al. Trimethylamine-N-oxide (TMAO)-induced atherosclerosis is associated with bile acid metabolism. *Lipids Health Dis.* 2018;17(1):286. https://doi.org/10.1186/s12944-018-0939-6.
194. Wang Z, Roberts AB, Buffa JA, et al. Non-lethal inhibition of gut microbial trimethylamine production for the treatment of atherosclerosis. *Cell.* 2015;163(7):1585–1595. https://doi.org/10.1016/j.cell.2015.11.055.
195. Wu P, Chen J, Chen J, et al. Trimethylamine N-oxide promotes apoE-/- mice atherosclerosis by inducing vascular endothelial cell pyroptosis via the SDHB/ROS pathway. *J Cell Physiol.* 2020;235(10):6582–6591. https://doi.org/10.1002/jcp.29518.
196. Zhu W, Wang Z, Tang WHW, Hazen SL. Gut microbe-generated trimethylamine N-oxide from dietary choline is prothrombotic in subjects. *Circulation.* 2017;135(17):1671–1673. https://doi.org/10.1161/CIRCULATIONAHA.116.025338.

197. Shih DM, Zhu W, Schugar RC, et al. Genetic deficiency of flavin-containing monooxygenase 3 (Fmo3) protects against thrombosis but has only a minor effect on plasma lipid levels—brief report. *Arterioscler Thromb Vasc Biol.* 2019;39(6):1045–1054. https://doi.org/10.1161/ATVBAHA.119.312592.
198. Suzuki H, Kurihara Y, Takeya M, et al. A role for macrophage scavenger receptors in atherosclerosis and susceptibility to infection. *Nature.* 1997;386(6622):292–296. https://doi.org/10.1038/386292a0.
199. Warrier M, Shih DM, Burrows AC, et al. The TMAO-generating enzyme flavin monooxygenase 3 is a central regulator of cholesterol balance. *Cell Rep.* 2015;10(3):326–338. https://doi.org/10.1016/j.celrep.2014.12.036.
200. Li X, Geng J, Zhao J, et al. Trimethylamine N-oxide exacerbates cardiac fibrosis via activating the NLRP3 inflammasome. *Front Physiol.* 2019;10:866. https://doi.org/10.3389/fphys.2019.00866.
201. Ma G, Pan B, Chen Y, et al. Trimethylamine N-oxide in atherogenesis: impairing endothelial self-repair capacity and enhancing monocyte adhesion. *Biosci Rep.* 2017;37(2). BSR20160244. https://doi.org/10.1042/BSR20160244.
202. Raggi P, Genest J, Giles JT, et al. Role of inflammation in the pathogenesis of atherosclerosis and therapeutic interventions. *Atherosclerosis.* 2018;276:98–108. https://doi.org/10.1016/j.atherosclerosis.2018.07.014.
203. Ke Y, Li D, Zhao M, et al. Gut flora-dependent metabolite trimethylamine-N-oxide accelerates endothelial cell senescence and vascular aging through oxidative stress. *Free Radic Biol Med.* 2018;116:88–100. https://doi.org/10.1016/j.freeradbiomed.2018.01.007.
204. Boini KM, Hussain T, Li PL, Koka S. Trimethylamine-N-oxide instigates NLRP3 inflammasome activation and endothelial dysfunction. *Cell Physiol Biochem.* 2017;44(1):152–162. https://doi.org/10.1159/000484623.
205. Sun X, Jiao X, Ma Y, et al. Trimethylamine N-oxide induces inflammation and endothelial dysfunction in human umbilical vein endothelial cells via activating ROS-TXNIP-NLRP3 inflammasome. *Biochem Biophys Res Commun.* 2016;481(1-2):63–70. https://doi.org/10.1016/j.bbrc.2016.11.017.
206. Brunt VE, Gioscia-Ryan RA, Richey JJ, et al. Suppression of the gut microbiome ameliorates age-related arterial dysfunction and oxidative stress in mice. *J Physiol.* 2019;597(9):2361–2378. https://doi.org/10.1113/JP277336.
207. Grover SP, Mackman N. Tissue factor: an essential mediator of hemostasis and trigger of thrombosis. *Arterioscler Thromb Vasc Biol.* 2018;38(4):709–725. https://doi.org/10.1161/ATVBAHA.117.309846.
208. Cheng X, Qiu X, Liu Y, Yuan C, Yang X. Trimethylamine N-oxide promotes tissue factor expression and activity in vascular endothelial cells: a new link between trimethylamine N-oxide and atherosclerotic thrombosis. *Thromb Res.* 2019;177:110–116. https://doi.org/10.1016/j.thromres.2019.02.028.
209. Jäckel S, Kiouptsi K, Lillich M, et al. Gut microbiota regulate hepatic von Willebrand factor synthesis and arterial thrombus formation via Toll-like receptor-2. *Blood.* 2017;130(4):542–553. https://doi.org/10.1182/blood-2016-11-754416.
210. Kamo T, Akazawa H, Suda W, et al. Dysbiosis and compositional alterations with aging in the gut microbiota of patients with heart failure. *PLoS One.* 2017;12(3):e0174099. https://doi.org/10.1371/journal.pone.0174099.
211. Katsimichas T, Ohtani T, Motooka D, et al. Non-ischemic heart failure with reduced ejection fraction is associated with altered intestinal microbiota. *Circ J.* 2018;82(6):1640–1650. https://doi.org/10.1253/circj.CJ-17-1285.

212. Hayashi T, Yamashita T, Watanabe H, et al. Gut microbiome and plasma microbiome-related metabolites in patients with decompensated and compensated heart failure. *Circ J.* 2018;83(1):182–192. https://doi.org/10.1253/circj.CJ-18-0468.
213. Wei H, Zhao M, Huang M, et al. FMO3-TMAO axis modulates the clinical outcome in chronic heart-failure patients with reduced ejection fraction: evidence from an Asian population. *Front Med.* 2021. Published online June 22. https://doi.org/10.1007/s11684-021-0857-2.
214. Kinugasa Y, Nakamura K, Kamitani H, et al. Trimethylamine N-oxide and outcomes in patients hospitalized with acute heart failure and preserved ejection fraction. *ESC Heart Fail.* 2021;8(3):2103–2110. https://doi.org/10.1002/ehf2.13290.
215. Dong Z, Zheng S, Shen Z, Luo Y, Hai X. Trimethylamine N-oxide is associated with heart failure risk in patients with preserved ejection fraction. *Lab Med.* 2021;52(4):346–351. https://doi.org/10.1093/labmed/lmaa075.
216. Videja M, Vilskersts R, Korzh S, et al. Microbiota-derived metabolite trimethylamine N-oxide protects mitochondrial energy metabolism and cardiac functionality in a rat model of right ventricle heart failure. *Front Cell Dev Biol.* 2020;8:622741. https://doi.org/10.3389/fcell.2020.622741.
217. Collins HL, Drazul-Schrader D, Sulpizio AC, et al. L-Carnitine intake and high trimethylamine N-oxide plasma levels correlate with low aortic lesions in ApoE(-/-) transgenic mice expressing CETP. *Atherosclerosis.* 2016;244:29–37. https://doi.org/10.1016/j.atherosclerosis.2015.10.108.
218. He K, Song Y, Daviglus ML, et al. Accumulated evidence on fish consumption and coronary heart disease mortality: a meta-analysis of cohort studies. *Circulation.* 2004;109(22):2705–2711. https://doi.org/10.1161/01.CIR.0000132503.19410.6B.
219. Parthasarathy A, Cross PJ, Dobson RCJ, Adams LE, Savka MA, Hudson AO. A three-ring circus: metabolism of the three proteogenic aromatic amino acids and their role in the health of plants and animals. *Front Mol Biosci.* 2018;5:29. https://doi.org/10.3389/fmolb.2018.00029.
220. Roager HM, Licht TR. Microbial tryptophan catabolites in health and disease. *Nat Commun.* 2018;9(1):3294. https://doi.org/10.1038/s41467-018-05470-4.
221. Fernstrom JD, Fernstrom MH. Tyrosine, phenylalanine, and catecholamine synthesis and function in the brain. *J Nutr.* 2007;137(6 Suppl 1):1539S–1547S. discussion 1548S https://doi.org/10.1093/jn/137.6.1539S.
222. Shishehbor MH, Aviles RJ, Brennan ML, et al. Association of nitrotyrosine levels with cardiovascular disease and modulation by statin therapy. *JAMA.* 2003;289(13):1675–1680. https://doi.org/10.1001/jama.289.13.1675.
223. Badawy AAB. Kynurenine pathway and human systems. *Exp Gerontol.* 2020;129:110770. https://doi.org/10.1016/j.exger.2019.110770.
224. Hubbard TD, Murray IA, Perdew GH. Indole and tryptophan metabolism: endogenous and dietary routes to Ah receptor activation. *Drug Metab Dispos.* 2015;43(10):1522–1535. https://doi.org/10.1124/dmd.115.064246.
225. Sonowal R, Swimm A, Sahoo A, et al. Indoles from commensal bacteria extend healthspan. *Proc Natl Acad Sci U S A.* 2017;114(36):E7506–E7515. https://doi.org/10.1073/pnas.1706464114.
226. Duranton F, Cohen G, De Smet R, et al. Normal and pathologic concentrations of uremic toxins. *J Am Soc Nephrol.* 2012;23(7):1258–1270. https://doi.org/10.1681/ASN.2011121175.
227. Barreto FC, Barreto DV, Liabeuf S, et al. Serum indoxyl sulfate is associated with vascular disease and mortality in chronic kidney disease patients. *Clin J Am Soc Nephrol.* 2009;4(10):1551–1558. https://doi.org/10.2215/CJN.03980609.
228. Niwa T. Role of indoxyl sulfate in the progression of chronic kidney disease and cardiovascular disease: experimental and clinical effects of oral sorbent AST-120. *Ther Apher Dial.* 2011;15(2):120–124. https://doi.org/10.1111/j.1744-9987.2010.00882.x.

229. Yamagami F, Tajiri K, Doki K, et al. Indoxyl sulphate is associated with atrial fibrillation recurrence after catheter ablation. *Sci Rep*. 2018;8(1):17276. https://doi.org/10.1038/s41598-018-35226-5.
230. Huć T, Nowinski A, Drapala A, Konopelski P, Ufnal M. Indole and indoxyl sulfate, gut bacteria metabolites of tryptophan, change arterial blood pressure via peripheral and central mechanisms in rats. *Pharmacol Res*. 2018;130:172–179. https://doi.org/10.1016/j.phrs.2017.12.025.
231. Metghalchi S, Ponnuswamy P, Simon T, et al. Indoleamine 2,3-dioxygenase fine-tunes immune homeostasis in atherosclerosis and colitis through repression of interleukin-10 production. *Cell Metab*. 2015;22(3):460–471. https://doi.org/10.1016/j.cmet.2015.07.004.
232. Eussen SJPM, Ueland PM, Vollset SE, et al. Kynurenines as predictors of acute coronary events in the Hordaland Health Study. *Int J Cardiol*. 2015;189:18–24. https://doi.org/10.1016/j.ijcard.2015.03.413.
233. Niinisalo P, Raitala A, Pertovaara M, et al. Indoleamine 2,3-dioxygenase activity associates with cardiovascular risk factors: the Health 2000 study. *Scand J Clin Lab Invest*. 2008;68(8):767–770. https://doi.org/10.1080/00365510802245685.
234. Pedersen ER, Tuseth N, Eussen SJPM, et al. Associations of plasma kynurenines with risk of acute myocardial infarction in patients with stable angina pectoris. *Arterioscler Thromb Vasc Biol*. 2015;35(2):455–462. https://doi.org/10.1161/ATVBAHA.114.304674.
235. Pertovaara M, Raitala A, Juonala M, et al. Indoleamine 2,3-dioxygenase enzyme activity correlates with risk factors for atherosclerosis: the Cardiovascular Risk in Young Finns Study. *Clin Exp Immunol*. 2007;148(1):106–111. https://doi.org/10.1111/j.1365-2249.2007.03325.x.
236. Alexeev EE, Lanis JM, Kao DJ, et al. Microbiota-derived indole metabolites promote human and murine intestinal homeostasis through regulation of interleukin-10 receptor. *Am J Pathol*. 2018;188(5):1183–1194. https://doi.org/10.1016/j.ajpath.2018.01.011.
237. Dodd D, Spitzer MH, Van Treuren W, et al. A gut bacterial pathway metabolizes aromatic amino acids into nine circulating metabolites. *Nature*. 2017;551(7682):648–652. https://doi.org/10.1038/nature24661.
238. Galligan JJ. Beneficial actions of microbiota-derived tryptophan metabolites. *Neurogastroenterol Motil*. 2018;30(2). https://doi.org/10.1111/nmo.13283.
239. Lamas B, Richard ML, Leducq V, et al. CARD9 impacts colitis by altering gut microbiota metabolism of tryptophan into aryl hydrocarbon receptor ligands. *Nat Med*. 2016;22(6):598–605. https://doi.org/10.1038/nm.4102.
240. Wikoff WR, Anfora AT, Liu J, et al. Metabolomics analysis reveals large effects of gut microflora on mammalian blood metabolites. *Proc Natl Acad Sci U S A*. 2009;106(10):3698–3703. https://doi.org/10.1073/pnas.0812874106.
241. Sun CY, Lin CJ, Pan HC, et al. Clinical association between the metabolite of healthy gut microbiota, 3-indolepropionic acid and chronic kidney disease. *Clin Nutr*. 2019;38(6):2945–2948. https://doi.org/10.1016/j.clnu.2018.11.029.
242. Konopelski P, Konop M, Gawrys-Kopczynska M, Podsadni P, Szczepanska A, Ufnal M. Indole-3-propionic acid, a tryptophan-derived bacterial metabolite, reduces weight gain in rats. *Nutrients*. 2019;11(3):E591. https://doi.org/10.3390/nu11030591.
243. Zhao ZH, Xin FZ, Xue Y, et al. Indole-3-propionic acid inhibits gut dysbiosis and endotoxin leakage to attenuate steatohepatitis in rats. *Exp Mol Med*. 2019;51(9):1–14. https://doi.org/10.1038/s12276-019-0304-5.
244. Venkatesh M, Mukherjee S, Wang H, et al. Symbiotic bacterial metabolites regulate gastrointestinal barrier function via the xenobiotic sensor PXR and Toll-like receptor 4. *Immunity*. 2014;41(2):296–310. https://doi.org/10.1016/j.immuni.2014.06.014.

245. de Mello VD, Paananen J, Lindström J, et al. Indolepropionic acid and novel lipid metabolites are associated with a lower risk of type 2 diabetes in the Finnish Diabetes Prevention Study. *Sci Rep.* 2017;7:46337. https://doi.org/10.1038/srep46337.
246. Tuomainen M, Lindström J, Lehtonen M, et al. Associations of serum indolepropionic acid, a gut microbiota metabolite, with type 2 diabetes and low-grade inflammation in high-risk individuals. *Nutr Diabetes.* 2018;8(1):35. https://doi.org/10.1038/s41387-018-0046-9.
247. Cason CA, Dolan KT, Sharma G, et al. Plasma microbiome-modulated indole- and phenyl-derived metabolites associate with advanced atherosclerosis and postoperative outcomes. *J Vasc Surg.* 2018;68(5):1552–1562.e7. https://doi.org/10.1016/j.jvs.2017.09.029.
248. Kappel BA, De Angelis L, Heiser M, et al. Cross-omics analysis revealed gut microbiome-related metabolic pathways underlying atherosclerosis development after antibiotics treatment. *Mol Metab.* 2020;36:100976. https://doi.org/10.1016/j.molmet.2020.100976.
249. Gesper M, Nonnast ABH, Kumowski N, et al. Gut-derived metabolite indole-3-propionic acid modulates mitochondrial function in cardiomyocytes and alters cardiac function. *Front Med (Lausanne).* 2021;8:648259. https://doi.org/10.3389/fmed.2021.648259.
250. Li M, Wang B, Zhang M, et al. Symbiotic gut microbes modulate human metabolic phenotypes. *Proc Natl Acad Sci U S A.* 2008;105(6):2117–2122. https://doi.org/10.1073/pnas.0712038105.
251. Aronov PA, Luo FJG, Plummer NS, et al. Colonic contribution to uremic solutes. *J Am Soc Nephrol.* 2011;22(9):1769–1776. https://doi.org/10.1681/ASN.2010121220.
252. Barrios C, Beaumont M, Pallister T, et al. Gut-microbiota-metabolite axis in early renal function decline. *PLoS One.* 2015;10(8):e0134311. https://doi.org/10.1371/journal.pone.0134311.
253. Price ND, Magis AT, Earls JC, et al. A wellness study of 108 individuals using personal, dense, dynamic data clouds. *Nat Biotechnol.* 2017;35(8):747–756. https://doi.org/10.1038/nbt.3870.
254. Pallister T, Jackson MA, Martin TC, et al. Hippurate as a metabolomic marker of gut microbiome diversity: modulation by diet and relationship to metabolic syndrome. *Sci Rep.* 2017;7(1):13670. https://doi.org/10.1038/s41598-017-13722-4.
255. Poesen R, Claes K, Evenepoel P, et al. Microbiota-derived phenylacetylglutamine associates with overall mortality and cardiovascular disease in patients with CKD. *J Am Soc Nephrol.* 2016;27(11):3479–3487. https://doi.org/10.1681/ASN.2015121302.
256. Elliott P, Posma JM, Chan Q, et al. Urinary metabolic signatures of human adiposity. *Sci Transl Med.* 2015;7(285). 285ra62. https://doi.org/10.1126/scitranslmed.aaa5680.
257. Loo RL, Zou X, Appel LJ, Nicholson JK, Holmes E. Characterization of metabolic responses to healthy diets and association with blood pressure: application to the Optimal Macronutrient Intake Trial for Heart Health (OmniHeart), a randomized controlled study. *Am J Clin Nutr.* 2018;107(3):323–334. https://doi.org/10.1093/ajcn/nqx072.
258. Urpi-Sarda M, Almanza-Aguilera E, Llorach R, et al. Non-targeted metabolomic biomarkers and metabotypes of type 2 diabetes: a cross-sectional study of PREDIMED trial participants. *Diabetes Metab.* 2019;45(2):167–174. https://doi.org/10.1016/j.diabet.2018.02.006.
259. Nemet I, Saha PP, Gupta N, et al. A cardiovascular disease-linked gut microbial metabolite acts via adrenergic receptors. *Cell.* 2020;180(5):862–877.e22. https://doi.org/10.1016/j.cell.2020.02.016.

260. Ottosson F, Brunkwall L, Smith E, et al. The gut microbiota-related metabolite phenylacetylglutamine associates with increased risk of incident coronary artery disease. *J Hypertens*. 2020;38(12):2427–2434. https://doi.org/10.1097/HJH.0000000000002569.
261. Molinaro A, Bel Lassen P, Henricsson M, et al. Imidazole propionate is increased in diabetes and associated with dietary patterns and altered microbial ecology. *Nat Commun*. 2020;11(1):5881. https://doi.org/10.1038/s41467-020-19589-w.
262. Sirich TL, Aronov PA, Plummer NS, Hostetter TH, Meyer TW. Numerous protein-bound solutes are cleared by the kidney with high efficiency. *Kidney Int*. 2013;84(3):585–590. https://doi.org/10.1038/ki.2013.154.
263. Gryp T, Vanholder R, Vaneechoutte M, Glorieux G. p-Cresyl sulfate. *Toxins (Basel)*. 2017;9(2):E52. https://doi.org/10.3390/toxins9020052.
264. Vanholder R, Schepers E, Pletinck A, Nagler EV, Glorieux G. The uremic toxicity of indoxyl sulfate and p-cresyl sulfate: a systematic review. *J Am Soc Nephrol*. 2014;25(9):1897–1907. https://doi.org/10.1681/ASN.2013101062.
265. Wang CP, Lu LF, Yu TH, et al. Serum levels of total p-cresylsulphate are associated with angiographic coronary atherosclerosis severity in stable angina patients with early stage of renal failure. *Atherosclerosis*. 2010;211(2):579–583. https://doi.org/10.1016/j.atherosclerosis.2010.03.036.
266. Lin CJ, Chuang CK, Jayakumar T, et al. Serum p-cresyl sulfate predicts cardiovascular disease and mortality in elderly hemodialysis patients. *Arch Med Sci*. 2013;9(4):662–668. https://doi.org/10.5114/aoms.2013.36901.
267. Lin CJ, Wu V, Wu PC, Wu CJ. Meta-analysis of the associations of p-cresyl sulfate (PCS) and indoxyl sulfate (IS) with cardiovascular events and all-cause mortality in patients with chronic renal failure. *PLoS One*. 2015;10(7):e0132589. https://doi.org/10.1371/journal.pone.0132589.
268. Shafi T, Meyer TW, Hostetter TH, et al. Free levels of selected organic solutes and cardiovascular morbidity and mortality in hemodialysis patients: results from the Retained Organic Solutes and Clinical Outcomes (ROSCO) Investigators. *PLoS One*. 2015;10(5):e0126048. https://doi.org/10.1371/journal.pone.0126048.
269. Wu IW, Hsu KH, Hsu HJ, et al. Serum free p-cresyl sulfate levels predict cardiovascular and all-cause mortality in elderly hemodialysis patients—a prospective cohort study. *Nephrol Dial Transplant*. 2012;27(3):1169–1175. https://doi.org/10.1093/ndt/gfr453.
270. Wang CP, Lu LF, Yu TH, et al. Associations among chronic kidney disease, high total p-cresylsulfate and major adverse cardiac events. *J Nephrol*. 2013;26(1):111–118. https://doi.org/10.5301/jn.5000111.
271. Meijers BKI, Bammens B, De Moor B, Verbeke K, Vanrenterghem Y, Evenepoel P. Free p-cresol is associated with cardiovascular disease in hemodialysis patients. *Kidney Int*. 2008;73(10):1174–1180. https://doi.org/10.1038/ki.2008.31.
272. Meijers BKI, Claes K, Bammens B, et al. p-Cresol and cardiovascular risk in mild-to-moderate kidney disease. *Clin J Am Soc Nephrol*. 2010;5(7):1182–1189. https://doi.org/10.2215/CJN.07971109.
273. Poesen R, Viaene L, Verbeke K, et al. Cardiovascular disease relates to intestinal uptake of p-cresol in patients with chronic kidney disease. *BMC Nephrol*. 2014;15:87. https://doi.org/10.1186/1471-2369-15-87.
274. Han H, Zhu J, Zhu Z, et al. p-Cresyl sulfate aggravates cardiac dysfunction associated with chronic kidney disease by enhancing apoptosis of cardiomyocytes. *J Am Heart Assoc*. 2015;4(6):e001852. https://doi.org/10.1161/JAHA.115.001852.
275. Erridge C, Bennett-Guerrero E, Poxton IR. Structure and function of lipopolysaccharides. *Microbes Infect*. 2002;4(8):837–851. https://doi.org/10.1016/S1286-4579(02)01604-0.

276. Bertani B, Ruiz N. Function and biogenesis of lipopolysaccharides. *EcoSal Plus*. 2018;8 (1). https://doi.org/10.1128/ecosalplus.ESP-0001-2018.
277. Goraca A, Piechota A, Huk-Kolega H. Effect of alpha-lipoic acid on LPS-induced oxidative stress in the heart. *J Physiol Pharmacol*. 2009;60(1):61–68.
278. Lu YC, Yeh WC, Ohashi PS. LPS/TLR4 signal transduction pathway. *Cytokine*. 2008;42(2):145–151. https://doi.org/10.1016/j.cyto.2008.01.006.
279. Zhang Y, Huang Z, Li H. Insights into innate immune signalling in controlling cardiac remodelling. *Cardiovasc Res*. 2017;113(13):1538–1550. https://doi.org/10.1093/cvr/cvx130.
280. Niebauer J, Volk HD, Kemp M, et al. Endotoxin and immune activation in chronic heart failure: a prospective cohort study. *Lancet*. 1999;353(9167):1838–1842. https://doi.org/10.1016/S0140-6736(98)09286-1.
281. Lorenzatti AJ, Servato ML. New evidence on the role of inflammation in CVD risk. *Curr Opin Cardiol*. 2019;34(4):418–423. https://doi.org/10.1097/HCO.0000000000000625.
282. Molteni M, Gemma S, Rossetti C. The role of toll-like receptor 4 in infectious and noninfectious inflammation. *Mediat Inflamm*. 2016;2016:6978936. https://doi.org/10.1155/2016/6978936.
283. Ruparelia N, Chai JT, Fisher EA, Choudhury RP. Inflammatory processes in cardiovascular disease: a route to targeted therapies. *Nat Rev Cardiol*. 2017;14(5):314. https://doi.org/10.1038/nrcardio.2017.33.
284. Asci H, Ozmen O, Erzurumlu Y, Sofu A, Icten P, Kaynak M. Agomelatine protects heart and aorta against lipopolysaccharide-induced cardiovascular toxicity via inhibition of NF-kβ phosphorylation. *Drug Chem Toxicol*. 2019;1–10. Published online September 13. https://doi.org/10.1080/01480545.2019.1663209.
285. Su Q, Yao J, Sheng C. Geniposide attenuates LPS-induced injury via up-regulation of miR-145 in H9c2 cells. *Inflammation*. 2018;41(4):1229–1237. https://doi.org/10.1007/s10753-018-0769-8.
286. Suliman HB, Welty-Wolf KE, Carraway M, Tatro L, Piantadosi CA. Lipopolysaccharide induces oxidative cardiac mitochondrial damage and biogenesis. *Cardiovasc Res*. 2004;64(2):279–288. https://doi.org/10.1016/j.cardiores.2004.07.005.
287. Peschel T, Schönauer M, Thiele H, Anker SD, Schuler G, Niebauer J. Invasive assessment of bacterial endotoxin and inflammatory cytokines in patients with acute heart failure. *Eur J Heart Fail*. 2003;5(5):609–614. https://doi.org/10.1016/s1388-9842(03)00104-1.
288. Pastori D, Carnevale R, Nocella C, et al. Gut-derived serum lipopolysaccharide is associated with enhanced risk of major adverse cardiovascular events in atrial fibrillation: effect of adherence to mediterranean diet. *J Am Heart Assoc*. 2017;6(6): e005784. https://doi.org/10.1161/JAHA.117.005784.
289. Pussinen PJ, Havulinna AS, Lehto M, Sundvall J, Salomaa V. Endotoxemia is associated with an increased risk of incident diabetes. *Diabetes Care*. 2011;34(2):392–397. https://doi.org/10.2337/dc10-1676.
290. Szeto CC, Kwan BCH, Chow KM, et al. Endotoxemia is related to systemic inflammation and atherosclerosis in peritoneal dialysis patients. *Clin J Am Soc Nephrol*. 2008;3(2):431–436. https://doi.org/10.2215/CJN.03600807.
291. Ahmadabady S, Beheshti F, Shahidpour F, Khordad E, Hosseini M. A protective effect of curcumin on cardiovascular oxidative stress indicators in systemic inflammation induced by lipopolysaccharide in rats. *Biochem Biophys Rep*. 2021;25:100908. https://doi.org/10.1016/j.bbrep.2021.100908.
292. Cani PD, Amar J, Iglesias MA, et al. Metabolic endotoxemia initiates obesity and insulin resistance. *Diabetes*. 2007;56(7):1761–1772. https://doi.org/10.2337/db06-1491.

293. Neves AL, Coelho J, Couto L, Leite-Moreira A, Roncon-Albuquerque R. Metabolic endotoxemia: a molecular link between obesity and cardiovascular risk. *J Mol Endocrinol.* 2013;51(2):R51–R64. https://doi.org/10.1530/JME-13-0079.
294. Horseman MA, Surani S, Bowman JD. Endotoxin, toll-like receptor-4, and atherosclerotic heart disease. *Curr Cardiol Rev.* 2017;13(2):86–93. https://doi.org/10.2174/1573403X12666160901145313.
295. Hsu CC, Wei TS, Huang CC, Chen YM. Endotoxemia is associated with acute coronary syndrome in patients with end stage kidney disease. *BMC Nephrol.* 2017;18(1):235. https://doi.org/10.1186/s12882-017-0652-0.
296. Zhang J, Wang G, Liu J, et al. Gut microbiota-derived endotoxin enhanced the incidence of cardia bifida during cardiogenesis. *J Cell Physiol.* 2018;233(12):9271–9283. https://doi.org/10.1002/jcp.26175.
297. Matsuura M. Structural modifications of bacterial lipopolysaccharide that facilitate gram-negative bacteria evasion of host innate immunity. *Front Immunol.* 2013;4:109. https://doi.org/10.3389/fimmu.2013.00109.
298. Steimle A, Autenrieth IB, Frick JS. Structure and function: lipid A modifications in commensals and pathogens. *Int J Med Microbiol.* 2016;306(5):290–301. https://doi.org/10.1016/j.ijmm.2016.03.001.
299. Slocum C, Coats SR, Hua N, et al. Distinct lipid a moieties contribute to pathogen-induced site-specific vascular inflammation. *PLoS Pathog.* 2014;10(7):e1004215. https://doi.org/10.1371/journal.ppat.1004215.
300. Zuo K, Li J, Li K, et al. Disordered gut microbiota and alterations in metabolic patterns are associated with atrial fibrillation. *Gigascience.* 2019;8(6):giz058. https://doi.org/10.1093/gigascience/giz058.
301. Tabata T, Yamashita T, Hosomi K, et al. Gut microbial composition in patients with atrial fibrillation: effects of diet and drugs. *Heart Vessel.* 2021;36(1):105–114. https://doi.org/10.1007/s00380-020-01669-y.
302. Yoshida N, Emoto T, Yamashita T, et al. Bacteroides vulgatus and Bacteroides dorei reduce gut microbial lipopolysaccharide production and inhibit atherosclerosis. *Circulation.* 2018;138(22):2486–2498. https://doi.org/10.1161/CIRCULATIONAHA.118.033714.
303. Yoshida N, Yamashita T, Kishino S, et al. A possible beneficial effect of Bacteroides on faecal lipopolysaccharide activity and cardiovascular diseases. *Sci Rep.* 2020;10(1):13009. https://doi.org/10.1038/s41598-020-69983-z.
304. Li J, Lin S, Vanhoutte PM, Woo CW, Xu A. Akkermansia muciniphila protects against atherosclerosis by preventing metabolic endotoxemia-induced inflammation in Apoe-/- mice. *Circulation.* 2016;133(24):2434–2446. https://doi.org/10.1161/circulationaha.115.019645.
305. Depommier C, Everard A, Druart C, et al. Supplementation with Akkermansia muciniphila in overweight and obese human volunteers: a proof-of-concept exploratory study. *Nat Med.* 2019;25(7):1096–1103. https://doi.org/10.1038/s41591-019-0495-2.
306. Sommer F, Bäckhed F. The gut microbiota — masters of host development and physiology. *Nat Rev Microbiol.* 2013;11(4):227–238. https://doi.org/10.1038/nrmicro2974.
307. Chen J, Guo Y, Gui Y, Xu D. Physical exercise, gut, gut microbiota, and atherosclerotic cardiovascular diseases. *Lipids Health Dis.* 2018;17:17. https://doi.org/10.1186/s12944-017-0653-9.
308. Khan TJ, Ahmed YM, Zamzami MA, et al. Effect of atorvastatin on the gut microbiota of high fat diet-induced hypercholesterolemic rats. *Sci Rep.* 2018;8(1):662. https://doi.org/10.1038/s41598-017-19013-2.

309. Romano KA, Vivas EI, Amador-Noguez D, Rey FE. Intestinal microbiota composition modulates choline bioavailability from diet and accumulation of the proatherogenic metabolite trimethylamine-N-oxide. *MBio*. 2015;6(2):e02481-14. https://doi.org/10.1128/mBio.02481-14.
310. Mayerhofer CCK, Awoyemi AO, Moscavitch SD, et al. Design of the GutHeart—targeting gut microbiota to treat heart failure—trial: a Phase II, randomized clinical trial. *ESC Heart Fail*. 2018;5(5):977–984. https://doi.org/10.1002/ehf2.12332.
311. Costanza AC, Moscavitch SD, Faria Neto HCC, Mesquita ET. Probiotic therapy with Saccharomyces boulardii for heart failure patients: a randomized, double-blind, placebo-controlled pilot trial. *Int J Cardiol*. 2015;179:348–350. https://doi.org/10.1016/j.ijcard.2014.11.034.
312. Gan XT, Ettinger G, Huang CX, et al. Probiotic administration attenuates myocardial hypertrophy and heart failure after myocardial infarction in the rat. *Circ Heart Fail*. 2014;7(3):491–499. https://doi.org/10.1161/CIRCHEARTFAILURE.113.000978.
313. Wilck N, Matus MG, Kearney SM, et al. Salt-responsive gut commensal modulates TH17 axis and disease. *Nature*. 2017;551(7682):585–589. https://doi.org/10.1038/nature24628.
314. Lin PP, Hsieh YM, Kuo WW, et al. Probiotic-fermented purple sweet potato yogurt activates compensatory IGF-IR/PI3K/Akt survival pathways and attenuates cardiac apoptosis in the hearts of spontaneously hypertensive rats. *Int J Mol Med*. 2013;32(6):1319–1328. https://doi.org/10.3892/ijmm.2013.1524.
315. Marques FZ, Nelson E, Chu PY, et al. High-fiber diet and acetate supplementation change the gut microbiota and prevent the development of hypertension and heart failure in hypertensive mice. *Circulation*. 2017;135(10):964–977. https://doi.org/10.1161/CIRCULATIONAHA.116.024545.
316. Koeth RA, Levison BS, Culley MK, et al. γ-Butyrobetaine is a pro-atherogenic intermediate in gut microbial metabolism of L-carnitine to TMAO. *Cell Metab*. 2014;20(5):799–812. https://doi.org/10.1016/j.cmet.2014.10.006.
317. Chan YK, Brar MS, Kirjavainen PV, et al. High fat diet induced atherosclerosis is accompanied with low colonic bacterial diversity and altered abundances that correlates with plaque size, plasma A-FABP and cholesterol: a pilot study of high fat diet and its intervention with Lactobacillus rhamnosus GG (LGG) or telmisartan in ApoE-/- mice. *BMC Microbiol*. 2016;16(1):264. https://doi.org/10.1186/s12866-016-0883-4.
318. Karlsson FH, Fåk F, Nookaew I, et al. Symptomatic atherosclerosis is associated with an altered gut metagenome. *Nat Commun*. 2012;3(1):1–8. https://doi.org/10.1038/ncomms2266.
319. Kasahara K, Krautkramer KA, Org E, et al. Interactions between Roseburia intestinalis and diet modulate atherogenesis in a murine model. *Nat Microbiol*. 2018;3(12):1461–1471. https://doi.org/10.1038/s41564-018-0272-x.
320. Molin G. Probiotics in foods not containing milk or milk constituents, with special reference to Lactobacillus plantarum 299v. *Am J Clin Nutr*. 2001;73(2 Suppl):380S–385S. https://doi.org/10.1093/ajcn/73.2.380s.
321. Wang B, Huang Q, Zhang W, Li N, Li J. Lactobacillus plantarum prevents bacterial translocation in rats following ischemia and reperfusion injury. *Dig Dis Sci*. 2011;56(11):3187–3194. https://doi.org/10.1007/s10620-011-1747-2.
322. Naruszewicz M, Johansson ML, Zapolska-Downar D, Bukowska H. Effect of Lactobacillus plantarum 299v on cardiovascular disease risk factors in smokers. *Am J Clin Nutr*. 2002;76(6):1249–1255. https://doi.org/10.1093/ajcn/76.6.1249.
323. Chan YK, El-Nezami H, Chen Y, Kinnunen K, Kirjavainen PV. Probiotic mixture VSL#3 reduce high fat diet induced vascular inflammation and atherosclerosis in ApoE−/− mice. *AMB Express*. 2016;6(1):61. https://doi.org/10.1186/s13568-016-0229-5.

324. Mencarelli A, Cipriani S, Renga B, et al. VSL#3 resets insulin signaling and protects against NASH and atherosclerosis in a model of genetic dyslipidemia and intestinal inflammation. *PLoS One.* 2012;7(9):e45425. https://doi.org/10.1371/journal.pone.0045425.
325. Dao MC, Everard A, Aron-Wisnewsky J, et al. Akkermansia muciniphila and improved metabolic health during a dietary intervention in obesity: relationship with gut microbiome richness and ecology. *Gut.* 2016;65(3):426–436. https://doi.org/10.1136/gutjnl-2014-308778.
326. Everard A, Belzer C, Geurts L, et al. Cross-talk between Akkermansia muciniphila and intestinal epithelium controls diet-induced obesity. *Proc Natl Acad Sci U S A.* 2013;110(22):9066–9071. https://doi.org/10.1073/pnas.1219451110.
327. Plovier H, Everard A, Druart C, et al. A purified membrane protein from Akkermansia muciniphila or the pasteurized bacterium improves metabolism in obese and diabetic mice. *Nat Med.* 2017;23(1):107–113. https://doi.org/10.1038/nm.4236.
328. Organ CL, Li Z, Sharp TE, et al. Nonlethal inhibition of gut microbial trimethylamine N-oxide production improves cardiac function and remodeling in a murine model of heart failure. *J Am Heart Assoc.* 2020;9(10):e016223. https://doi.org/10.1161/JAHA.119.016223.
329. Halkjær SI, Christensen AH, Lo BZS, et al. Faecal microbiota transplantation alters gut microbiota in patients with irritable bowel syndrome: results from a randomised, double-blind placebo-controlled study. *Gut.* 2018;67(12):2107–2115. https://doi.org/10.1136/gutjnl-2018-316434.
330. Zhang F, Cui B, He X, et al. Microbiota transplantation: concept, methodology and strategy for its modernization. *Protein Cell.* 2018;9(5):462–473. https://doi.org/10.1007/s13238-018-0541-8.
331. Brandt LJ, Aroniadis OC, Mellow M, et al. Long-term follow-up of colonoscopic fecal microbiota transplant for recurrent Clostridium difficile infection. *Am J Gastroenterol.* 2012;107(7):1079–1087. https://doi.org/10.1038/ajg.2012.60.
332. van Nood E, Vrieze A, Nieuwdorp M, et al. Duodenal infusion of donor feces for recurrent Clostridium difficile. *N Engl J Med.* 2013;368(5):407–415. https://doi.org/10.1056/NEJMoa1205037.
333. Smits LP, Bouter KEC, de Vos WM, Borody TJ, Nieuwdorp M. Therapeutic potential of fecal microbiota transplantation. *Gastroenterology.* 2013;145(5):946–953. https://doi.org/10.1053/j.gastro.2013.08.058.
334. Zhang T, Li Q, Cheng L, Buch H, Zhang F. Akkermansia muciniphila is a promising probiotic. *Microb Biotechnol.* 2019;12(6):1109–1125. https://doi.org/10.1111/1751-7915.13410.
335. Vrieze A, Van Nood E, Holleman F, et al. Transfer of intestinal microbiota from lean donors increases insulin sensitivity in individuals with metabolic syndrome. *Gastroenterology.* 2012;143(4):913–916.e7. https://doi.org/10.1053/j.gastro.2012.06.031.
336. Cicero AFG, Colletti A. Role of phytochemicals in the management of metabolic syndrome. *Phytomedicine.* 2016;23(11):1134–1144. https://doi.org/10.1016/j.phymed.2015.11.009.
337. Shaito A, Thuan DTB, Phu HT, et al. Herbal medicine for cardiovascular diseases: efficacy, mechanisms, and safety. *Front Pharmacol.* 2020;11:422. https://doi.org/10.3389/fphar.2020.00422.

CHAPTER TEN

Human gut microbiota and Parkinson's disease

Archana Pant[†,††], Krishna Singh Bisht[††], Swati Aggarwal[††], and Tushar Kanti Maiti*

Regional Centre for Biotechnology, Faridabad, Haryana, India
*Corresponding author: e-mail address: tkmaiti@rcb.res.in

Contents

[†] Present Address: National Institute of Immunology, Delhi.
[††] All authors contribute equally to this manuscript.

Progress in Molecular Biology and Translational Science, Volume 192
ISSN 1877-1173
https://doi.org/10.1016/bs.pmbts.2022.08.004

Abstract

The bidirectional communication between the gut and the brain has come up very fascinating in recent years. Many studies have reported that the onset of gastrointestinal issues appears long before the actual manifestation of Parkinson's disease (PD) symptoms. Disturbances in the gut-brain axis have been found to be linked with PD. PD-linked neuropathological changes in the enteric nervous system and significant alteration of gut microbiota suggest a vital role of gut microbiota in PD pathogenesis. Studies have also suggested that aggregation of α-synuclein, one of the major proteins associated with PD neuropathology, might start from the gut and move to the central nervous system (CNS) through the vagus nerve and olfactory bulb. Inflammation in the gut has been suggested to be associated with PD initiation and progression. The flushing out of healthy gut microbiota and replacing with pathogens induces gut inflammation and promotes neuroinflammation in the CNS. Therefore, it is intriguing to understand the mechanism of gut-brain communications associated with the development of PD. This review sheds light on the PD pathology, the gut dysbiosis that is associated with PD and its medications, altered gene expression, pathways and microbial metabolites during PD.

1. Parkinson's Disease

Parkinson's Disease (PD) is the second most common neurodegenerative disorder which majorly affects the elderly population and it is marked by slow movements with tremors symptoms. A recent global burden analysis identified more than 6.1 million individuals are suffering from PD of whom 2.9 million are women and 3.2 million are men.[1] The disease prevalence increases with the age. Approximately 41 people per 100,000 at their age of 40s and 1900 people per 100,000 at their age of 80s or more are living with PD. Almost 4% of people with the age of 50 are diagnosed with PD The disease is more common among males when compared to females.

2. History

The first account of PD was initially provided by James Parkinson in the article titled *"An Essay on the Shaking Palsy"* published 200 years ago in 1817. In this monograph, James Parkinson described his observations on 6 subjects on the signs and symptoms of each individual case. He described it as a disease of progressive but gradual onset with a disabling course. Though he did not elaborate on the motor symptoms such as bradykinesia. However, he reported resting tremors, flexed posture and gait disturbances as an outcome of the disease.[2] Interestingly, James Parkinson also found that these patients show severe constipation problem, continuous saliva drooling, reduced speech and swallowing difficulties. He hinted patient with depression preceded by trembling.[3] James Parkinson died on the 21st of December 1824. Later, the neurological disorder highlighted by James as *"shaking palsy"* and it was characterized as Parkinson's Disease. Fritz Jakob Heinrich Lewy first described the characteristic inclusions in nerve cell bodies at different parts of the patient brains while studying brains of PD patients. These are identified as proteinaceous eosinophilic inclusions and mainly observed in the substantia nigra region of brain in PD patients. Later on Tretiakoff postulated that the degeneration of the nerve cells in the substantia nigra contributes rigidity and tremor. The abnormal inclusions are present not only in the nerve cell bodies but also in the nerve cell processes and these inclusions are considered as Lewy neurites.[4] It has been demonstrated that the PD associated Lewy neurites or Lewy body are immunoreactive for protein called α-synuclein.[5] After 200 years of its first published work in PD, multitude of clinical facets has exceeded breathtakingly and endorsed researcher to scrutinize the disease etiology, risk factor analysis, molecular mechanism of disease progression and finally the therapeutic interventions.

3. Etiology

Parkinson's Disease is characterized by progressive loss of dopaminergic neurons in the Substantia Nigra pars compacta (SNc) region of mid brain. The degeneration of neurons is majorly due to the deposition of α-synuclein positive proteinaceous aggregates called Lewy Bodies.[5] Substantia nigra helps to initiate and control movement patterns by sending the signals to a region called basal ganglia through the nerve fibers. Dopamine is a neurotransmitter produced by the neurons in the substantia nigra which function as an inhibitor of nerve impulses and thus suppress the unintended movement. When neurons are degenerated or damaged in substantia nigra, there is an imbalance of excitatory (acetylcholine) and inhibitory (dopamine) neurotransmitters in this region disrupting the normal signaling system. The altered neurotransmitter signaling disturbs the movement control, mood, behavior, thinking and sensation in the PD patients.[6]

4. Symptoms

Symptoms of PD have been classified majorly as motor and non-motor types and the motor symptoms include rigidity, tremor, akinesia, bradykinesia and postural instability whereas non-motor symptoms include anxiety, depression, insomnia, psychosis, rapid-eye-movement (REM) sleep behavior disorder, fatigue, constipation, dysautonomia, and hyposmia.[7] Many PD patients show a gastrointestinal (GI) impairments and more than 30% of patients with GI symptoms display constipation, dysphagia and gastroparesis. Some patients also display swallowing difficulties with unintended mouth opening. Constipation is the most common and the most prevalent prodromal GI symptom in patients with PD[8]. In more advanced cases loss of smell and taste, reduction of appetite, poor diet intake leading to weight loss are the common features (Fig. 1).[9]

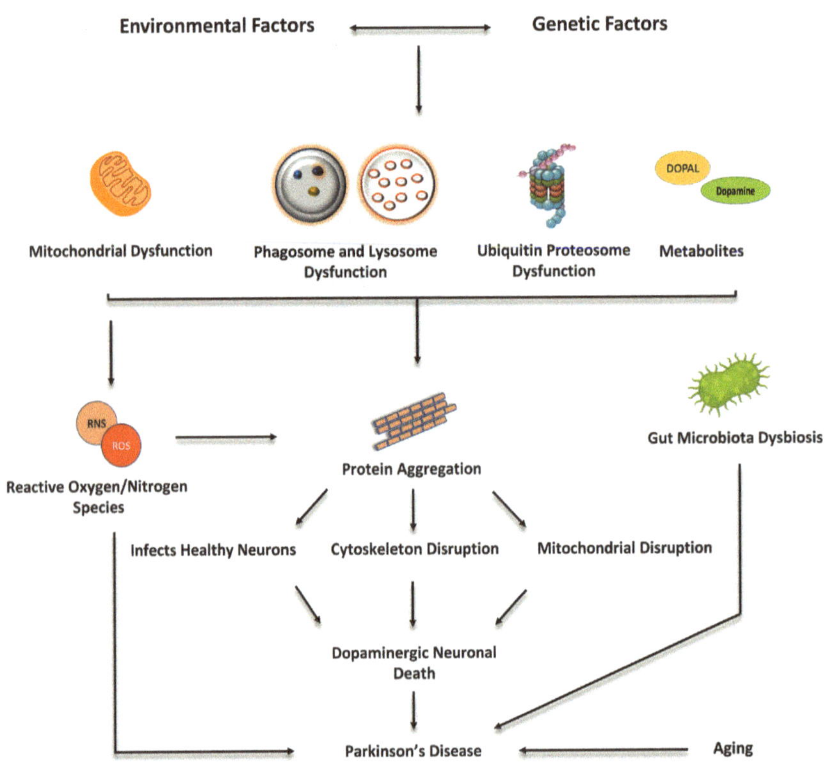

Fig. 1 Mechanistic insights of Parkinson's Disease development. Environmental and genetic risk factors lead to mitochondrial dysfunction generating oxidative stress, collapsing of protein homeostasis machinery (phagosome and lysosome dysfunction, ubiquitin proteosome dysfunction) and metabolites generation. These phenomena may cause protein aggregation which affects healthy neurons leading to cytoskeleton and organelle disruption and ultimately neuronal cell death. In addition, alterations in the gut microbiota, aging also induce the pathogenesis and development of Parkinson's Disease.

5. Risk factors

The current view of PD pathogenesis revolves around the complex interplay of sporadic and genetic factors. The sporadic PD cases are more common and only 10% of the patients show familial PD[10]. However, clinical and epidemiological studies strongly support a complex systemic pathophysiology accompanying PD-associated cognitive impairment which develop decades ago prior to the clinical manifestation. In 1997, the first PD associated missense mutation (Ala53Thr) in *SNCA gene* encoding α-synuclein protein has been mapped and identified. Subsequently, Ala30Pro and Glu46Lys mutations have been identified in families with PD and DLB.[4] After nearly 15 years of the first identification of genetic factor, 28 distinct PD related chromosomal regions have been mapped. However, the heritable monogenic PD manifests due to the mutation of genes that reside on the six of these specific regions. Mutations in the genes SNCA (PARK1) and LRRK2 (PARK8) indicate an autosomal dominant form of PD. However, the mutations in *Parkin (PARK2), DJ-1 (PARK7), PINK1 (PARK6)* and *ATP13A2 (PARK9)* genes are associated with the autosomal recessive form of PD[11].

5.1 SNCA (PARK)

In 1997, α-synuclein encoding gene *SNCA* was the first gene to show mutation as A53T mutant that caused autosomal dominant PD[12]. Many other missense mutations in α-synuclein including A30P, E46K, H50Q, A53E, G51D that have been found in the PD condition and are associated with early-onset of PD with typical motor symptoms.[13] α-Synuclein is a cytosolically abundant 140-amino acid protein consisting three distinct domains including (i) N or amino-terminal region (1–61 amino acids) containing 7 imperfect KTKEGV motifs (ii) a central hydrophobic non-amyloid component (NAC) of β amyloid (61–95 amino acids) and (iii) a negatively charged carboxy-terminal domain (96–140 amino acids). The N-terminal domain resides all missense mutations reported so far. α-Synuclein is a natively unfolded protein with almost no secondary structure and adopts α-helical rich structures upon phospholipid membranes binding through its N-terminal.[14] The point mutations in α-Synuclein tend to induce stable β sheets structure and thus aggravate the formation of species like, toxic oligomers, protofibrils and fibrils.[15] These atypical conformers upon generation, act as "seeds" that via unknown mechanisms convert normal proteins into a pathological ones.[16] Therefore, it is postulated that toxic gain of function mutations in *SNCA* cause PD.

5.2 LRRK2 (PARK8)

LRRK2 gene encodes for a cytoplasmic GTPase dependent kinase called leucine-rich repeat kinase 2 (LRRK2). However, its exact physiological function remains largely unknown.[17] LRRK2 gene mutations are also reported in late-onset autosomal dominant PD. The frequency of mutations varies from 2% to 40% in different populations. Several mutations have been reported in LRRK2 including I2020T, G2019S, Y1699C, R1441H, R1441G and R114C and they are found to be pathogenic. These pathogenic mutations are populated mainly in the C-terminal region. Among all mutations, G2019S has been studied extensively and it is prevalent in the most PD cases.[11]

5.3 Parkin (PARK2), PINK1 (PARK6) and DJ-1 (PARK7)

The mutations in *Parkin*, *PINK1*, and *DJ-1* are also linked with early onset PD (EOPD) with indistinguishable clinical phenotype. *Parkin* is second PD gene to be identified and the first to cause autosomal recessive PD. In case of juvenile PD, the homozygous mutations of *Parkin* have been observed most frequently. It is an E3 ubiquitin ligase which regulates the ubiquitination of proteins and determines the cellular fate. Parkin is also known to be involved in mitochondrial maintenance.[17]

PINK1 is a ubiquitously expressed protein kinase consisting of N-terminal conserved serine-threonine kinase domain and an autoregulatory C-terminal domain. Loss-of-function due to mutations majorly affects the PINK1 kinase activity and demonstrates its role in the development of PD. PINK1 and Parkin participate in a process to clear damaged mitochondria. PINK1 recruits Parkin from cytosol and potentiates Parkin to be enzymatically active. Activated Parkin facilitates the autophagic clearance of damaged mitochondria via mitophagy.[18]

Mutations in gene encoding protein DJ-1 is associated in 1–2% of autosomal recessive EOPD cases. DJ-1 is a 189 amino acid protein which acts as a redox sensor and it is present as a dimer under physiologic conditions. However, the mutations drive protein to unstable and misfolded condition which is eliminated by the proteasome system and reduces its function in maintaining neuronal homeostasis and antioxidant activity (Fig. 2).[19]

5.4 Other risk factors

Male individuals are more prone to develop PD than females and the gender is considered as established risk factor for PD. Another risk factor is the ethnicity which contributes significantly to the disease. However, age is the

Fig. 2 Risk factors associated with the development of Parkinson's Disease.

major risk factor of PD. The occurrence of disease progression increases with age and shows a highest peak at around 80 years of age. Due to an increase of life expectancy, aging population provides an important public health trend where more than 50% enhancement of PD cases are expected by 2030.[20] Environmental exposures also account for risk factors for Parkinson's disease. The factors that increase risk are pesticide exposure, β-blocker uses, agricultural occupation, prior head injury, rural living. The drinking of coffee, use of calcium-ion channel blockers and non-steroidal anti-inflammatory drugs, tobacco smoking, and alcohol consumption also add significantly on the disease etiology. Epidemiologic studies have also suggested that use of antipsychotics drugs like benzamides, haloperidol, phenothiazines and risperidone by elderly people also enhance risk of Parkinson's disease.[21]

The pathological findings of α-synuclein aggregated in Lewy body and Lewy neurites in dopaminergic neurons are now established as a hallmark of PD. However, the accumulated α-synuclein is not only observed in the central dopaminergic neurons but also identified in the gastrointestinal (GI) tract. Braak and co-workers hypothesize that α-synuclein migrates trans synaptically cell to cell from gut to brain. According to his hypothesis, pathogens or toxins enter in the host system mostly through upper Gastrointestinal (GI) tract and stimulate the inflammation. The aggregated α-synuclein in the Enteric Nervous System (ENS) of the GI tract transmit to the Central Nervous System (CNS) through vagus nerve leading to loss of dopaminergic neurons in the substantia nigra and progression of PD signs and symptoms. This hypothesis nicely corroborates with the fact that the PD patients develop GI symptoms prior to motor and CNS symptoms.[22] Animal studies have also supported that α-synuclein disseminates from the gut to the brain through vagus nerve. The vagus nerve resection (vagotomy) has been shown to restrict the transmission successfully.[8]

6. Gut brain axis and gut microbiota

The communication between CNS and the ENS (Gut-brain axis) is a bidirectional and this communication connects the peripheral intestinal functions with the cognitive centers of the brain (Fig. 3). The information from the brain is commuted through both sympathetic and parasympathetic arms of autonomous nervous system and neuro endocrine signaling. However, the feedback from the gut is mainly governed by the afferent spinal and sensory vagal nerves. The communication of gut and brain is influenced by the microbiota either directly affecting vagus nerve or through metabolites released, like short-chain fatty acids (SCFAs) or peptidoglycan by microbiota.[23] The signaling from brain and gut-microbiota has been established by the means of endocrine, neural, immune and humoral links.[24] The gut microbiota constitutes a complex ecological community comprising of trillions of microbes. Their presence in the gut determines both normal physiology and susceptibility to disease through metabolic activity and host interactions. The colonization of gut microbiota is established at birth and changes throughout the lifetime. The gut microbiota composition is rather stable. However, it can be altered by internal and external factors including lifestyle changes, stress, antibiotics, bacterial infections, diet, surgery, personal hygiene and environmental exposures. Majority of gut microbes are beneficial or harmless and they protect the host against pathogens. Gut microbiota

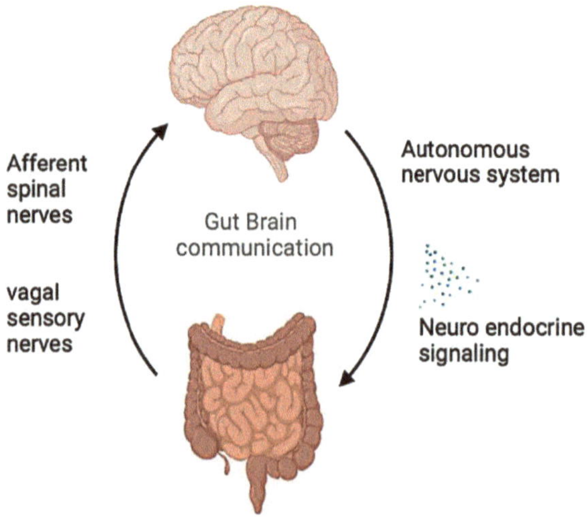

Fig. 3 Mode of communication between gut and brain.

metabolites also contribute many cellular function including normal immune functions. The gut microbiota is not identical within and among different populations. The variability is observed in the microbiome of healthy individuals vs people with different ailments where the intestinal dysbiosis is reported thus confirming its role in health and disease.[25] Gut microbiota activates the vagus nerve that plays a critical role in brain function and behavior. For example, the *Lactobacillus johnsonii* La1 strain intraduodenally incorporation reduces the renal sympathetic nerve activity and enhances the gastric vagal nerve activity in urethane-anesthetized rats.[26]

7. Gut microbiota dysbiosis in PD

Gut microbiota of an individual is now considered as second genome as it carries a wide range of genes and has the capability to modulate the pathways, thereby playing an essential role in individual's well-being. Gut microbiota is also involved in intestinal immunity, permeability, motility, and secretion. Its role in gut-brain communications and signaling to CNS is also being studied.[27] Recent studies suggests that the gut microbiota also regulates the growth and development functions of both the CNS and ENS.[28–30] Dysbiosis, which is defined as alterations within the microbial composition of human gut microbiota, has been reported in patients diagnosed with multiple neurological disoders.[31] For example, Gut microbes associated with feces and mucosa differ between Parkinson's disease patients and healthy controls.[32] Studies have shown that depletion of resident gut microbiota by the use of broad-spectrum antibiotics in specific pathogen free (SPF) mice evidently compromises the shape and maturation of microglial cells. However, restoration of the complex microbiota had somewhat restored microglia features.[33,34] Also, recent studies have revealed that changes in diet with high sugar or high energy diet are associated with gut microbiota dysbiosis, inflamed gut, microglia activation and interruption in vagus nerve associated gut–brain communication.[35] Many studies establish an interconnection between gut microbiota composition and Parkinson's disease.[36,37] The symptoms that often precede motor issues are gut dysfunctions, motility, constipation, dysfunctional defecation and weight loss.[38] This suggested that gut dysfunction is a possible pre-symptomatic phase in PD pathogenesis.[39]

Multiple studies have reported altered gut microbiota composition in PD patients compared to controls. Gut microbiota analysis of 89 PD patients Siberian population shows a significant differences in comparison with the healthy control. There is a decrease in the taxonomical diversity in

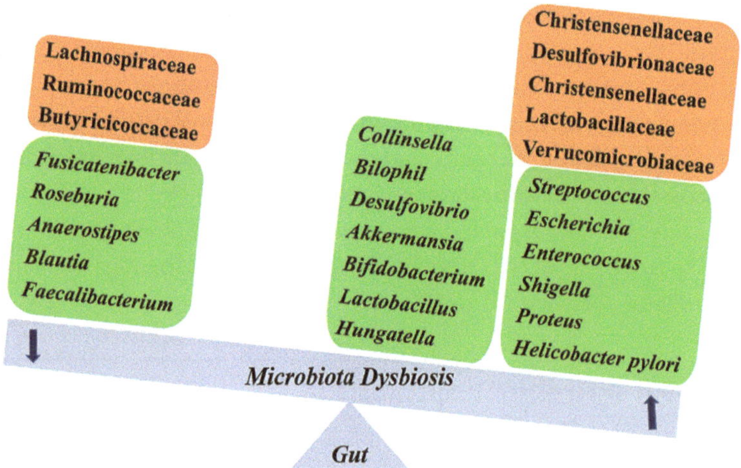

Fig. 4 Representation of gut microbiota dysbiosis in PD patients. Upside arrow: abundance, Downside arrow: depletion, Genera in green boxes, Families in orange boxes.

PD microbiota along with difference in 9 genera and 15 species.[40] Many studies suggest an enrichment of genus *Akkermansia* and *Verrucomicrobiaceae* family and depletion of *Lachnospiraceae* family and conflicting results for *Prevotellaceae* family in PD individuals. Different cohorts also show inconsistencies in their findings like Chinese cohorts never suggested enriched *Lactobacillaceae* family, however, western cohorts do (Fig. 4).[41–43]

A study by Romano and his group has provided a meta-analysis of 10 studies with individual patient data and described the probable gut microbiota composition associated with PD. Their analysis suggests that the PD gut microbiome differ significantly from controls. Consistent to previous studies, they also suggest a raise in abundance of rare taxa and depletion in the loads of the most abundant species.[44,45] Interestingly, the PD patients have shown a similar microbiota depletion profiles like to other neurodegenerative, neuro-inflammatory and gastrointestinal disorders.[46,47] A lower abundance of the genera *Fusicatenibacter*, *Roseburia*, *Anaerostipes*, *Blautia* and *Faecalibacterium* and Ruminococcaceae, Lachnospiraceae families along with Butyricicoccaceae family (highly depleted) is also observed in PD patients and the Butyriciccocaceae family is enriched in controls. *Fusicatenibacter* and *Faecalibacterium* not only have significant depletion in PD patients but also correlate to the degree of gut inflammation.[48] A study has shown that the recurrent infection with *Helicobacter pylori* promotes PD[49] and in Indians, almost 50% of the PD patients are seropositive to *H. pylori*[50]. A study has also

Table 1 Keystone taxa alteration during PD and their association with PD.

Genus	Alteration in PD pathology	Effects
Akkermansia	Increase	Degradation of mucin, increasing the gut permeability Induce ROS generation by increasing mitochondrial uptake of Ca^{++} in enteroendocrine cells[109]
Bifidobacterium	Increase	Though these are probiotic in nature but impact of their abundance in PD progression in needed to elucidate
Lactobacillus	Increase	
H. pylori	Increase	Hinders the absorption of levodopa, a primary drug used for PD management[110] Increases IL-6 and TNF-alfa in serum which may aggravate blood brain barrier[111]
Prevotella	Decrease	Decrease leads to less production of small fatty acids along with thiamine and folate[110]
Faecalibacterium	Decrease	
Lachnospiraceae	Decrease	

reported an increased abundance of pathogens that are belong to genus *Streptococcus*, *Escherichia*, *Enterococcus*, *Shigella* and *Proteus* in PD patients compared to the controls. The genera *Akkermansia*, *Bifidobacterium*, *Lactobacillus* and *Hungatella,* which are considered as beneficial bacteria, show highly prevalent in PD patient's gut (Table 1). The Lactobacillaceae and Christensenellaceae are the most strongly enriched taxa in PD.[40,43,51,52] In a study by Murros and coworkers, have identified the presence, a Gram-negative sulfate-reducing bacteria of genus *Desulfovibrio*, in the fecal samples PD patients. The abundance of *Desulfovibrio* species is very much correlated with the PD disease severity. The bacteria like *D. piger, D. fairfieldensis and D. desulfuricans* are also significantly common in PD samples than in control.[53] There are various bacteria which play a protective role toward PD pathology including *Bacillus subtilis* and *Clostridium butyricum*[23]. In the PD patients, high relative prevalence of *Christensenellaceae, Desulfovibrionaceae, Collinsella* and *Bilophila* and low presence of *Lachnospiraceae* (family level) have also been documented. In a Chinese cohort, PD gut is enriched with genera *Butyricicoccus, Clostridium IV* and *XVIII, Holdemania, Aquabacterium, Anaerotruncus and Sphingomonas* with reduction in *Sediminibacterium* and *Lactobacillus* population. The study also suggested a negative correlation between genera *Escherichia/ Shigella* and disease progression.[41]

Apart from bacterial component, the non-bacterial members may also contribute in PD development. Fungi, which is omnipresent may have a role similar to bacteria. Limited researches have accessed the role of fungi in PD. Saccharomyces was found to be dominant in both PD as well as controls.[54] The genera abundant in PD group were Kazachstania, Hanseniaspora, Penicillium and uncultured Tremellaceae.[54] Another study suggested the enhanced relative abundance of family Dipodascaceae and its genus Geotrichum and a reduced relative abundances of Aspergillus/ Penicillium and unidentified Opisthokonta in the PD samples vs controls.[55] Neuroinflammation presents as the most striking features of PD, however, the trigger leading to it remains unidentified. The systemic microbial infection in these patients may induce inflammation and different clinical symptoms. Effect of fungal infection on PD is not well explored. Seborrheic dermatitis (SD), a skin infection, caused by Malassezia, is associated to PD[56]. A recent study had suggested the presence of multiple microbial infections and found the presence of fungal structure in the CNS of PD patients. The study supported the presence of fungal species belonging to genera Candida, Botrytis, Malassezia and Fusarium.[57]

8. Neuroinflammation and gut microbiota in Parkinson's disease

Neuroinflammation is defined as an immune response within the nervous system that is accompanied by the production of various cytokines, chemokines, and reactive oxygen species. These molecules are produced by CNS glia (microglia and astrocytes), endothelial cells and peripherally-derived immune cells. These neuroinflammatory responses contribute to many physiological and biochemical outcome including immune cell recruitment, tissue damage, edema and ultimately cell death. However, the degree of neuroinflammation depends on the condition, duration and type of stimulus.[58]

Increasing evidence confirms that altering the gut microbiota composition can alter intestinal barrier function and intestinal permeability. This alteration affects and activate immune system at both GI epithelial cells and the ENS. The change of gut microbiota is linked with gut inflammation which aids to the initiation of α-synuclein misfolding.[59] Molecular mimicry is an interesting concept that proposes the involvement of the microbiota in neurodegeneration. In this process, bacterial proteins drive cross seeding mediated protein misfolding and aggregation, inflammation and oxidative stress which induce the neuronal toxicity leading to the development of PD[60]. Neuroinflammation in PD is associated with the induction of Toll-like receptors (TLRs), particularly TLR2 and TLR4 signaling. TLRs play an important role in regulating intestinal barrier integrity. Activation of TLRs has also been shown to activate glial cells in the CNS. Peripheral immune responses, characterized by the production of proinflammatory cytokines such as TNF-α, IL-1β and IL-8 by immune cells, also disrupt the blood-brain barrier and promote microglia-mediated inflammation and neurotoxicity. Bacterial amyloid also induces the pathways of molecular mimicry with the involvement of TLR2/4 and NFκB pathways. Therefore, activation of peripheral immune responses by the gut microbiota may enhance inflammatory responses mediated by brain amyloids such as α-synuclein. It has also been suggested that cerebral amyloid mimics bacterial or viral infection, leading to TLR-mediated glial cell activation. Furthermore, pro-inflammatory factors associated with chronic gastrointestinal disorders induce neuroinflammation leading to dopaminergic neuron death and subsequent PD[61].

9. PD medications and the gut microbiota

Carbidopa-levadopa, dopamine agonists including ropinirole, pramipexole, rotigotine and apomorphine, MAO B inhibitors such as rasagiline, selegiline and safinamide, catechol *O*-methyltransferase (COMT) inhibitors like opicapone and entacapone, the anticholinergics like benztropine and trihexyphenidyl are the majorly prescribed drugs in the PD treatment. These medications have been shown to affect the gut microbiota. The microbiota may alter the bioactivity or bioavailability of these medicine. L-Dopa is degraded by the gut flora and causes a reduced drug effectiveness or adverse drug response.[62] Tyrosine decarboxylases (tyrDCs) from *Enterococcus* species like *Enterococcus faecalis* and *Lactobacillus* species are also involved in its metabolism and reducing its efficacy.[63] However, both these genera have significant efficiency differences in L-dopa metabolism.[64] Dopamine-dehydroxylating enzymes within *E. lenta* and 10 other related species convert bacterial-derived dopamine to a toxic m-tyramine metabolite and induce hypertensive crisis.[63]

A study by Weis and the group has shown an increase in abundance of *Finegoldia* and *Peptoniphilus* is associated with a decrease in abundance of *Faecalibacterium* in PD patients treated with L-dopa. However, the abundance change in the bacterial species is not replicated in PD patients if they are given other dopaminergic drugs. Entacapone, the frequently used

COMT inhibitor is also metabolized by several gut bacteria. *E. faecalis* effectively metabolizes both entacapone and levodopa with an efficacy of 98.9%.[65] Patients who have been treated with entacapone show a significant difference in relative abundances of number of families and genera like *Faecalibacterium*, *Bifidobacterium*, *Blautia*, Lachnospiraceae and Enterobacteriaceae.[32,48,66] The entacapone intake specifically reduces the relative abundance of *Faecalibacterium prausnitzii* along with a reduced level of butyrate. However other COMT inhibitors like opicapone and tolcapone do not show this diferences[44,48,66]. On the other hand Entacapone treated patients also show a relative increase of *Sellimonas*, *Anaerococcus* and *Enterococcus*. The relative abundance of *Bifidobacteriaceae* is also significantly high in male PD patients treated with entacapone.[48]

In contrast, a short-term longitudinal study by Palacios and co-workers have inspected if L-dopa and gut microbiota of PD patients show biphasic interaction and their study confirms with no significant alterations in gut diversity with L-dopa. They observed a marginal level of *Clostridium* group IV in patients with a good motor response to Levodopa compared to those with no response. Surprisingly these results do not corroborate well with the previous studies which correlate the abundance of *Dorea*, *Phascolarctobacterium*[41] and *Bacillaceae*[44] species upon L-dopa use.[44]

10. Microbial metabolites in Parkinson's disease

The gut microbiota influences the PD pathogenesis in multiple ways including metabolites release, immune response modulation through cytokines release and amyloidogenic proteins. These substances either affect the blood brain barrier and induce neuro inflammation or directly play role in accumulation of amyloids. PD is accompanied by altered gut microbiota along with changes in microbial metabolites. *Streptococcus*, a pathobiont, has been found in PD gut abundantly and produces neurotoxins like streptokinase and streptomycin, which permanently damage the nervous system. Calprotectin, an inflammatory marker has been found to be higher in the fecal sample of PD patients than healthy controls.[67] An opportunistic pathogen like *Finegoldia magna* infect immunocompromised people[68] and degrades defensive proteins produced by the host.[69] PD subjects may have raised gut inflammation and permeability. The reason for this is thought to be lower SCFAs in the gut. Bacterial fermentation of dietary fiber produces SCFAs that are important for fueling colonocytes and maintaining colonic epithelial integrity. Butyrate and other SCFA also impact the ENS and microglial function. These metabolites also induce the systemic anti-inflammatory response leading to the epigenetic alteration in the CNS.[25] Low SCFA levels have been attributed to the reduced abundance of SCFA-producing taxa in PD patients. Many beneficial bacteria that ferment dietary fiber and produce SCFA have been shown with decrease in abundance during PD. Some bacteria like *Clostridium* group IV have also shown to induce T regulatory cells accumulation and maintain immune homeostasis in mice colon.[70] Enhanced levels of different proinflammatory cytokines in both serum and colon are reported in PD subjects.[71] The gut microbiota also secretes many neuroactive substances. Examples of serotonin producers are *Candida, Escherichia coli, Enterococci* and *Streptococci*. Similarly, *Bifidobacterium* and *Lactobacillus* produce gamma amino-butyric acid (GABA), *Lactobacillus* produce acetylcholine, *Bacillus* and *Serratia* produce dopamine, while *Escherichia* and *Saccharomyces* secrete norepinephrine. Gut microbiota regulate the expression of the large numbers of factors including 5-hydroxytryptamine receptor (5-HT1A), brain-derived neurotropic factor (BDNF) and NMDA receptor subunit 2 (NR2A) along with enhanced enteric and circulating serotonin production.[72] But how they directly influence the brain is not completely known yet. Decreased levels of thiamine, folic acid biosynthesis, SCFAs, and ghrelin are also associated with decreased frequency of Prevotella in PD.[23] Direct protective effect of *Prevotella* by releasing H_2S on dopaminergic neurons has also been demonstrated.[73]

Although less studied, the metabolites which may be linked to fungi can also be differentially associated with PD and controls. A study had found that fungal metabolites like 1-octen-3-ol can cause neurological damage in Drosophila by directly disrupting dopamine packaging.[74] In another study, there was a marked difference in six volatile organic compounds (VOCs) in stool samples of PD vs control groups. Two VOCs, 2,6,6-trimethylbicyclo [3.1.1]hept-2-ene and 1,3-ditert-butylbenzene were linked to PD and four were linked to controls of which three were SCFAs. An abundant species in PD, Torulaspora (OTU-314) was strongly correlated with two VOCs: 2,6,6-trimethylbicyclo[3.1.1]hept-2-ene (β-pinene) and 6,6-dimethyl-2-methylene-bicyclo [3.1.1]heptane.[42]

11. Altered gene expression and associated pathways in Parkinson's disease patient's gut

The family *Christensenellaceae*, being one of the most abundant taxa in PD, can help in *Methanobrevibacter* proliferation, through H_2 production.[75]

This archaebacterium is the major methane producer in the human gut. The positive correlation of these two taxa in PD patients supports metabolic pathways involved in methanogenesis from CO_2 and H_2. Also, the pathways involved in synthesizing coenzyme B, M and Factor 420 (key co-factors participating in methanogenesis) are also abundant in PD subjects than in control group. A healthy gut microbiota helps to generate different neurotransmitters like acetylcholine, serotonin, melatonin, catecholamine, GABA and histamine. These neuroactive molecules influence gastrointestinal motility and the immune system, nutritional absorption, gut microbiota and blood flow regulation. The enhanced glutamate degradation pathway and reduced glutamate/ glutamine biosynthesis pathway in PD patients advocate differential enteric production of neurotransmitters. The menaquinone (vitamin K2) and ubiquinone (CoQ) biosynthesis pathways are also enriched in PD associated microbiota. Previous studies have suggested that both vitamin K and CoQ are important to prevent mitochondrial dysfunctions associated with PD.[76,77] A study has reported that fecal samples from PD patients show a positive for the *Desulfovibrio*-specific [Fe-Fe]-hydrogenase gene[78] which reduces the soluble ferric iron to ferrous ion.[79] *Desulfovibrio* generates hydrogen sulfide which removes the iron from ferritin and enhances iron levels in the cytosol.[80,81] Both ferrous and ferric ions are capable of inducing α-synuclein aggregation and considered as neuropathologic factor of PD.[82]

12. Changes in nutrients profile in Parkinson's disease patients

The increased homocysteine level also causes neurotoxicity. Certain vitamins like Vitamin B6, B9, and B12 play an major role in homocysteine metabolism, thereby acting as neuroprotectors.[83] Vitamin C is abundant in CNS is considered as a neuroprotective vitamin. However, its neuroprotection mechanism is not confirmed.[79] Vitamin E has been found to decrease the PD risk.[84,85] Vitamin D also involves in calcium regulation which indirectly affects the neuronal loss within the substantia nigra. Vitamin B12 and vitamin D deficiencies are detected in PD patients and their supplementation have shown to improve motor symptoms.[86–88] The abundance of N-3 polyunsaturated fatty acids (PUFAs) within the brain is considered to have a neuroprotective effect.[89] The α-linolenic acid, a fatty acid precursor of PUFAs has shown a reduced PD risk.[90] Micro and macronutrients are considered to be very important for growth and normal body functions and also their levels are greatly associated with the composition of gut microbiota. However, this understanding remains unexplored in PD patients.

13. Models to study microbiota brain axis

13.1 Germ-free models

Germ free mice give the opportunity to perform study in completely sterile gut, i.e., free from any detectable bacteria, virus or any other microbe in the gut. These models give the platform to study the impact of exclusively introduced known microbial colonies.[78] To create the GF mouse model, caesarean sections are carefully performed to avoid contact with microorganisms present in both the vagina and skin of the pup and mother. These pups are then raised in sterile isolators, colonies are maintained on sterile feed, bedding and water and fecal samples are continuously monitored for gut microbiota.[91]

13.2 Antibiotics

Despite of large number of advantages, it is not possible to generate germ free mice model for every research group due to high demand resource and expertise. he advent of antibiotic-based mouse models has removed hurdles as they offer much greater temporal flexibility and specificity compared to GF models.[78] To create these models, subtherapeutic doses of a single antibiotic or a cocktail of antibiotics are administered to eliminate the entire microbiota. The use of non-absorbable antibiotics [vancomycin, neomycin, bacitracin] is necessary to avoid potential systemic and central nervous system effects and to be able to directly assess the effects of microbiota loss on the brain.[91]

13.3 Fecal microbiota transplant

Fecal microbiota transplantation (FMT) is a mode of direct introduction of gut microbiota from one individual to other by administration of fecal suspension from donor to acceptor.[92] Transfer of intestinal microbiota is commonly performed via oral administration of fecal material through gavaging to the rodents or colonoscopy in humans. Studies have already shown that the transfer of various behavioral phenotypes constitute through FMT including anxiety-like behavior and depressive symptomatology. However, the study of fecal material in GF mice or non-GF mice needs a priory broad spectrum antibiotics treatment.[91] FMT is being investigated as a potential treatment for various diseases, including PD. Improvements in both motor and non-motor symptoms upon FMT are documented in both the PD patients and rodent model.[23,93] Fecal microbiota from the PD patients transplanted in GF-ASO (allele-specific oligonucleotide) mice shows significant early motor dysfunction as compared to GF-ASO transplanted with fecal microbiota of healthy individual.[94] Sun and colleagues also showed that FMT decreased signaling of the TLR4/TNF-α pathway in the brain, as well as decreased activation of microglia and astrocytes.[31]

14. Implications of gut microbiota on brain

14.1 Role of gut microbiota in blood brain barrier

The blood-brain barrier (BBB) or intestinal epithelial barrier is primarily composed of capillary endothelial cells (EC). The integrity is maintained by the tight junction sealing where the mode of transport is only paracellular. A healthy and stable gut microbiota community plays an important role in maintaining the homeostatic balance between barrier integrity and reduced permeability. Gut Permeability is also influenced by the specific gut microbiota population. The decrease in the *Prevotella* species in PD has been shown to increase gut permeability by affecting the mucin synthesis.[95] An increased permeability of BBB in GF mice than the pathogen free mice has been documented. Recently, Hoban et al. reported altered cortical myelination and impaired BBB functions in GF mice.[96] This abnormality may be due to reduced tight junction proteins (occludin and claudin-5) expression which are necessary for BBB functions.[23] In addition, permeability is also reduced by colonization of the gut microbiota and administration of SCFAs such as butyrate, which are primarily released by SCFA-releasing bacteria in the gut.[97] Gut microbiota is also found to govern the brain immune system by regulating the activation of microglia.

14.2 Role of gut flora in neuro inflammation

The gut microbiota has been shown to influence the immune system either by stimulating it or by releasing numerous cytokines.[34] Gut bacteria has shown to affect differentiation and activation of immune cells in the intestine, periphery as well as brain.[98] Role of gut microbiota in activation of brain immune cells, microglia and astrocytes have been already demonstrated by various groups. Lack of complex microbiota in germ free mice is enough for global defect in microglia including immature phenotype and altered cell proportion as compared to SCF. The partial reversal of phenotype by introduction of complex microbiota in germ free mice has also been repored.[33] This observation may be supported by the fact that these mice show altered circulating cytokine profile. The ratio of pro-inflammatory and anti-inflammatory circulating cytokines has been found to be highly affected by the gut microbiota, especially in case of diseases with intestinal barrier including PD.[99] Choi et al. has demonstrated role of *Proteus mirabilis* derived LPS induces early PD progression with motor deficit and loss of dopaminergic neurons.[100] Systemic mitigation of microbial products like LPS have been shown to activate TLR4 on the macrophages with increase in level of pro- inflammatory cytokines.[93] TLRs are also activated by the biological active signaling molecule released by the gut microbiota like serotonin, purines as well as many neurotropic factors (Fig. 5).[101]

14.3 Gut microbiota and oxidative stress

Gut microbiota has been shown to affect the oxidative stress locally as well as in CNS. The gut microbiota and CNS crosstalk may influence antioxidant system of CNS or enhances the oxidative stress directly through the activation of microglia. This oxidative stress may result in higher levels of toxic hydroxyl radical and plays major role in PD progression.[102] Recently the role of gut flora, in activation of innate immune system mediated by inflammasome, has also focused in neurodegenerative disorders. Inflammasomes are the complexes that activate and assemble in response

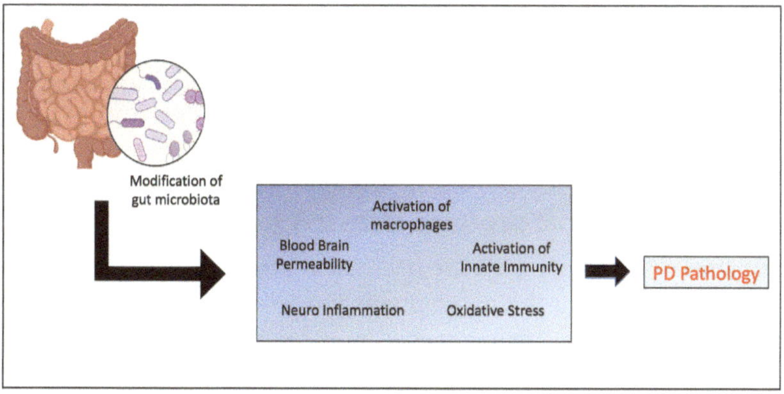

Fig. 5 Representation of effect of gut microbiota on various factors playing important in PD pathology.

to pathogens or danger signals leading to the pro-inflammatory cytokines, such as IL-18 and IL-1b, mediated cell death.[99] The NLRP3 inflammasome in MPTP induced PD mice model has already been demonstrated.[103]

15. Gut microbiota induced PD progression

Gut microbes have been shown to promote motility and gastrointestinal disorders in a Thy1-α-synuclein mouse model.[94] The PF-ASO (pathogen free) animals exhibit early progressive motor deficit as well as less fluid containing feces as compared GF-ASO littermates. Increase of region specific particularly substantia nigra α-synuclein agregation has been found in PF-ASO compared to GF-ASO. Braak first hypothesized that the etiology of PD is of gut origin, beginning with the accumulation of α-synuclein in the gut, which is transmitted to the brain via the vagus nerve in a prion-like manner.[104] α-Synuclein inclusion has been observed in the ENS specifically in the vagus and glossopharyngeal nerve.[105] Importantly, in addition to neurons many cell types, including enteroendocrine cells (EECs) of the intestinal epithelium, express α-synuclein. Because gut has more neurons than spinal cord, the ENS can act as a pathway for pathology to spread to her CNS. The bidirectional route between gut and brain facilitates the transport of many microbial agents into the brain without entering circulatory system. This is considered as one of ways of prion proteins spread into the CNS in synucleinopathies. This notion is also supported by the fact that the enteric nervous system also harbors a Lewy body (LB) at early stages of PD.[106] Animal model studies have also confirmed that the introduction of preformed α-synuclein amyloid fibrils in GI tract induces LB formation in the brain. Oral consumption of other compounds which are used intracranially for development of Parkinson models like rotenone, paraquat with lectin shows an elevated α-synuclein expression leading to early motor deficit. Transfer of gut microbiota mainly *Proteus mirabilis* of MPTP induced PD model to wild type mice displays a PD-like symptoms.[106]

15.1 Amyloidogenic proteins released by the bacteria

Other than α-synuclein, there are many food products which are not been digested completely in the gastrointestinal tract. The proteins from many food materials display amyloidogenic property and play role in initiation of PD pathology. Many of them are also directly transferred to blood or absorbed in ENS. A few well reported food proteins having amyloidogenic properties are β-parvalbumin (from fish), casein proteins (from milk), ovalbumin (from egg), gluten (from wheat). The amyloidogenic proteins released by the microbiota are mainly involved in biofilm formation and stabilization. For example, *E. coli* endotoxin has been shown to induce Aβ fibril formation in vitro, suggesting its role in the pathogenesis of (AD).[107] Nonhuman amyloidogenic protein also acts as a seed for other amyloidogenic proteins aggregation initiation in vivo. Some food proteins that act as allergens adopt the amyloid state, conferring protection from gastrointestinal digestion and allowing absorption into the body. It triggers amyloid formation upon cross-reacting with α-synuclein in neighboring cells, which is then transmitted to the brain. On the other hand, bacterial or dietary amyloid itself also travels to the brain via the blood or ENS and interacts with α-synuclein polypeptides, causing aggregation of α-synuclein. α-synuclein overexpression has been reported in response to viral and bacterial infection.[106] CsgA is the extracellular protein that is secreted by *E. coli* and assembles to amyloids called curli fibers. These amyloid polymers with cross-beta-sheets help bacterial cell binding to form biofilms and conferring resistance for destruction. Due to its amyloidogenic property its role in neurodegenerative disorders has also been investigated. The feeding of curli-producing *E. coli* to rats develops PD pathology with α-synuclein amyloid formation. However, curli-deficient *E. coli* does not show PD phenotype. CsgA also directly enhances the α-synuclein aggregation in vitro using biochemical assays. Curli has been also shown to enhance α-synuclein production both brain and gut. The curli exposed rat displays enhanced neuroinflammation along with over expression of IL6, TNF, TLR2 and microgliosis.[108] FapC is another protein which is synthesized by *Pseudomonas aeruginosa* and it provides the stability to the biofilm.[106] It also possess amyloidogenic properties and plays an important role in cell to cell attachment. As *Pseudomonas* is highly abundant in gut, the role of FapC in α-synuclein amyloid formation has been studied. Apart from these many other members of the human microbiome are capable of assembling extracellular amyloids, including *Bacillus, Streptococcus, Salmonella, Staphylococcus, Mycobacteria, Citrobacter and Klebsiella* species[103] and they need detailed investigation (Table 1).

16. Knowledge gaps, conclusions and future prospects

Although PD and its pathophysiology have been studied extensively. However, little is known about the factors that influence the initiation and progression of the disease. Similarly, the etiology of different PD phenotypes and disease severity make the problem complex. Recent evidences have shown that apart from metabolic and immune functions, gut microbiota has important role in bidirectional interactions between the gut and the nervous system. The interaction with CNS is maintained by regulating the brain chemistry and influencing associated neuro-endocrine system. The altered gut bacteria impair the gut brain barrier and disturbs the gut-immune-brain axis leading to gut inflammation thus contributing to neuroinflammation and neurodegeneration in the CNS. The diversity and abundance of microbiota also impact the function of both ENS and CNS. Also, more research is required to gain better understanding of linkage between geographical dependencies or inter-individual variations in gut microbiota in PD progression. Hence, a better understanding of gut microbiota-metabolites-brain axis may help in early diagnosing, preventing and curing the PD pathogenesis. In this chapter, we have compiled the current understanding of gut-brain communication through gut microbiota and regulation of PD disease initiation and progression. Diet, exercises, meditation, prebiotics, probiotics and symbiotics have been shown to regulate disease pathology and needs more detailed studies to circumvent the complex problem.

Acknowledgments

A.P. would like to thank Department of Science and Technology (DST)—SERB-National Postdoctoral Fellowship (N-PDF) (Grant Number PDF/2020/000421) and DST, Innovation in Science Pursuit for Inspired Research (INSPIRE)-Faculty Fellowship (Grant Number DST/INSPIRE/04/2020/001246) for providing fellowship and research grant.

K.S.B. would like to thank DST for INSPIRE fellowship for his PhD. S.A. would like to thank Council of Scientific & Industrial Research (CSIR) for her fellowship during PhD. Authors are also thankful to Regional Centre for Biotechnology, Faridabad, Haryana for intramural research support.

References

1. Ray Dorsey E, et al. Global, regional, and national burden of Parkinson's disease, 1990–2016: a systematic analysis for the global burden of disease study 2016. *Lancet Neurol*. 2018;17:939–953.
2. Jost WH, Reichmann H. "An essay on the shaking palsy" 200 years old. *J Neural Transm*. 2017;124:899–900.
3. Lees A. Dorsal column grey matter an essay on the shaking palsy. *Brain*. 2017;140:843–848.
4. Goedert M, Spillantini MG, Del Tredici K, Braak H. 100 years of Lewy pathology. *Nat Rev Neurol*. 2013;9:13–24.
5. Osterhaus A, et al. α-Synuclein in Lewy bodies. *Nature*. 1997;388:839–840.

6. Abruzzo A, Damiano G, Buscemi S, Gulotta L, Buscemi G. Reoperation for persistent or recurrent secondary hyperparathyroidism. Surgical treatment of renal hyperparathyroidism. *Acta Biomed Ateneo Parm.* 2017;88:325.
7. Balestrino R, Schapira AHV. Parkinson disease. *Eur J Neurol.* 2020;27:27–42.
8. Chiang H-L, Lin C-H. Altered gut microbiome and intestinal pathology in Parkinson's disease. *J Mov Disord.* 2019;12:67–83.
9. Hawkes CH, Del Tredici K, Braak H. Parkinson's disease: a dual-hit hypothesis. *Neuropathol Appl Neurobiol.* 2007;33(6):599–614. https://doi.org/10.1111/j.1365-2990.2007.00874.x. PMID: 17961138; PMCID: PMC7194308.
10. Thomas B, Flint Beal M. Parkinson's disease. *Hum Mol Genet.* 2007;16:183–194.
11. Huang Y, Chan P, Halliday G. Genetics of Parkinson's disease. In: *Oxidative Stress and Neurodegenerative Disorders*; 2007:663–697. https://doi.org/10.1016/B978-044452809-4/50169-1.
12. Polymeropoulos MH, Lavedan C, Leroy E, et al. Mutation in the alpha-synuclein gene identified in families with Parkinson's disease. *Science.* 1997;276(5321):2045–2047. https://doi.org/10.1126/science.276.5321.2045. PMID: 9197268.
13. Zhang PL, Chen Y, Zhang CH, Wang YX, Fernandez-Funez P. Genetics of Parkinson's disease and related disorders. *J Med Genet.* 2018;55:73–80.
14. Giasson BI, Murray IVJ, Trojanowski JQ, Lee VM. A hydrophobic stretch of 12 amino acid residues in the middle of α-synuclein is essential for filament assembly. *J Biol Chem.* 2001;276:2380–2386.
15. Bertoncini CW, Fernandez CO, Griesinger C, Jovin TM, Zweckstetter M. Familial mutants of α-synuclein with increased neurotoxicity have a destabilized conformation. *J Biol Chem.* 2005;280:30649–30652.
16. Katsnelson A, De Strooper B, Zoghbi HY. Neurodegeneration: from cellular concepts to clinical applications. *Sci Transl Med.* 2016;8:1–6.
17. Nuytemans K, Theuns J, Cruts M, Van Broeckhoven C. Genetic etiology of Parkinson disease associated with mutations in the SNCA, PARK2, PINK1, PARK7, and LRRK2 genes: a mutation update. *Hum Mutat.* 2010;31(7):763–780. https://doi.org/10.1002/humu.21277.
18. Youle RJ, Narendra DP. Mechanisms of mitophagy. *Nat Rev Mol Cell Biol.* 2011;12:9–14.
19. Anderson PC, Daggett V. Molecular basis for the structural instability of human DJ-1 induced by the L166P mutation associated with Parkinson's disease. *Biochemistry.* 2008;47:9380–9393.
20. Dorsey ER, et al. Projected number of people with Parkinson disease in the most populous nations, 2005 through 2030. *Neurology.* 2007;68:384–386.
21. Kalia LV, Lang AE. Parkinson's disease. *Lancet.* 2015;386:896–912.
22. Braak H, Del Tredici K. Neuropathological staging of brain pathology in sporadic Parkinson's disease: separating the wheat from the chaff. *J Parkinsons Dis.* 2017;7: S73–S87.
23. Socała K, et al. The role of microbiota-gut-brain axis in neuropsychiatric and neurological disorders. *Pharmacol Res.* 2021;172, 105840.
24. Carabotti M, Scirocco A, Maselli MA, Severi C. The gut-brain axis: interactions between enteric microbiota, central and enteric nervous systems. *Ann Gastroenterol.* 2015;28:203–209.
25. Keshavarzian A, Engen P, Bonvegna S, Cilia R. The gut microbiome in Parkinson's disease: a culprit or a bystander? *Prog Brain Res.* 2020;252:357–450. https://doi.org/10.1016/bs.pbr.2020.01.004. PMID: 32247371.
26. Tanida M, et al. Effects of intraduodenal injection of Lactobacillus johnsonii La1 on renal sympathetic nerve activity and blood pressure in urethane-anesthetized rats. *Neurosci Lett.* 2005;389:109–114.

27. Wiley NC, Dinan TG, Ross RP, Stanton C, Clarke G, Cryan JF. The microbiota-gut-brain axis as a key regulator of neural function and the stress response: implications for human and animal health. *J Anim Sci.* 2017;95(7):3225–3246.
28. Hyland NP, Cryan JF. Microbe-host interactions: influence of the gut microbiota on the enteric nervous system. *Dev Biol.* 2016;417:182–187.
29. Sharon G, Sampson TR, Geschwind DH, Mazmanian SK. The central nervous system and the gut microbiome. *Cell.* 2016;167(4):915–932. https://doi.org/10.1016/j.cell.2016.10.027. PMID: 27814521; PMCID: PMC5127403.
30. Ma Q, Xing C, Long W, Wang HY, Liu Q, Wang R-F. Impact of microbiota on central nervous system and neurological diseases: the gutbrain axis. *J Neuroinflammation.* 2019;16:53.
31. Schroeder BO, Bäckhed F. Signals from the gut microbiota to distant organs in physiology and disease. *Nat Med.* 2016;22:1079–1089.
32. Scheperjans F, et al. Gut microbiota are related to Parkinson's disease and clinical phenotype. *Mov Disord.* 2015;30:350–358.
33. Erny D, et al. Host microbiota constantly control maturation and function of microglia in the CNS. *Nat Neurosci.* 2015;18:965–977.
34. Sun M-F, Shen Y-Q. Dysbiosis of gut microbiota and microbial metabolites in Parkinson's disease. *Ageing Res Rev.* 2018;45:53–61.
35. Sen T, Cawthon CR, Ihde BT, et al. Diet-driven microbiota dysbiosis is associated with vagal remodeling and obesity. *Physiol Behav.* 2017;173:305–317.
36. Bedarf JR, Hildebrand F, Coelho LP, et al. Functional implications of microbial and viral gut metagenome changes in early stage L-DOPA-naïve Parkinson's disease patients. *Genome Med.* 2017;9(1):39.
37. Heintz-Buschart A, Pandey U, Wicke T, et al. The nasal and gut microbiome in Parkinson's disease and idiopathic rapid eye movement sleep behavior disorder. *Mov Disord.* 2018;33(1):88–98.
38. Cloud LJ, Greene JG. Gastrointestinal features of Parkinson's disease. *Curr Neurol Neurosci Rep.* 2011;11(4):379–384.
39. Tereshchenko LV, Anisimov VN, Shul'govsky VV, Latanov AV. Early changes in saccadic eye movement in hemiparkinsonian MPTP-treated monkeys. *Perception.* 2015;44 (8–9):1054–1063.
40. Petrov VA, et al. Analysis of gut microbiota in patients with Parkinson's disease. *Bull Exp Biol Med.* 2017;162:734–737.
41. Qian Y, et al. Alteration of the fecal microbiota in Chinese patients with Parkinson's disease. *Brain Behav Immun.* 2018;70:194–202.
42. Li C, et al. Gut microbiota differs between Parkinson's disease patients and healthy controls in Northeast China. *Front Mol Neurosci.* 2019;12:171.
43. Jin M, et al. Analysis of the gut microflora in patients with Parkinson's disease. *Front Neurosci.* 2019;13:1184.
44. Keshavarzian A, et al. Colonic bacterial composition in Parkinson's disease: colonic microbiota in Parkinson's disease. *Mov Disord.* 2015;30:1351–1360.
45. Barichella M, et al. Unraveling gut microbiota in Parkinson's disease and atypical parkinsonism. *Mov Disord.* 2019;34:396–405.
46. Shi Z, et al. Dysbiosis of gut microbiota in patients with neuromyelitis optica spectrum disorders: a cross sectional study. *J Neuroimmunol.* 2020;339, 577126.
47. Eeckhaut V, et al. Butyricicoccus pullicaecorum in inflammatory bowel disease. *Gut.* 2013;62:1745–1752.
48. Weis S, et al. Effect of Parkinson's disease and related medications on the composition of the fecal bacterial microbiota. *NPJ Park Dis.* 2019;5:28.
49. Charlett A, Dobbs RJ, Dobbs SM, Weller C, Brady P, Peterson DW. Parkinsonism: siblings share helicobacter pylori seropositivity and facets of syndrome. *Acta Neurol Scand.* 1999;99:26–35.

50. Mridula KR, et al. Association of *Helicobacter pylori* with Parkinson's disease. *J Clin Neurol.* 2017;13(2):181–186.
51. Hill-Burns EM, et al. Parkinson's disease and Parkinson's disease medications have distinct signatures of the gut microbiome: PD, medications, and gut microbiome. *Mov Disord.* 2017;32:739–749.
52. Baldini F, et al. Parkinson's disease-associated alterations of the gut microbiome predict disease-relevant changes in metabolic functions. *BMC Biol.* 2020;18:62. https://doi.org/10.1186/s12915-020-00775-7.
53. Murros KE, Huynh VA, Takala TM, Saris PEJ. Desulfovibrio bacteria are associated with Parkinson's disease. *Front Cell Infect Microbiol.* 2021;11, 652617. https://doi.org/10.3389/fcimb.(2021.652617.
54. Fernandez EDP, Gebeyehu GG, Flain L, et al. The faecal metabolome and mycobiome in Parkinson's disease. *Parkinsonism Relat Disord.* 2022;95:65–69. https://doi.org/10.1016/j.parkreldis.2022.01.005 [Epub 2022 Jan 13. PMID: 35045378].
55. Weis S, Meisner A, Schwiertz A, et al. Association between Parkinson's disease and the faecal eukaryotic microbiota. *NPJ Parkinsons Dis.* 2021;7(1):101. https://doi.org/10.1038/s41531-021-00244-0. PMID: 34795317; PMCID: PMC8602383.
56. Ravn AH, et al. Skin disorders in Parkinson's disease: potential biomarkers and risk factors. *Clin Cosmet Investig Dermatol.* 2017;10:87–92.
57. Pisa D, et al. Parkinson's disease: a comprehensive analysis of fungi and bacteria in brain tissue. *Int J Biol Sci.* 2020;16:1135–1152.
58. DiSabato DJ, Quan N, Godbout JP. Neuroinflammation: the devil is in the details. *J Neurochem.* 2016;139:136–153.
59. Olanow CW, Wakeman DR, Kordower JH. Peripheral alpha-synuclein and Parkinson's disease. *Mov Disord.* 2014;29:963–966.
60. Friedland RP. Mechanisms of molecular mimicry involving the microbiota in neurodegeneration. *J Alzheimer's Dis.* 2015;45:349–362.
61. Morais LH, Schreiber HL, Mazmanian SK. The gut microbiota–brain axis in behaviour and brain disorders. *Nat Rev Microbiol.* 2021;19:241–255.
62. Weersma RK, Zhernakova A, Fu J. Interaction between drugs and the gut microbiome. *Gut.* 2020;69:1510–1519.
63. Maini Rekdal V, Bess EN, Bisanz JE, Turnbaugh PJ, Balskus EP. Discovery and inhibition of an interspecies gut bacterial pathway for levodopa metabolism. *Science.* 2019;364(6445):eaau6323. https://doi.org/10.1126/science.aau6323. PMID: 31196984; PMCID: PMC7745125.
64. van Kessel SP, Frye AK, El-Gendy AO, et al. Gut bacterial tyrosine decarboxylases restrict levels of levodopa in the treatment of Parkinson's disease. *Nat Commun.* 2019;10:310.
65. Zimmermann M, Zimmermann-Kogadeeva M, Wegmann R, et al. Mapping human microbiome drug metabolism by gut bacteria and their genes. *Nature.* 2019;570: 462–467.
66. Unger MM, et al. Short chain fatty acids and gut microbiota differ between patients with Parkinson's disease and age-matched controls. *Parkinsonism Relat Disord.* 2016;32:66–72.
67. Schwiertz A, Spiegel J, Dillmann U, et al. Fecal markers of intestinal inflammation and intestinal permeability are elevated in Parkinson's disease. *Parkinsonism Relat Disord.* 2018;50:104–107. https://doi.org/10.1016/j.parkreldis.2018.02.022. PMID: 29454662.
68. Higaki S, Morohashi M. Characteristics of anaerobes from skin specimens. *Drugs Exp Clin Res.* 2003;29:153–155.
69. Frick I-M, et al. Constitutive and inflammation-dependent antimicrobial peptides produced by epithelium are differentially processed and inactivated by the commensal Finegoldia magna and the pathogen Streptococcus pyogenes. *J Immunol.* 2011;187: 4300–4309.

70. Atarashi K, Tanoue T, Shima T, et al. Induction of colonic regulatory T cells by indigenous Clostridium species. *Science*. 2011;331:337–341. https://doi.org/10.1126/science.1198469 Science.
71. Elfil M, Kamel S, Kandil M, Koo BB, Schaefer SM. Implications of the gut microbiome in Parkinson's disease. *Mov Disord*. 2020;35(6):921–933. https://doi.org/10.1002/mds.28004. PMID: 32092186.
72. Heijtz RD, et al. Normal gut microbiota modulates brain development and behavior. *Proc Natl Acad Sci U S A*. 2011;108:3047–3052.
73. Cakmak YO. Prevotella derived hydrogen sulfide, constipation, and neuroprotection in *Parkinson's disease. Mov Disord*. 2015;30(8):1151.
74. Inamdar AA, et al. Fungal-derived semiochemical 1-octen-3-ol disrupts dopamine packaging and causes neurodegeneration. *Proc Natl Acad Sci U S A*. 2013;110:19561–19566.
75. Ruaud A, et al. Syntrophy via interspecies H2 transfer between Christensenella and Methanobrevibacter underlies their global cooccurrence in the human gut. *MBio*. 2020;11. e03235–19.
76. Vos M, et al. Vitamin K2 is a mitochondrial electron carrier that rescues Pink1 deficiency. *Science*. 2012;336:1306–1310.
77. Shults CW. Therapeutic role of coenzyme Q10 in Parkinson's disease. *Pharmacol Ther*. 2005;107:120–130.
78. Kennedy EA, King KY, Baldridge MT. Mouse microbiota models: comparing germ-free mice and antibiotics treatment as tools for modifying gut bacteria. *Front Physiol*. 2018;9:1534.
79. Park HS, Lin S, Voordouw G. Ferric iron reduction by desulfovibrio Vulgaris Hildenborough wild type and energy metabolism mutants. *Antonie Van Leeuwenhoek*. 2008;93(1–2):79–85. https://doi.org/10.1007/s10482-007-9181-3.
80. Cassanelli S, Moulis J. Sulfide is an efficient iron releasing agent for mammalian ferritins. *Biochim Biophys Acta*. 2001;1547(1):174–182. https://doi.org/10.1016/s0167-4838(01)00182-0.
81. Hälldin J, Land T. Sulfide increases labile iron pool in RD4 cells. *Biometals*. 2008;21(2):127–131. https://doi.org/10.1007/s10534-007-9099-2.
82. Joppe K, Roser A-E, Maass F, Lingor P. The contribution of Iron to protein aggregation disorders in the central nervous system. *Front Neurosci*. 2019;13:15. https://doi.org/10.3389/fnins.2019.00015.
83. Agim ZS, Cannon JR. Dietary factors in the etiology of Parkinson's disease. *Biomed Res Int*. 2015;672838.
84. Etminan M, Gill SS, Samii A. Intake of vitamin E, vitamin C, and carotenoids and the risk of Parkinson's disease: a meta-analysis. *Lancet Neurol*. 2005;4:362–365.
85. Hantikainen E, Trolle Lagerros Y, Ye W, et al. Dietary antioxidants and the risk of Parkinson disease: The Swedish national march cohort. *Neurology*. 2021;96:e895–e903.
86. Suzuki M, Yoshioka M, Hashimoto M, Murakami M, Noya M, Takahashi D. Urashima M randomized,double-blind, placebo-controlled trial of vitamin D supplementation in Parkinson disease. *Am J Clin Nutr*. 2013;97:1004–1013.
87. Luo X, Ou R, Dutta R, Tian Y, Xiong H, Shang H. Association between serum vitamin D levels and Parkinson's disease: a systematic review and meta-analysis. *Front Neurol*. 2018;9:909.
88. XieY FH, Peng S, Xiao J, Zhang J. Association of plasma homocysteine, vitamin B12 and folate levels with cognitive function in Parkinson's disease: a metaanalysis. *Neurosci Lett*. 2017;636:190–195.
89. Bazinet RP, Laye S. Polyunsaturated fatty acids and their metabolites in brain function and disease. *Nat Rev Neurosci*. 2014;15:771–785.
90. Kamel F, Goldman SM, Umbach DM, et al. Dietary fat intake, pesticide use, and Parkinson's disease. *Parkinsonism Relat Disord*. 2014;20:82–87.

91. Cryan JF, et al. The microbiota-gut-brain axis. *Physiol Rev.* 2019;99:1877–2013.
92. Gupta S, Allen-Vercoe E, Petrof EO. Fecal microbiota transplantation: in perspective. *Therap Adv Gastroenterol.* 2016;9:229–239.
93. Sun M-F, et al. Neuroprotective effects of fecal microbiota transplantation on MPTP-induced Parkinson's disease mice: gut microbiota, glial reaction and TLR4/TNF-α signaling pathway. *Brain Behav Immun.* 2018;70:48–60.
94. Sampson TR, et al. 5718049: gut microbiota regulate motor deficits and neuroinflammation in a model of Parkinson's disease. *Cell.* 2016;167(6), e12. 469–1480.
95. Hasegawa S, et al. Intestinal dysbiosis and lowered serum lipopolysaccharide-binding protein in Parkinson's disease. *PLOS ONE.* 2015;10:e0142164.
96. Hoban AE, Stilling RM, Ryan FJ, et al. Regulation of prefrontal cortex myelination by the microbiota. *Transl Psychiatry.* 2016;6(266):e774. https://doi.org/10.1038/tp.2016.42 [Erratum in: Sci Transl Med. 2014; 6(266):266er7].
97. Braniste V, Al-Asmakh M, Kowal C, et al. The gut microbiota influences blood-brain barrier permeability in mice. *Sci Transl Med.* 2014;6.
98. Richards PJ, et al. Galacto-oligosaccharides modulate the juvenile gut microbiome and innate immunity to improve broiler chicken performance. *mSystems.* 2020;5. e00827–19.
99. Rutsch A, Kantsjö JB, Ronchi F. The gut-brain axis: how microbiota and host inflammasome influence brain physiology and pathology. *Front Immunol.* 2020;11:3237.
100. Choi JG, Kim N, Ju IG, et al. Oral administration of *Proteus mirabilis* damages dopaminergic neurons and motor functions in mice. *Sci Rep.* 2018;8:1275. https://doi.org/10.1038/s41598-018-19646-x.
101. Miller RL, James-Kracke M, Sun GY, Sun AY. Oxidative and inflammatory pathways in Parkinson's disease. *Neurochem Res.* 2009;34:55–65.
102. Dumitrescu L, Popescu-Olaru I, Cozma L, et al. Oxidative stress and the microbiota-gut-brain axis. *Oxid Med Cell Longev.* 2018;2018:2406594. https://doi.org/10.1155/2018/2406594. PMID: 30622664; PMCID: PMC6304899.
103. Yan Y, Jiang W, Liu L, et al. Dopamine controls systemic inflammation through inhibition of NLRP3 inflammasome. *Cell.* 2015;160(1–2):62–73. https://doi.org/10.1016/j.cell.2014.11.047. PMID: 25594175.
104. del Tredici K, Braak H. *A Not Entirely Benign Procedure: Progression of Parkinson's Disease*; 2008.
105. Shannon KM, Keshavarzian A, Dodiya HB, Jakate S, Kordower JH. Is alpha-synuclein in the colon a biomarker for premotor Parkinson's disease? Evidence from 3 cases. *Mov Disord.* 2012;27:716–719.
106. Wittung-Stafshede P. Gut power: modulation of human amyloid formation by amyloidogenic proteins in the gastrointestinal tract. *Curr Opin Struct Biol.* 2022;72:33–38.
107. Asti A, Gioglio L. Can a bacterial endotoxin be a key factor in the kinetics of amyloid fibril formation? *J Alzheimers Dis.* 2014;39:169–179.
108. Friedland RP. Mechanisms of molecular mimicry involving the microbiota in neurodegeneration. *J Alzheimers Dis.* 2015;45:349–362.
109. Neto DPA, et al. *Akkermansia muciniphila* induces mitochondrial calcium overload and α-synuclein aggregation in an enteroendocrine cell line. *Iscience.* 2022;25(3):103908.
110. Romano S, et al. Meta-analysis of the Parkinson's disease gut microbiome suggests alterations linked to intestinal inflammation. *NPJ Parkinson's Dis.* 2021;7(1):1–13.
111. Cannon T, Gruenheid S. Microbes and Parkinson's disease: from associations to mechanisms. *Trends Microbiol.* 2022;30(8):749–760.

CHAPTER ELEVEN

Vaginal microbiome dysbiosis in preterm birth

Taruna Ahrodia, J.R. Yodhaanjali, and Bhabatosh Das*
Molecular Genetics Laboratory, Infection and Immunology Division, Translational Health Science and Technology Institute, NCR Biotech Science Cluster, Faridabad, India
*Corresponding author: e-mail address: bhabatosh@thsti.res.in

Contents

Abstract

All the environmentally exposed surfaces of the human body harbor ecologically distinct microbial communities with a mutualistic beneficial relationship. Depending on the body sites, microbes may provide metabolic functions, protection against pathogens, and signaling molecules to modulate host physiology and reduce disease susceptibility. Our recent understanding of the vaginal microbiome based on culture-independent 16S rRNA gene sequencing indicates that *Lactobacillus*-dominated microbial communities of healthy women play an important role in decreasing susceptibility to several urogenital diseases, including bacterial, fungal and viral infections. The findings of shotgun sequencing of the vaginal microbiome suggest that microbial-derived lactic acid, bacteriostatic, bactericidal molecules, and lower vaginal pH mediate such protections and regulations. Bacterial species, the dominant component of the vaginal microbiome, also play a key role in determining the gestation period and birth outcomes of reproductive-age women. The presence of *Lactobacillus crispatus* species in the vaginal milieu reduces the risk of preterm delivery in women of Asian ancestry. A deeper knowledge of the vaginal microbiota's role in the succession and development of newborn gut bacteria would also be beneficial. The microbiome of the mother changes throughout pregnancy and is linked to the microbiome of the newborn. This chapter highlights updated information and new opportunities for human microbiome research, focusing on the assessment of the risk of preterm birth.

Progress in Molecular Biology and Translational Science, Volume 192
ISSN 1877-1173
https://doi.org/10.1016/bs.pmbts.2022.08.001

Abbreviations

BV bacterial vaginosis
BVAB bacterial vaginosis associated bacterium
CSTs community state types
PTB preterm birth
SCFAs short-chain fatty acids
STDs sexually transmitted disease

1. Introduction

Bacteria, archaea, fungi, and viruses associated with different parts of the human body play a key role in an individual's health, with the bacterial component playing the most important function. The coevolution of the human-microbiome association with distinct structural and functional signatures in different body parts indicates a mutualistic relationship that has provided humans with incredible benefits. The community of microorganisms that live on and in the human body contains up to 10–100 trillion cells, outnumbering human cells by a factor of 1.3/cell and having 100 times more genes than the human genome.[1,2] The human body harbor's different bacterial species that vary with the differential sites and with the geo-ethnicity,[3] and are highly dependent on the existing environmental conditions and host factors.[4,5] To further discern the useful function of the microbiome in human health and diseases, the NIH has launched HMP2, now termed the integrative HMP or iHMP, which initiates the 2nd phase of the study and promises a more in-depth and better "multi-omics" method to explore host–bacterial interactions and network dynamics in the context of human health and disease.

The skin, mouth, nasal cavity, stomach, reproductive system, and perhaps the placenta, for example, are home to diverse microbial populations.[6] The microbiome of the skin, oral, and mucosal membranes plays an important role in the first line of defense by secreting several antimicrobial peptides against invading pathogens or by creating a nutrition-depleted environment that limits pathogen growth, thereby enhancing innate immunity. They have also been known to aid digestion and release certain metabolites like vitamin B12, thiamine, and riboflavin, which are beneficial for humans. Amid various human microbiomes being identified, one of the diverse bacterial populations is identified in the vaginal microbiome. Historically, investigations of vaginal ecosystem components depended first on microscopic examination and subsequently on culture techniques to identify individual bacteria.[7] Because 80% of the bacterial community that makes up the human microbiome is non-culturable, the microbiome is not being investigated effectively due to culture-dependent technology.[8] Recent advances in culture-free approaches such as 16s ribosomal RNA gene sequencing, pyrosequencing, and WGS (whole genome sequencing) have provided us with a more defined and clearer image of the human microbiome and have allowed us to identify, characterize, and distinguish the bacterial populations that were dwelling in a confined community. Furthermore, several kinds of research are being conducted to better understand the impact of the microbiome on birth outcomes.[9] As the maternal microbiome is recognized as a crucial driver of a wide range of important maternal and child health outcomes, it influences the newborn microbiome in combination with perinatal factors. The maternal microbiome serves as a source for the neonatal microbiota throughout pregnancy. Until the mid-twentieth century, the placenta and uterus were thought to be germ-free or sterile, thanks to vaginal secretions. By indulging advanced techniques for extensive classification of the placental microbiome in over 300 healthy, term, and preterm associated pregnancies, this study detected a metabolically enriched microbiome in the placenta.[10] The placenta provides a network between the developing fetus and the womb for the exchange of gases, providing the requisite nutrition and also protecting the embryonic membrane from pathogenic bacterial infections. The vaginal microbiota is notably different between pregnant and non-pregnant women. The distinctness is observed based on the microbial composition and physiological changes, as during pregnancy the microbiota is more stable and dominated by *Lactobacillus* species. Such stability in the microbial population can be attributed to a high hormonal concentration of estrogen, absence of menstruation or changes in vaginal secretions [11]. The vaginal microbiome, in particular, plays a key role in the establishment of the newborn microbiome during normal deliveries. Furthermore, several studies were undertaken in which C-section-born individuals were subjected to vaginal seeding or micro-birthing, where a swab obtained from the vaginal tract could be applied to the kid during a caesarean for improved growth and development.[12] Also, recent studies[13] show that the initial inoculating microbial community has a significant impact on the community composition and the health of the offspring at later phases. The understanding of diverse microbial communities residing in the vaginal milieu provides a very promising tool to determine the health status of the newborn population. Furthermore, there are studies, which show the effect of early exposure to the microbes associated with the disease during adulthood.[14] Therefore, it's important to look into the vaginal milieu and their interactions with newborns and also to develop certain predictive models on the basis of the vaginal microbiome to predict the frequency of preterm birth in the population and, therefore, provide possible intervention strategies to attain homeostasis.

This chapter reported the role of the maternal microbiome in pregnancy outcomes that affect neonatal and infant health. In addition, we also reported several other factors such as vaginal dysbiosis or inflammation caused by a bacterial infection, which is attributed to BV and chorioamnionitis. Therefore, it eventually leads to preterm birth.

2. Normal vaginal microbiota

The vaginal microbiome is the least diverse compared to the complexity of microbiomes of other body habitats in the human body; it is mostly dominated by *Lactobacillus* species, which maintain acidic pH in the vaginal milieu and inhibit colonization and multiplication of allochthonous bacterial and fungal species (Table 1). The *Lactobacillus* species catabolize the glycogen by fermenting it into lactic acid, which maintains the vaginal pH at around 4.4–4.6. One of the comparative studies performed in-vitro has shown the

Table 1 Vaginal microbial signatures associated with homeostatic and dysbiotic states.

Conditions	Microbiota	Role of microbiota	References
Term birth	*L. crispatus* *Lactobacillus gasseri* *Lactobacillus jensenii*	Bacteriocins Protective Protective	18,19
Pre-term birth	*Gardnerella vaginalis* *Lactobacillus iners* *Megasphera* sp. *Sneathia sanguinegens*	Harmful Harmful Harmful Harmful	18,19
Bacterial vaginosis	*G. vaginalis* *Prevotella*	Produces bacteriocins active in-vitro against lactobacilli Harmful	67
Chorioamnionitis	*G. vaginalis, E. coli, Group B Streptococcus, Fusobacterium, Sneathia, Leptotrichia*	Bacterial endotoxins and exotoxins production	56 57

addition of lactic acid to peripheral blood mononuclear cells (PBMCs) up-regulates the secretion of IL-23 compared to the lipopolysaccharide-induced cytokine secretion. In contrast, in the absence of microbial lipopolysaccharide, lactic acid doesn't impel the production of IL-23 beyond the normal levels,[15] signifying the importance of lactic acid in maintaining vaginal health. The vaginal microbiome should be made up of bacterial species, with *Lactobacillus* spp. dominating. One of the important characterizations of these species is maintaining biofilms in the vagina, which are the colonies of microorganisms that adhere to the vaginal epithelial surface and prevent the growth of harmful pathogenic bacteria. The typical vaginal microbiome comprises four main *Lactobacillus* spp. exhibiting conserved signatures, which are categorized into four CSTs (community state types).[16] According to Ravel *et al.*, CST I is dominated by *Lactobacillus crispatus*, CST II is associated with *Lactobacillus gasseri*, CST III is composed of *Lactobacillus iners*, and CST IV is associated with loss of diversity and the presence of anaerobic bacterial species, which is linked to the dysbiosis of healthy vaginal flora. The latter is further classified into CST IV-A and CST IV-B; the latter contains high levels of *Atopobiumvaginae, Megasphera,* and others, while the CST IV-A contains high levels of *Aerocooccus, Peptoniphillus, Prevotella,* and *Streptococcus.*[17] CST V is dominated by *Lactobacillus jensenii.*[18,19] The vaginal microbiome is highly dynamic, with changes in composition related to pregnancy, menstruation, and illness. The microbial community's stability declines during menstruation and coincides with estrogen levels. In terms of bacterial makeup, *Gardnerella vaginalis* abundance increases during menstruation, while *Lactobacillus* species, except *L. iners*, decrease.[20] The vaginal microbiome plays a vital function in the female body, acting as the first line of defense against pathogens that try to invade via the vaginal route. Once inside the body, pathogens cause dysbiosis, which eventually results in diseased states. In addition, various bacteria are found to be associated with the most prevalent vaginal infection, called bacterial vaginosis. One of the major consequences of BV in pregnant women is preterm birth, and further, it is also associated with affecting the placenta and fetus, such as in chorioamnionitis. Depletion of *Lactobacillus is a potential risk* of dysbiosis of the vaginal microbiome and becomes more susceptible to the colonization of diverse, anaerobic species such as *Megasphera, Gardnerella, Candida,* and *Sneathia*, etc.[18,19] Thus, it is important to understand the microbial signatures and their association with healthy and diseased states.

3. Variation in vaginal microbiome among ethnicities

Both genetic and environmental variables are thought to influence the composition and function of the microbiome, with socioeconomic status being a prominent environmental element. Significant disparities in the microbiome makeup of healthy people of various races and ethnicities have been discovered in recent studies. There is a progressive fluctuation in the composition and diversity of the microbiome, which varies with nutrition, lifestyle, and geographic representation. Also, recent studies discovered a major disruption in the microbiome as a result of westernization,[21] which leads to more harmful and dysbiotic states.[22] Apart from nutrition, lifestyle, and regional representation, Estrogen, a steroid chemical implicated in reproductive development, has a substantial influence on the microbial makeup of the vaginal environment.[23] In this section, we attempt to give insights into the microbiome variance between ethnicities, as well as the likely cause of this heterogeneity.

As discussed earlier, based on an analysis of healthy American women from four ethnic groups (white, black, Hispanic, and Asian)[16] divided the human vaginal microbiome into five major community state types (CSTs), with CSTs I, II, III, and V dominated by *Lactobacillus*, and is more common in Asian and white women, while CST IV, containing lower levels of lactic acid-producing bacteria and larger proportions of species that are composed of facultative or strictly anaerobic bacteria, such as *Gardnerella, Atopobium, Prevotella, Sneathia*, and *Streptococcus* species, is more prevalent in healthy black and Hispanic women than *Lactobacillus*.[16] Zhou *et al.* discovered substantial differences in vaginal microbiota composition between healthy Caucasian and black women of reproductive age in North America. In addition, *Lactobacillus* species were shown to be common in microbial communities in Caucasian women, although *Atopobium* and a wide range of Clostridiales phylotypes predominated in black women.[24] Furthermore, a study on apparently healthy Japanese women in Tokyo, discovered that the proportion of non-*Lactobacillus* species in vaginal communities expanded gradually from white to Japanese to black populations.[25] According to Fettweis *et al.*, the European-American vaginal microbiome is low in bacterial diversity and dominated by *Lactobacillus* spp., while the African-American vaginal microbiome is more diverse and dominated by *G. vaginalis* and the uncultivated bacterial vaginosis associated bacterium-1 (BVAB1).[26] It's uncertain if ethnic diversity is a product of inheritance or the environment.

According to Zhou *et al.*,[24] host genetic variables, such as the innate and adaptive immune systems, may be more relevant than cultural and behavioral variations between ethnic groups in influencing the makeup of vaginal microbiota.[24] Furthermore, Mehta *et al.* observed the vaginal microbial signatures of the reproductive-aged Indian women and found them more complicated, with *Lactobacillus* signals predominating. However, as compared to African American and Hispanic women, the non-*Lactobacillus* milieu is shown to be uncommon in Indian women[27].

4. Structure and functions of microbiome with birth outcomes

During gestation, the microbiome plays a critical function in regulating the immune system and metabolism, also providing benefits to the female body by resisting infections. The female body goes through a lot of changes throughout pregnancy, and the microbiome changes as well, and these modifications in the community, in particular, the vaginal microbiome, are thought to be responsible for the diverse birth outcomes. As vaginal delivery remains the most common method of birth and the infants born via caesarean compared to the vaginally developed are more likely to develop immune reactions. Therefore, in this section, we're mainly going to focus on pre-term birth, bacterial vaginosis, and preeclampsia because they're all connected to the vaginal microbiota and are a serious concern nowadays (Table 2).

4.1 Vaginal microbiota linked with preterm birth

Preterm birth (PTB), or preterm labor, is characterized by delivery before completing 37 weeks of gestation. It is a global health challenge affecting more than 12.9 million infants worldwide. PTB affects 1 out of every 10 infants born and is the leading cause of death in children under 5 years of age. According to WHO (2018), across 184 countries, PTB ranges from 5% to 18% and is considered the second most dominant cause of neonatal death worldwide.[28] The outcomes of PTB are divided into six categories, which include: congenital malformations; iatrogenic PTB (caused due to clinical infection, abruption of the placenta, placenta praevia, preeclampsia, etc); intrauterine fetal death; multiple gestations; spontaneous PTB; and preterm premature rupture of membranes. Many factors, such as socio-economic conditions, medical complications, obstetric and fatal conditions, and environmental factors, are considered risk factors for spontaneous PTB. Nowadays,

Table 2 Immunological response and the risks associated with infectious conditions.

Infectious condition	Immunological response	Birth outcome	Associated health disorders	References
Bacterial vaginosis	Activation of macrophage inflammatory protein (MIP)-1α, MIP-1, IL-6, IL-7, IL-17α. Inhibition of IgA and IL-8 induced neutrophil infiltration. Degradation of mucosal sialic acid.	Premature rupture of membranes, low birth weight, amniotic fluid infection, postpartum endometritis, chorioamnionitis	Preterm birth (PTB), spontaneous preterm birth (sPTB), fetal growth restriction (FGR), preterm rupture of membranes (PPROM), failure of in-vitro fertilization, miscarriages, preeclampsia (PE), Pelvic inflammatory disease (PID), increased risk of HIV	46 53 49
Chorio-amnionitis	Activation of IL-6, IL-1α, IL-1β, IL-8 and TNF-α; release of prostaglandins and neutrophil infiltration	Premature rupturing of fetal membrane and preterm birth	Fetal inflammatory response syndrome which causes neonatal morbidity	62 61 68
Preterm birth	Intrauterine inflammation and maternal effector/activated T cells	Increased risk of early-onset sepsis and T cell activation caused by maternal hypothermia, bradycardia, systemic inflammation, cervical dilation, intra-amniotic inflammation, and fetal growth restriction	Newborn respiratory disease, preterm labor, and adverse neonatal outcomes	69 70

the dysbiotic microbiome is being considered as one of the potential key factors of PTB after knowing its importance in diseased states.[29] During pregnancy, changes in the woman's body provide an opportunity to establish pathogenic bacteria and remold well-established commensal microorganisms in the vaginal environment. A number of studies in the last few years based on culture-dependent as well as culture-independent data have increased knowledge about the normal and abnormal vaginal microbiome. A longitudinal diversity study of the vaginal microbiome between non-pregnant and pregnant women reveals lower richness and diversity, but a higher bacterial load is found as compared to non-pregnant women.[30] As most studies have revealed that the *Lactobacillus* genus is the most dominant bacteria in vaginal flora, some other pathogenic or non-pathogenic bacterial genera reside in the vagina viz. *Gardnerella, Prevotella, Atopobium, Bifidobacterium, Chlamydia, Neisseria, Mobiluncus, Megasphaera, Sneathia, Bacteroides, Blautia*, BVAB1, BAVB2, BVAB3, *Streptococcus, Corynebacterium, Porphyromonas, Anaerococcus, Campylobacter, Finegoldia Peptoniphilus, Eggerthella, Leptotrichia, Staphylococcus*, and *Escherichia* spp. have been also identified.[31–34] The diversity and relative abundance of these bacterial species depend on subjects' health, socioeconomic status, and ethnicity. The various species of *Lactobacillus* play a protective role by producing lactic acid (pH = 3.5–4.5) as their fermentation by-product. Such a lower pH is beneficial to the host vagina and inhibits the growth of many other pathogenic and invading microbes.[35] Additionally, lactobacilli also produce a bactericidal peptide known as bacteriocin. Hydrogen peroxide (H_2O_2) is another defensive factor produced by *Lactobacillus* though its role is controversial.[36,37] Two evolutionary hypotheses have been proposed about the dominancy of *Lactobacillus* spp. and the uniqueness of the human vaginal microbiome among mammals: (i) the "obstetric protection hypothesis" suggests that the selection of a lactobacilli-rich environment is due to microbial complications associated with pregnancy and childbirth, and (ii) the "disease risk hypothesis" in the vaginal microbiome proposes that humans are at higher risk of getting STD infections than non-human mammal, the STD may have a significant impact on the host. Therefore, the lactobacilli-dominated environment is needed because of its protective role.[38,39] Many longitudinal and spatial studies have shown an association between the risk of preterm birth and vaginal microbial composition, but the key species have not yet been identified. Only relative abundance has been reported.[40,41] The longitudinal study of the diversity and structure of the vaginal microbiome at two points in gestation (before 16 weeks or between 20 and 24 weeks) of pregnant nulliparous African-American women shows that women who deliver preterm have lower abundance and diversity of *Prevotella, Sneathia, Coriobacteriaceae* and *Aerococcus* compared to term-delivery women.[42] Another study of Irish populations shows a distinct abundance pattern of lactobacilli populations in vaginal microbiota between preterm and term-delivered women. *L. crispatus* (*P*-value = 0.03) and *Lactobacillus reuteri* (*P*-value = 0.04) were found to be significantly lower in the preterm birth group. A non-Lactobacillus species, *Aldercreutzia equolifaciens* was also associated with preterm birth (*P*-value = 0.004). Two distinct community state types, CST-IIIa and CST-V, dominated by *L. iners* and *L. jensenii* respectively, were clustered within the preterm birth group. This study shows a positive association between *L. iners* and *L. jensenii* with preterm birth.[43] A different association pattern of the vaginal microbiome with preterm and term delivery was observed in a Canadian population. A relative abundance of *L. crispatus/Lactobacillus acidophilus*, *L. gasseri/Lactobacillus johnsonii*, *L. iners/Ralstonia solanacearum*, and *Bifidobacterium longum/Bifidobacterium breve* may be associated with a decreased risk of early spontaneous preterm birth.[44] The term prevalence ratio (term/preterm) of bacteria in 216 Canadian women in which term (*n* = 170) preterm (*n* = 46) shows that *Lactobacillus delbrueckii* subsp. lactis, *Pseudovibrio* sp., *Sphingobium yanoikuyae*, *Clostridium innocuum*, and *Massilia timonae* were greater than 6 times in term as compared to the preterm birth. Whereas, there are species, *Facklamia hominis, Corynebacterium amycolatum, Parvimonas micra, Mobiluncus curtsii* subsp. *curtsii, Brevibacterium linens, Streptococcus thermophilus, Dethiobacter alkaliphilus, Brevibacterium massiliense, Eremococcus coleocola, Anaeromusa acidaminophila*, and *Arthrobacter globiformis*, were identified, whose prevalence was found to be greater than 6 times greater in preterm than in term subjects.[30] In Asian countries, only a handful of studies of the association of the vaginal microbiome with preterm and term-delivered women have been performed. A recent study from India has observed that the vaginal microbiome of both term and preterm samples reveals similar alpha diversity indices. However, there was a significantly higher abundance of L. *iners* (*P*-value < 0.02) in preterm during all trimesters. In the first trimester, *Megasphaera* sp. (*P*-value 0.05), *G. vaginalis* (*P*-value = 0.01), and *Sneathia sanguinegens* (*P*-value 0.0001) were identified as significantly higher in preterm than term. Only L. *gasseri* was significantly high in abundance in the 3rd trimester (*P*-value = 0.010) in terms of.[18,19] The genomes of L. *crispatus* and L. *gasseri* are enriched with horizontally acquired genetic elements and peptides potentially linked with antimicrobial functions.

4.2 Bacterial vaginosis (BV)

BV or vaginal dysbiosis is an infection seen in the lower genital tract of reproductive-age women worldwide, with an average global prevalence rate of 25% across different regions such as 29% in Southern Asia, 27% in Northern America, 25% in Sub-saharan Africa, Northern Africa, and the

Middle East, 24% in East Asia and Pacific, 24% Latin America and the Caribbean, 23% East and Central Asia.[45] Asymptomatic BV or BV with only malodorous vaginal discharge without acute inflammation is the most common form of BV, thus, the previous name "vaginitis" (inflammation of the vagina) was replaced with "vaginosis." In severe cases of BV, the characteristic symptoms are (i) itchy or non-itchy homogeneous white or gray milky vaginal discharge, (ii) Increase in vaginal pH >4.5, (iii) An amine, fishy malodor which gives a positive whiff test, and (iv) Identification of clue cells (attachment of BV-related bacteria to exfoliated vaginal epithelial cells) and the destruction of the epithelial cells by the secretion of bacterial lytic enzymes on a saline wet mount through microscopy.[46,47] The probability of homosexual transmission of BV-related bacteria in women is higher than in heterosexual transfer.[48] Thus, BV is often categorized as a sexually enhanced or sexually amplified disease rather than a sexually transmitted disorder. The leading reproductive health outcomes of BV comprise conditions such as preterm birth (PTB), low birth weight, fetal growth restriction (FGR), amniotic fluid infection, preterm rupture of membranes (PROM), failure of in-vitro fertilization, miscarriages, preeclampsia (PE), pelvic inflammatory disease (PID), postpartum endometritis, chorioamnionitis, increased risk of HIV and other sexually transmitted diseases.[49] BV is characterized by an increase in pathogenic gram variable facultative anaerobes resulting in the loss of beneficial *Lactobacillus* dominant microbiome. It's a polymicrobial disease generating a niche in the host with the aid of two or more bacterial species suppressing the growth of *Lactobacillus* strains.[50] *G. vaginalis* is the primary colonizer, which provides a protective environment to other anaerobes against the resident *Lactobacillus* spp. Other than *Gardnerella* some major commensal organisms or contributors of BV are *Prevotella, Ureaplasma, Bacteroides, Mobiluncus, Veillonellaceae bacterium, Mycoplasma, Peptostreptococcus, Anaerococcus, Sneathia, Clostridium, Leptotrichia,* and newly discovered *Atopobium, Megaphaera,* BV-associated bacterium 1 (BVAB1) to BVAB3, etc.[51] The infestation of BV-related species can tolerate the normal vaginal pH of 4–5 and increases it to 6–7 pH. This increase in the pH allows the growth of other anaerobic bacteria and is caused by increases in the concentration of short-chain fatty acids (SCFA) such as acetates, butyrate, isobutyrate, formate, propionate, succinate; amines such as putrescine, cadaverine, and trimethylamine.[46] Studies have also shown that 2.7% of the total normal healthy vaginal microbiome consists of *G. vaginalis* in comparison with the BV-associated vaginal microbiome.[16] This suggests that *G. vaginalis* is a virulent, opportunistic organism that can survive at very low pH and manifests asymptomatic BV.

4.2.1 Pathogenesis of bacterial vaginosis (BV)

BV is diagnosed by Amsel's criteria based on the symptoms exhibited by the host and by gram staining of vaginal smears to detect and quantify clue cells using a scoring system in the range 0–10 called the Nugent score. This led to the new definition of BV as a synergic poly-bacterial infection (Fig. 1). A recent hypothesis explains the etiology of BV has four main steps which are: (i) Initiation of BV by the invasion of "early or primary colonizer species," e.g., *Gardnerella* spp. and *Prevotella* spp. which can adhere to the vaginal microbiome and alters the pH from 4.5 to 6–7, (ii) Characteristic polymicrobial biofilm formation which helps the pathogenic bacteria against antibacterial effects of the normal *Lactobacillus* dominant spp., (iii) Maturation of the biofilm and coaggregation of "secondary colonizers" such as *Atopobium* spp., *Megasphaera* spp., *Mycoplasma* spp., etc. (iv) dispersion or deterioration of vaginal epithelial cells covered with pathogenic bacteria (Clue cells), and vaginal discharge induced by hydrolytic enzymes produced by the pathogenic bacteria, and (v) propagation by invading other parts of the vaginal environment which marks the beginning of the next cycle (Fig. 1).[46,52] A biofilm is a systematic community of organisms embedded in a self-made extracellular matrix adherent to the vaginal epithelial layer. BV-associated biofilm facilitates the growth of other pathogenic bacteria; a collaborative relationship between these organisms and *Gardnerella* spp. supports the growth and maturation of the biofilm. *G. vaginalis* associated biofilm adheres to vaginal epithelial cells and provides the needs of the microorganisms such as maintaining an optimal pH gradient and providing nutrients and oxygen. They also protect the organisms from antibacterial agents like H_2O_2, lactic acid, and other bacteriocins secreted by lactobacilli, tolerance to host immune response, and provide antibacterial resistance thus promoting the recurrence of BV.[50] *G. vaginalis* and *Prevotella* spp. can tolerate the high redox potential of a healthy lactobacillus-dominated vagina and displaces the resident lactobacilli by lowering the redox potential. An increase in strict anaerobic BV-related species is a result of low redox potential. These pathogens exhibit a metabolic interdependency between them to fulfill their nutritional demands. As the primary colonizer, *G. vaginalis* deplete the protein-rich vaginal epithelial layer and release various peptides and amino acids, which are taken up by *Prevotella* spp., *Atopobium* spp., and other fastidious BV-associated pathogens. *Prevotella* spp. and *Atopobium* spp. utilize this peptide and produce ammonia, which in turn is used up by *Gardnerella* spp. as a growth factor. This symbiotic relationship between BV-associated bacteria makes the infection more complex and might be responsible for the rate of recurrence of the disease (Fig. 2).

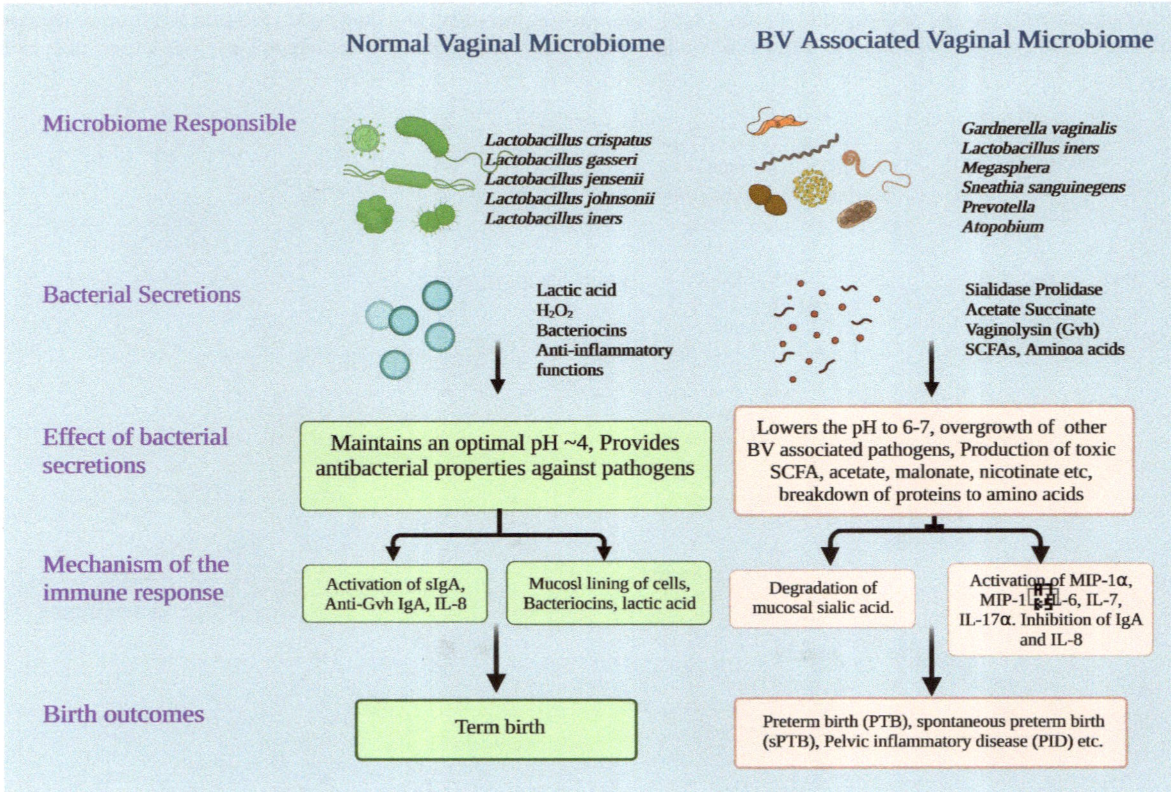

Fig. 1 Sequential mechanisms in different birth etiologies in the normal vs. BV associated microbiome.

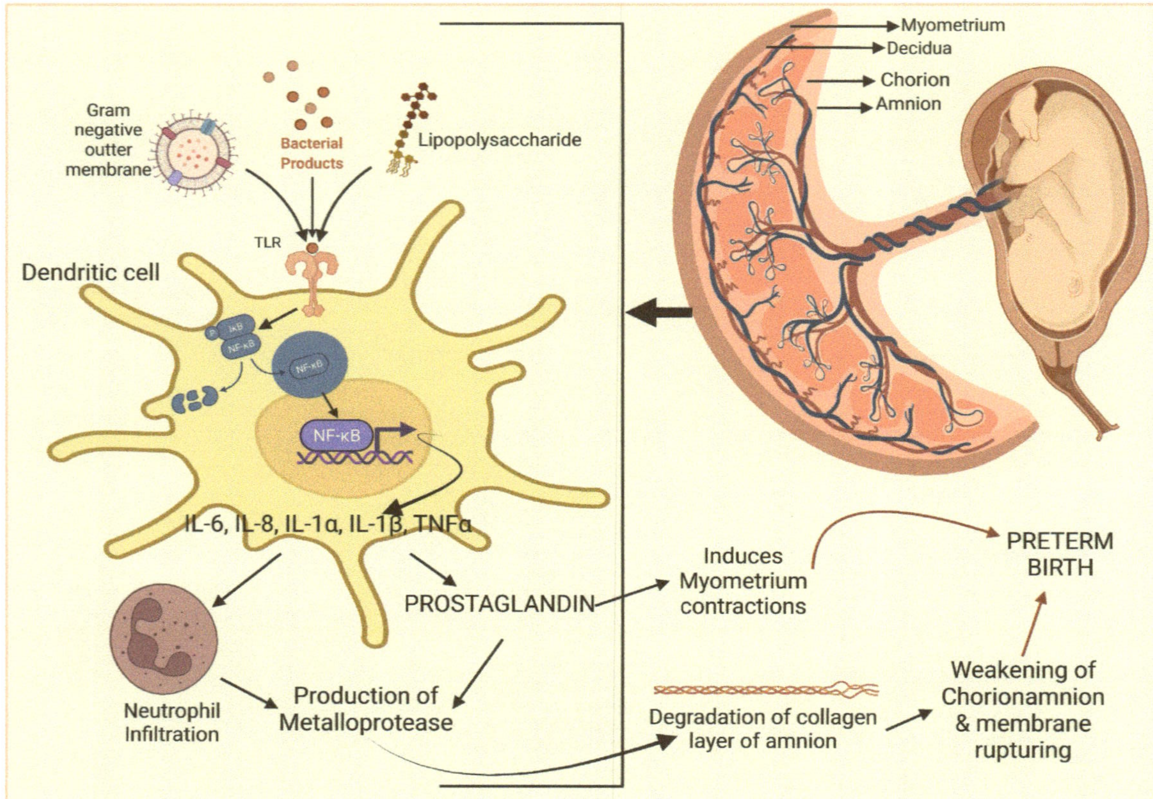

Fig. 2 Molecular and Immunological Mechanisms during Chorioamnionitis: Toll-like-receptors of dendritic cells residing in the chorion decidua detect the bacterial endotoxins and exotoxins which activate the downstream NF-κB signaling pathways, NF-κB induces transcription and translation of genes to produce IL-6, IL-8, IL-1α, IL-1β, and TNFα. These pro-inflammatory cytokines stimulate the production of prostaglandin and neutrophil infiltration. Neutrophils produce a secondary inflammatory response by releasing metalloproteases. Prostaglandin induces myometrium contraction and metalloprotease catalysis, the degradation of the collagen layer, which further leads to chorion amnion rupture, which is associated with preterm birth.

4.2.2 Bacterial vaginosis and preterm birth

There is strong evidence to believe that BV has a substantial association with preterm birth. The pathophysiologic pathways linking BV and preterm birth are still unclear but a recent hypothesis explains. Studies show that BV might lead to spontaneous preterm birth, preterm birth, preterm rupture of membranes, etc. However, it is not clear why BV resolves in some women spontaneously and not in others. Several studies suggest that BV is one of the leading causes of poor perinatal outcomes especially preterm birth.[53–55] The vaginal microbiome plays a consequential role in the maintenance of gestational health by either inhibiting preterm birth or contributing to preterm birth. The vaginal environment of a healthy pregnant woman is enriched with circulatory estrogen in the vaginal tissues, and blood as it increases glycogen levels, which in turn enhances the growth of *Lactobacillus* spp. in the progression of the gestational period. The glycogen present in the vaginal epithelial cells and lumen is used by matrix metalloproteinase—8 (MM-8), MM-9, and hyaluronidase-1 during pregnancy and facilitates the host α-amylase to produce maltose. The vaginal Lactobacilli use maltose for further growth and proliferation. On the other hand, causes dysbiosis, stress, immune dysregulation, uterine hemorrhage, etc., results in the loss of vaginal *Lactobacillus* spp. and an increase in the level of BV pathogens which doubles the risk of premature preterm birth.

4.3 Chorioamnionitis

It is an intrauterine inflammation affecting the chorion and amniotic membrane and it is manifested as polymicrobial bacterial infections which cause early rupture of the membrane, leading to preterm birth or neonatal inflammation.[15] The microorganisms involved in chorioamnionitis are *G. vaginalis*, *E. coli*, *Bacteroides*, and group B *Streptococcus*.[56] Other studies have shown a broad range of bacterial colonies such as *Fusobacterium, Sneathia*, and *Leptotrichia* in the amniotic cavity. Furthermore, distinguishable differences in the bacterial species exist between women with preterm delivery without premature rupture of the fetal membrane and women that deliver preterm with premature rupture of the membrane.[57] Some of these bacteria are commonly found in association with BV. Therefore, there might be a chance of pregnant women with BV having a high risk of chorioamnionitis. Chorioamnionitis is classified as clinical, which has symptoms like uterine tenderness, genital discharge, tachycardia, and histological Chorioamnionitis is the most prevalent acute microbial infection with inflammation of the chorion, amnion, and placenta with or without premature rupture of the fetal membrane.[58] Acute chorioamnionitis is distinguished as a neutrophilic infiltration and inflammation at the maternal-fetal interface.[59] The amniotic sac is formed of various membranes such as decidua, which is characterized as maternal tissue and fetal tissues such as chorion, which is composed of reticular layer, basement membrane, trophoblasts, and amniotic membranes. The amnion encapsulates a layer of epithelial cells, underlying mesenchymal cells, and collagen. On the other hand, decidua is in direct contact with chorion and comprises maternal immune cells.[60] Understanding the membrane architecture provides an insight into the intrauterine space and microbial invasion pathways. The major significance of bacterial infection is the release of bacterial endotoxins and exotoxins. These are recognized by the toll-like receptors of leukocytes and dendritic cells residing in the chorion decidua and fetal membrane. This interaction activates the downstream signaling molecules such as NF-B and AP-1. These transcription factors stimulate transcription and translation of cytokines such as IL-6, IL-1, IL-1, IL-8, and TNF-α. These inflammatory cytokines induce the release of prostaglandins and cause neutrophil infiltration.[61,62]

Infiltration of neutrophils into chorion decidua and amnion is quite distinct as the neutrophils in chorion decidua are of maternal origin and those in amniotic fluid are of fetal origin.[63] During intrauterine infections, there is a notable increment in certain neutrophils such as $CD15^+$ in the amniotic fluid and other inflammatory cytokines produced by neutrophils of chorion decidua are TNF-α, IL-8, and MIP-1 ([64]; Roberto[65]). The neutrophils of decidua inhibit pathogen growth by producing antimicrobial enzymes such as neutrophil elastase and myeloperoxidase. They also trigger the production of prostaglandin, which activates uterine contractions and the metalloprotease, causing cervical ripening and degrading the collagen layer of the chorioamniotic membrane, resulting in preterm rupture.[65] Under normal conditions, chorionic tissue produces prostaglandin dehydrogenase, which catalyzes the inactivation of prostaglandins produced in the amnion. Thus, it intercepts the prostaglandin from reaching the myometrium and initiating uterine contractions.[66]

Other risks associated with chorioamnionitis are developmental damages caused to the fetus; the newly born baby gets the microbial components from the mother as soon as the amnion bursts. These bacteria further localize to the site-specific regions and induce gene expression in the intestinal epithelial cells. Microbes residing in the intestinal region are critical for the development of GALT-gut-associated lymphoid tissue. Hence, a mother with intrauterine infection and the microbes associated with this infection may have an adverse effect on the growth and development of the infant, as the infant may lack the necessary microbes that impart good health.

5. Conclusion

Microbial communities living in the reproductive tract are crucial for reproductive health and shaping the gut microbiome of newborns. There are significant differences in the composition and functional potency of the vaginal microbiome in different ethnic groups. The dominance of different species of *Lactobacillus* in the vaginal microbiome is very common across the globe; although non-Lactobacillus dominated diverse microbial communities are also reported in healthy African and American black women. The dynamics of the vaginal microbiome and changes in community compositions over time within and between subjects are clearly demonstrated. Community composition, functional potency, and dynamics of the vaginal microbiome significantly contribute to the pathophysiology of preterm birth. *L. crispatus* dominated vaginal microbiota reduces the risk of preterm delivery; while increasing the abundance of *G. vaginalis* and *S. sanguinegens* in a *L. iners* dominated community potentially increases the risk of preterm delivery. Quantitative measurement of the abundances of these bacteria in the vaginal milieu of pregnant women can potentially be used for risk assessment and diagnosis of preterm delivery. Moreover, recent research emphasized the significance of the vaginal microbiome in in-vitro fertilization, suggesting a potential role of a healthy microbiome in the successful implantation of the embryos. Understanding the vaginal microbiota may therefore be utilized to determine pregnancy outcomes as well as to develop certain predictive models for the prediction of any birth defects that can be addressed utilizing prospective treatment strategies. Furthermore, it is also feasible to establish intervention measures based on the vaginal microbiome to help pregnant women reverse dysbiosis toward homeostasis, which will eventually aid in the growth and development of unborn children.

Acknowledgment

We thank all the members of the Molecular Genetics Laboratory (MGL), THSTI for their valuable feedback. We acknowledge the support of the Executive Director, THSTI. TA and BD acknowledge the Department of Biotechnology (DBT), Govt. of India for research fellowship and funding (No.BT/PR9983/MED/97/194/2013).

References

1. Sender R, Fuchs S, Milo R. Revised estimates for the number of human and bacteria cells in the body. *PLoS Biol.* 2016;14(8), e1002533.
2. Turnbaugh PJ, Ley RE, Hamady M, Fraser-Liggett CM, Knight R, Gordon JI. The human microbiome project. *Nature.* 2007;449(7164):804–810.
3. Jayaraman S. Of ethnicity, environment, and microbiota. *Cell Mol Immunol.* 2019;16 (2):106–108.
4. Costello EK, Lauber CL, Hamady M, Fierer N, Gordon JI, Knight RJ. Bacterial community variation in human body habitats across space and time. *Science.* 2009;326 (5960):1694–1697.
5. Ehrlich SD, Consortium M. *Metagenomics of the Human Body.* Springer New York; 2011.
6. Gomez-Arango LF, Barrett HL, McIntyre HD, Callaway LK, Morrison M, Nitert MD. Contributions of the maternal oral and gut microbiome to placental microbial colonization in overweight and obese pregnant women. *Sci Rep.* 2017;7(1):2860.
7. Larsen B, Monif GRJCID. Understanding the bacterial flora of the female genital tract. *Clin Infect Dis.* 2001;32(4):e69–e77.
8. Lagier JC, Dubourg G, Million M, et al. Culturing the human microbiota and culturomics. *Nat Rev Microbiol.* 2018;16:540–550.
9. Dunlop AL, Mulle JG, Ferranti EP, Edwards S, Dunn AB, Corwin EJ. Maternal microbiome and pregnancy outcomes that impact infant health: a review. *Adv Neonatal Care.* 2015;15(6):377–385.
10. Aagaard K, Ma J, Antony KM, Ganu R, Petrosino J, Versalovic J. The placenta harbors a unique microbiome. *Sci Transl Med.* 2014;6(237):237ra265.
11. Romero R, Hassan SS, Gajer P, et al. The composition and stability of the vaginal microbiota of normal pregnant women is different from that of non-pregnant women. *Microbiome.* 2014;2(1):4.
12. Lokugamage AU, Pathberiya SDC. The microbiome seeding debate—let's frame it around women-centred care. *Reprod Health.* 2019;16(1):91.
13. Mueller NT, Dominguez-Bello MG, Appel LJ, Hourigan SK. 'Vaginal seeding' after a caesarean section provides benefits to newborn children: FOR: does exposing caesarean-delivered newborns to the vaginal microbiome affect their chronic disease risk? The critical need for trials of 'vaginal seeding' during caesarean section. *BJOG.* 2020;127(2):301.
14. Paquette K, Coltin H, Boivin A, Amre D, Nuyt AM, Luu TM. Cancer risk in children and young adults born preterm: A systematic review and meta-analysis. *PLoS One.* 2019;14(1):e0210366.
15. Witkin SS, Alvi S, Bongiovanni AM, Linhares IM, Ledger WJ. Lactic acid stimulates interleukin-23 production by peripheral blood mononuclear cells exposed to bacterial lipopolysaccharide. *FEMS Immunol Med Microbiol.* 2011;61(2):153–158.
16. Ravel J, Gajer P, Abdo Z, et al. Vaginal microbiome of reproductive-age women. *Proc Natl Acad Sci U S A.* 2011;108(Suppl 1):4680–4687.
17. Ma ZS, Li L. Quantifying the human vaginal community state types (CSTs) with the species specificity index. *PeerJ.* 2017;5:e3366.
18. Kumar S, Kumari N, Talukdar D, et al. The vaginal microbial signatures of preterm birth delivery in Indian women. *Front Cell Infect Microbiol.* 2021;13:11. Original Research.
19. Kumar S, Kumari N, Talukdar D, et al. The vaginal microbial signatures of preterm birth delivery in Indian women. *Front Cell Infect Microbiol.* 2021;11:622474.
20. Srinivasan S, Liu C, Mitchell CM, et al. Temporal variability of human vaginal bacteria and relationship with bacterial vaginosis. *PLoS One.* 2010;5(4):e10197.
21. Gupta VK, Paul S, Dutta C. Geography, ethnicity or subsistence-specific variations in human microbiome composition and diversity. *Front Microbiol.* 2017;8:1162.

22. Nam YD, Jung MJ, Roh SW, Kim MS, Bae JW. Comparative analysis of Korean human gut microbiota by barcoded pyrosequencing. *PLoS One*. 2011;6(7):e22109.
23. Farage MA, Miller KW, Sobel JD. Dynamics of the vaginal ecosystem—hormonal influences. *Infect Dis Res Treat*. 2010;3: IDRT.S3903.
24. Zhou X, Brown CJ, Abdo Z, et al. Differences in the composition of vaginal microbial communities found in healthy Caucasian and black women. *ISME J*. 2007;1 (2):121–133.
25. Zhou X, Hansmann MA, Davis CC, et al. The vaginal bacterial communities of Japanese women resemble those of women in other racial groups. *FEMS Immunol Med Microbiol*. 2010;58(2):169–181.
26. Fettweis JM, Brooks JP, Serrano MG, et al. Differences in vaginal microbiome in African American women versus women of European ancestry. *Microbiology (Reading)*. 2014;160 (Pt 10):2272–2282.
27. Mehta O, Ghosh TS, Kothidar A, et al. Vaginal microbiome of pregnant Indian women: insights into the genome of dominant Lactobacillus species. *Microb Ecol*. 2020;80 (2):487–499.
28. Fettweis JM, Serrano MG, Brooks JP, et al. The vaginal microbiome and preterm birth. *Nat Med*. 2019;25(6):1012–1021.
29. Qin J, Li R, Raes J, et al. A human gut microbial gene catalogue established by metagenomic sequencing. *Nature*. 2010;464(7285):59–65.
30. Freitas AC, Chaban B, Bocking A, et al. The vaginal microbiome of pregnant women is less rich and diverse, with lower prevalence of Mollicutes, compared to non-pregnant women. *Sci Rep*. 2017;7(1):1–16.
31. Gupta P, Singh MP, Goyal K. Diversity of vaginal microbiome in pregnancy: deciphering the obscurity. *Front Public Health*. 2020;8:326.
32. Jespers V, Kyongo J, Joseph S, et al. A longitudinal analysis of the vaginal microbiota and vaginal immune mediators in women from sub-Saharan Africa. *Sci Rep*. 2017;7(1):1–13.
33. Kindinger LM, Bennett PR, Lee YS, et al. The interaction between vaginal microbiota, cervical length, and vaginal progesterone treatment for preterm birth risk. *Microbiome*. 2017;5(1):1–14.
34. Mitchell C, Srinivasan S, Zhan X, et al. Associations between serum estrogen, vaginal microbiota and vaginal glycogen in postmenopausal women. *Menopause*. 2016;215(6): S827.
35. Danielsson D, Teigen PK, Moi HJA. The genital econiche: focus on microbiota and bacterial vaginosis. *Ann NY Acad*. 2011;1230(1):48–58.
36. Dennerstein GJ, Ellis DHJA, Obstetrics NZJ. Oestrogen, glycogen and vaginal candidiasis. *Aust N Z J Obstet Gynaecol*. 2001;41(3):326–328.
37. O'Hanlon DE, Moench TR, Cone RA. Vaginal pH and microbicidal lactic acid when lactobacilli dominate the microbiota. *PLoS One*. 2013;8(11):e80074.
38. Amabebe E, Anumba DOC. The vaginal microenvironment: the physiologic role of lactobacilli. *Front Med (Lausanne)*. 2018;5:181.
39. Miller EA, Beasley DE, Dunn RR, Archie EA. Lactobacilli dominance and vaginal pH: why is the human vaginal microbiome unique? *Front Microbiol*. 2016;7:1936.
40. Bayar E, Bennett PR, Chan D, Sykes L, MacIntyre DA. The pregnancy microbiome and preterm birth. In: *Paper presented at the seminars in immunopathology*; 2020.
41. Bennett PR, Brown RG, MacIntyre DAJO, Clinics G. Vaginal microbiome in preterm rupture of membranes. *Obstet Gynecol Clin N Am*. 2020;47(4):503–521.
42. Nelson DB, Shin H, Wu J, Dominguez-Bello MGJA. The gestational vaginal microbiome and spontaneous preterm birth among nulliparous. *Afr Am Women*. 2016;33 (9):887–893.
43. Cruickshank R, Sharman A. The biology of the vagina in the human subject. *BJOG*. 1934;41(2):208–226.

44. Tabatabaei N, Eren A, Barreiro L, et al. Vaginal microbiome in early pregnancy and subsequent risk of spontaneous preterm birth: a case–control study. *BLOG.* 2019;126(3):349–358.
45. Peebles K, Velloza J, Balkus JE, McClelland RS, Barnabas RV. High global burden and costs of bacterial vaginosis: a systematic review and meta-analysis. *Sex Transm Dis.* 2019;46(5):304–311.
46. Amabebe E, Anumba DOC. A combination of cervicovaginal fluid glutamate, acetate and D-lactate identified asymptomatic low-risk women destined to deliver preterm: a prospective cohort study. *Reprod Sci.* 2022;29(3):915–922.
47. Amsel R, Totten PA, Spiegel CA, Chen KC, Eschenbach D, Holmes KKJT. Nonspecific vaginitis: diagnostic criteria and microbial and epidemiologic associations. *Am J Med.* 1983;74(1):14–22.
48. Chen X, Lu Y, Chen T, Li R. The female vaginal microbiome in health and bacterial vaginosis. *Front Cell Infect Microbiol.* 2021;11:631972.
49. Cho H, Shim S, Park H, Cha D. Influence of maternal microbiome and inflammatory response in preterm birth: recent aspects of the prevention of preterm birth. *Microbiol Res.* 2021;13:1–13.
50. Redelinghuys MJ, Geldenhuys J, Jung H, Kock MM. Bacterial vaginosis: current diagnostic avenues and future opportunities. *Front Cell Infect Microbiol.* 2020;10:354.
51. Morrill S, Gilbert NM, Lewis AL. Gardnerella vaginalis as a cause of bacterial vaginosis: appraisal of the evidence from in vivo models. *Front Cell Infect Microbiol.* 2020;10:168.
52. Almeida MO, Carmo FLRD, Gala-García A, et al. *Lactobacillus crispatus* protects against bacterial vaginosis. *Genet Mol Res.* 2019;18:GMR18475.
53. Basavaprabhu HN, Sonu KS, Prabha R. Mechanistic insights into the action of probiotics against bacterial vaginosis and its mediated preterm birth: an overview. *Microb Pathog.* 2020;141:104029.
54. Larsson P-G, Poutakidis G, Adolfsson A, Charonis G, Bauer P, Ekstrom L. Treatment of bacterial vaginosis in early pregnancy and its effect on spontaneous preterm delivery and preterm rupture of membranes. *Clin Microbiol Open Access.* 2016.
55. Viscardi RMJN. Evidence for role of genital mycoplasmas in preterm birth and neonatal lung injury. *NeoReviews.* 2018;19(2):e69–e77.
56. Sperling RS, Newton E, Gibbs RS. Intraamniotic infection in low-birth-weight infants. *J Infect Dis.* 1988;157(1):113–117.
57. DiGiulio DB. Diversity of microbes in amniotic fluid. *Semin Fetal Neonatal Med.* 2012;17(1):2–11.
58. Su BH. Histological chorioamnionitis and neonatal outcome in preterm infants. *Pediatr Neonatol.* 2014;55(2):154–155.
59. Cappelletti M, Presicce P, Kallapur SG. Immunobiology of acute chorioamnionitis. *Front Immunol.* 2020;11:649.
60. Menon R, Richardson LS, Lappas M. Fetal membrane architecture, aging and inflammation in pregnancy and parturition. *Placenta.* 2019;79:40–45.
61. Agrawal V, Hirsch E. Intrauterine infection and preterm labor. In: *Paper presented at the seminars in fetal and neonatal medicine*; 2012.
62. Peng CC, Chang JH, Lin HY, Cheng PJ, Su BH. Intrauterine inflammation, infection, or both (triple I): a new concept for chorioamnionitis. *Pediatr Neonatol.* 2018;59(3):231–237.
63. Sampson JE, Theve RP, Blatman RN, et al. Fetal origin of amniotic fluid polymorphonuclear leukocytes. *Am J Obstet Gynecol.* 1997;176(1 Pt 1):77–81.
64. Martinez-Varea A, Romero R, Xu Y, et al. Clinical chorioamnionitis at term VII: the amniotic fluid cellular immune response. *J Perinat Med.* 2017;45(5):523–538.
65. Romero R, Gomez R, Ghezzi F, et al. A fetal systemic inflammatory response is followed by the spontaneous onset of preterm parturition. *Am J Obstet Gynecol.* 1998;179(1):186–193.

66. Goldenberg RL, Hauth JC, Andrews WWJNE. Intrauterine infection and preterm delivery. *N Engl J Med.* 2000;342(20):1500–1507.
67. Teixeira G, Soares-Brandao K, Branco K, et al. Antagonism and synergism in *Gardnerella vaginalis* strains isolated from women with bacterial vaginosis. *J Med Microbiol.* 2010;59 (8):891–897.
68. Gotsch F, Romero R, Kusanovic JP, et al. The fetal inflammatory response syndrome. *Clin Obstet Gynecol.* 2007;50(3):652–683.
69. Melville JM, Moss TJM. The immune consequences of preterm birth. *Front Neurosci.* 2013;7:79.
70. Arenas-Hernandez M, Gomez-Lopez N, Garcia-Flores V, et al. Choriodecidual leukocytes display a unique gene expression signature in spontaneous labor at term. *Genes Immun.* 2019;20(1):56–68.

Index

Note: Page numbers followed by "*f*" indicate figures and "*t*" indicate tables.

U

V

W

Z

Kolibri 360 GmbH
Ferdinand-Jühlke-Straße 7
99095 Erfurt, Deutschland
produktsicherheit@kolibri360.de